Matrix Analysis of Electrica

Matrix Analysis of Electrical Machinery

SECOND EDITION

by

N. N. HANCOCK, B.SC., B.SC.(ENG.),
M.SC.TECH., PH.D., C.ENG., F.I.E.E.
SENIOR LECTURER IN ELECTRICAL ENGINEERING AT THE
UNIVERSITY OF MANCHESTER, INSTITUTE OF SCIENCE AND TECHNOLOGY

PERGAMON PRESS
OXFORD · NEW YORK · TORONTO · SYDNEY

Pergamon Press Ltd., Headington Hill Hall, Oxford
Pergamon Press Inc., Maxwell House, Fairview Park, Elmsford, New York 10523
Pergamon of Canada Ltd., 207 Queen's Quay West, Toronto 1
Pergamon Press (Aust.) Pty. Ltd., 19a Boundary Street, Rushcutters Bay, N.S.W. 2011, Australia

---

First edition 1964

Second edition 1974

Library of Congress Cataloging in Publication Data

Hancock, Norman Napoleon.
Matrix analysis of electrical machinery.
Bibliography: p.
1. Electric machinery. 2. Matrices. I. Title.
TK2211.H3 1974 621.31'04'201512943 74-3286
ISBN 0-08-017898-7
ISBN 0-08-017899-5 (flexicover)

Printed in Hungary

Contents

Preface to the Second Edition

After a further ten years the author remains convinced that it is educationally sound for the study of electrical machine analysis in terms of matrices to precede that in terms of tensors. In this order it is shown early that the method can be applied to practical problems and later the desirability of a formal statement of the principles upon which it is based may be revealed. As Dr. W. J. Gibbs pointed out, the situation is then delightfully reminiscent of that of M. Jourdain in Molière's *Le Bourgeois gentilhomme*, when he discovered that he had long been speaking prose without knowing that he was doing so.

This second edition, like the first, is intended for readers who have some knowledge of, or are concurrently studying, the physical nature of electrical machines. Whilst it is not assumed that the reader has a knowledge of "classical" machine theory, for the convenience of those who have, reference to it is sometimes made, in particular to the companion volume in the series: *Electrical Machines and their Applications* by J. Hindmarsh.

The changes from the first edition are of four kinds. Those in the first group are concerned with making more explicit some matters previously left to the reader and with greater precision of statement. Among these are the statements on units and the direction of torque and the complete inversion of a matrix by partitioning. The second group consists of the rectification of two former errors of judgement, namely, the omission of any reference to leakage in the synchronous machine in the interest of what proved to be oversimplification, and the scant reference to three-phase machines on the ground that the relation between three-phase and two-phase machines was so simple that it was sufficient to consider only the latter. The third group may be regarded as improved exposition, of which the principal example is the completely different and far more elegant treatment of the induction machine, omitting the dangerous use of half a transformation. The fourth group is comprised of additional material, in particular to illustrate interconnection and to provide an introduction to small oscillations. Also under this heading is the inclusion of hints and answers to the Exercises.

Some of the changes have resulted from the further experience of the author in teaching the topic to undergraduates at UMIST. Other changes have been prompted by the comments of correspondents, to whom the author is grateful. Once again the author is especially indebted to his colleague, Mr. J. Hindmarsh, whose pertinent questions revealed deficiencies by no means apparent to the author.

In conclusion the author would like to record his appreciation of the helpful advice given him by the General Editor of the series, Prof. P. Hammond, on several matters of policy.

Preface to the First Edition

THE systematic analysis of the performance of electrical machinery owes its origin to the inspiration of Gabriel Kron who used various approaches in his earlier writings on the subject according to whether the intended reader was a mathematician[1] or an engineer.[2] Even in the latter reference, however, Kron refers repeatedly to mathematical concepts which are outside the conscious experience of the average electrical engineer and student. It was pointed out by Karapetoff in America and Gibbs in England that much of the application of the technique to electrical machines could be derived without even mentioning the word "tensor". All that is required is a knowledge of the circuit equations, elementary matrix algebra, and the principle that the power of the system must be the same, irrespective of the terms in which it is expressed. This principle is not new to the electrical engineer, who uses it every time he "refers" secondary quantities to the primary in transformer or induction motor.

The author had the privilege of hearing Dr. W. J. Gibbs lecture on the subject in 1950 and was, in turn, inspired to study it. Consequently it has formed part of graduate courses at the Manchester College of Science and Technology for the last ten years and of undergraduate courses for a shorter time. It is largely on this experience that this book is based.

The author's view is that this is the best approach to electrical machine performance for both the non-specialist and the specialist, and that the latter will find in it a powerful tool when he is faced with more complicated performance problems. Machine design, which reduces to an attempt to produce a machine having the parameters which analysis shows lead to the required performance, is for the specialist only and is very dependent on the concept of magnetic flux which, however, figures little in the present work.

Space has not permitted the treatment of some topics, in particular small oscillations have not been considered and interconnection has been mentioned only briefly. Nevertheless, an attempt has been made to treat most of the machine performance normally covered in an undergraduate course and also to include one or two more advanced problems to indicate the power and the limitations of the method.

It would not be fitting to close this preface without expressing the author's debt to Dr. Gibbs, and also to his colleagues Mr. J. Hindmarsh and Mr. R. W. Whitehead, whose constructive criticism has frequently improved his exposition of the subject.

CHAPTER 1

Introduction

A CURSORY glance will show that, as a consequence of the inclusion of all the necessary material in the earlier chapters, specific consideration of machine performance starts with Chapter 9—about halfway through the book. Some readers will naturally be tempted to skip part of the first half in order to see what use is made of this matrix algebra. There is no serious objection to this. A return to the earlier part can be made later. It is suggested, however, that at least Chapters 1 and 2 and Chapter 5, pp. 53–55 and 65–76, be studied before leaping to Chapter 9. The more methodical student may prefer the less exciting procedure of following the development of the method from Chapter 2 on, omitting nothing, unless his lack of interest in transformers drives him to bypass Chapter 4.

Conventions

In all mathematical work conventions are of the utmost importance even though they have no intrinsic value. As far as possible consistent conventions have been maintained and it is appropriate to review them here.

(1) In the circuit diagrams of slip-ring machines, rotor windings are shown as coils, both when representing the actual windings and also when representing "stationary" transformed windings. For the latter, straight radial lines and brushes with or without radial lines have also been used by other writers. Commutator windings are indicated by brushes with or without radial lines. The latter are essential only when it is necessary to indicate the position of brushes not on the axes of symmetry.

For simplicity the rotor windings shown as coils are represented as occupying only half the diameter, and stator windings are shown on one side of the machine only. This serves to indicate the arbitrarily chosen positive directions of current and m.m.f., which are radially outwards in both cases. The actual machine windings, if balanced, occupy the whole periphery. Rotor phase windings are not necessarily connected together, although it may appear so from some diagrams.

(2) The positive direction of both current and m.m.f. is represented by the current arrows in the circuit diagrams. The terminal voltages are "rises" in the direction of the voltage arrows and all other voltages are "falls" in the direction of the current arrows.

(3) The positive direction of rotation of the rotor is taken as counter-clockwise, since this is the direction of measurement of positive angles. This sometimes appears to be in conflict with the convention of positive current and m.m.f., but in such cases a negative sign arises to show that the motoring torque is clockwise.

(4) Lower case v and i are used for instantaneous values of terminal voltage and current. To avoid a plethora of $\sqrt{2}$'s, $\hat{V}$ and $\hat{I}$ are used for the amplitudes of sinusoidal voltages and currents, leaving V and I for steady-state d.c. and r.m.s. a.c. values.

(5) Capital L and X are used for total (i.e. self-) inductance and the corresponding reactance respectively. The difference $(L-M)$ between the self-inductance and the mutual inductance is represented by the lower case letter l and, similarly, the difference $(X-X_M)$ between total reactance and mutual (magnetizing) reactance is represented by the lower case letter x. If all quantities are referred to the same base, l and x are respectively the leakage inductance and leakage reactance referred to that base.

The values of inductance used here may be regarded as actual values or referred values as preferred. Those of Kron and some other writers are referred to single-turn windings which, whilst simplifying the permeance–inductance relationships, causes complication when interconnection is involved.

(6) Matrices are distinguished from their elements by bold type. As a consequence it is not possible to use bold type for phasors which

may be distinguished from their magnitudes, where necessary, by the vertical parallel lines of the latter.

(7) Laplace transforms of voltage and current are indicated by a bar over the voltage and current symbols. The complex number of the Laplace transform is denoted by s, whereas the differential operator d/dt is denoted by p. This distinction is desirable because in some cases the differentiation is actually performed in the course of a linear transformation.

(8) The indices used to indicate stator winding circuits are capitals, those of rotor circuits are lower case.

(9) Since the difference between generators and motors is due solely to the different operating conditions, only "machines" are considered in general. Nevertheless, one has to choose between $V = E+IR$ and $V = E-IR$, and as there is no compromise between these, the former has been preferred on the ground that all motor windings and some generator windings are "sinks", and thus $V = E+IR$ is more appropriate for the majority of windings.

The E in this equation is thus the "back" e.m.f. of a motor. It appears in this book as a voltage determined by the triple product of a quantity having the dimensions of inductance, the angular velocity, and a current. When this product, with its appropriate sign, is positive, it represents a positive voltage drop in the same direction as the resistance drop IR, i.e. it is then a back e.m.f. There may be more than one such term in one voltage equation, in which case the total e.m.f. is their algebraic sum.

It can be seen that the difference between the motor and generator equations is essentially the opposite conventional directions, and hence opposite signs, of the armature currents. In the motor equation, used throughout this book, the generating condition will correspond to a negative value for the armature current in a d.c. machine and a negative value for the power component of the current in an a.c. machine.

Units

The units are not referred to in the text, but are assumed to be SI units throughout, namely: the metre, the kilogram, the second, and the common electrical units, ampere, volt, ohm, henry, and farad, together with the power unit the watt, the energy unit the joule, and the force unit the newton. Since these form a consistent set of units, no constants or conversion factors are required.

Angular velocity and angular frequency are, of course, in radians per second.

Parameters

A knowledge of the physical significance of a machine parameter may well suggest a method for its measurement. Equally, inspection of the circuit equations may well suggest a method of measurement which, conversely, reveals the physical significance of the parameter concerned. Again, one familiar with the standard tests[†] for the measurement of the parameters of "classical" machine theory, can analyse those tests by the circuit techniques described herein. This will show the relationship—usually identity—of the parameters used in the two different theoretical treatments. In practice it may not be possible to distinguish between these various approaches.

† See ref. 3.

CHAPTER 2

Elements of Matrix Algebra

Matrix Representation of Simultaneous Equations

If three variables y_1, y_2, y_3 are related to three other variables x_1, x_2, x_3 by three equations

$$y_1 = a_{11}x_1 + a_{12}x_2 + a_{13}x_3$$

$$y_2 = a_{21}x_1 + a_{22}x_2 + a_{23}x_3$$

$$y_3 = a_{31}x_1 + a_{32}x_2 + a_{33}x_3$$

where a_{11}, a_{12}, a_{13}, etc., are all constant, the three equations are said to form a set of linear equations with constant coefficients. There are three types of quantity involved: the variables y_1, y_2, y_3; the constant coefficients a_{11}, a_{12}, a_{13}, etc.; and the variables x_1, x_2, x_3. Moreover, the y's are formed of products of a's and x's. This suggests the possibility of writing the whole set of equations in a symbolic form as one equation $\mathbf{y} = \mathbf{A}\mathbf{x}$, where $\mathbf{y}$ represents all the y's, $\mathbf{A}$ represents all the a's and $\mathbf{x}$ represents all the x's. This equation $\mathbf{y} = \mathbf{A}\mathbf{x}$ may be written in full as

$$\begin{bmatrix} y_1 \\ y_2 \\ y_3 \end{bmatrix} = \begin{bmatrix} a_{11} & a_{12} & a_{13} \\ a_{21} & a_{22} & a_{23} \\ a_{31} & a_{32} & a_{33} \end{bmatrix} \begin{bmatrix} x_1 \\ x_2 \\ x_3 \end{bmatrix}$$

where

$$\mathbf{y} = \begin{bmatrix} y_1 \\ y_2 \\ y_3 \end{bmatrix} \quad \mathbf{A} = \begin{bmatrix} a_{11} & a_{12} & a_{13} \\ a_{21} & a_{22} & a_{23} \\ a_{31} & a_{32} & a_{33} \end{bmatrix} \quad \mathbf{x} = \begin{bmatrix} x_1 \\ x_2 \\ x_3 \end{bmatrix}$$

and where each array of elements in a square bracket is called a matrix and the equation $\mathbf{y} = \mathbf{Ax}$ is called a matrix equation. This equation must, of course, lead to the same values for the individual variables y_1, y_2, y_3 as do the original three equations with which we started, and the operation of "multiplying" $\mathbf{A}$ and $\mathbf{x}$ in their expanded form must be defined so that this is so.

Multiplication of Matrices

From the original equations

$$\mathbf{y} = \begin{bmatrix} y_1 \\ y_2 \\ y_3 \end{bmatrix} = \begin{bmatrix} a_{11}x_1 + a_{12}x_2 + a_{13}x_3 \\ a_{21}x_1 + a_{22}x_2 + a_{23}x_3 \\ a_{31}x_1 + a_{32}x_2 + a_{33}x_3 \end{bmatrix}$$

so that

$$\begin{bmatrix} a_{11} & a_{12} & a_{13} \\ a_{21} & a_{22} & a_{23} \\ a_{31} & a_{32} & a_{33} \end{bmatrix} \begin{bmatrix} x_1 \\ x_2 \\ x_3 \end{bmatrix} = \begin{bmatrix} a_{11}x_1 + a_{12}x_2 + a_{13}x_3 \\ a_{21}x_1 + a_{22}x_2 + a_{23}x_3 \\ a_{31}x_1 + a_{32}x_2 + a_{33}x_3 \end{bmatrix}$$

from which the rule for multiplication in such cases is seen to be:

The nth element of the column matrix $\mathbf{y}$ is equal to the sum of the product of the first element of the nth row of $\mathbf{A}$ and the first element of $\mathbf{x}$, the product of the second element of the nth row of $\mathbf{A}$ and the second element of $\mathbf{x}$, and so on to the product of the last element of the nth row of $\mathbf{A}$ and the last element of $\mathbf{x}$.

Thus, taking the second element y_2 of $\mathbf{y}$, this is given by the product of the first element of the second row of $\mathbf{A}$, namely a_{21} and the first element x_1 of $\mathbf{x}$, plus the product of the second element a_{22} of the second row of $\mathbf{A}$ and the second element x_2 of $\mathbf{x}$, plus the product of the third element a_{23} of the second row of $\mathbf{A}$ and the third element x_3 of $\mathbf{x}$, giving $y_2 = a_{21}x_1 + a_{22}x_2 + a_{23}x_3$ as required.

This is in fact a particular case of matrix multiplication in that the second matrix $\mathbf{x}$ of the product has only one column. The more general case is represented by

$$\mathbf{C} = \mathbf{AB}$$

where both **A** and **B** have more than one row and more than one column. Such a product can be considered as arising from two matrix equations

$$\mathbf{y} = \mathbf{Ax} \quad \text{and} \quad \mathbf{x} = \mathbf{Bz}$$

whence $\mathbf{y} = \mathbf{A}(\mathbf{x}) = \mathbf{A}(\mathbf{Bz}) = (\mathbf{AB})\mathbf{z}$. If it is desired to represent **y** more directly as **Cz**, it is necessary to define the process of multiplication in such a case. The rule can be derived by examining the matrix **C** obtained by the normal algebraic method of eliminating x_1, x_2, etc., from the two sets of equations represented in matrix form by $\mathbf{y} = \mathbf{Ax}$, $\mathbf{x} = \mathbf{Bz}$. It is found to be:

The *n*th element of the *m*th row of **C** is the sum of the product of the first element of the *m*th row of **A** and the first element of the *n*th column of **B**, the product of the second element of the *m*th row of **A** and the second element of the *n*th column of **B**, and so on.

Thus if

$$\mathbf{A} = \begin{bmatrix} a_{11} & a_{12} & a_{13} \\ a_{21} & a_{22} & a_{23} \\ a_{31} & a_{32} & a_{33} \end{bmatrix} \quad \mathbf{B} = \begin{bmatrix} b_{11} & b_{12} & b_{13} \\ b_{21} & b_{22} & b_{23} \\ b_{31} & b_{32} & b_{33} \end{bmatrix} \text{ and } \mathbf{C} = \begin{bmatrix} c_{11} & c_{12} & c_{13} \\ c_{21} & c_{22} & c_{23} \\ c_{31} & c_{32} & c_{33} \end{bmatrix}$$

then

$$c_{21} = a_{21}b_{11} + a_{22}b_{21} + a_{23}b_{31}$$

$$c_{22} = a_{21}b_{12} + a_{22}b_{22} + a_{23}b_{32}$$

and so on.

It is not necessary for matrices to be square in such products, but they must be "conformable", that is the number of elements in the rows of the first must be equal to the number of elements in the columns of the second. This is equivalent to saying that the number of columns of the first must be equal to the number of rows of the second. The product matrix has as many rows as the first and as many columns as the second matrix.

There may be more than two matrices in a product and in such cases matrix multiplication is associative, that is the order in which the operations of multiplication are performed does not matter:

$$\mathbf{ABC} = \mathbf{A}(\mathbf{BC}) = (\mathbf{AB})\mathbf{C}$$

Irrespective of the number of matrices in the product, however, multiplication is not commutative, that is, except in certain particular cases, the order of the matrix factors in the product may not be changed. It is easy to show, for example, that in general

$$\mathbf{AB} \neq \mathbf{BA}$$

by a simple case:

$$\begin{bmatrix} a_{11} & a_{12} \\ a_{21} & a_{22} \end{bmatrix} \begin{bmatrix} b_{11} & b_{12} \\ b_{21} & b_{22} \end{bmatrix} = \begin{bmatrix} a_{11}b_{11}+a_{12}b_{21} & a_{11}b_{12}+a_{12}b_{22} \\ a_{21}b_{11}+a_{22}b_{21} & a_{21}b_{12}+a_{22}b_{22} \end{bmatrix}$$

whereas

$$\begin{bmatrix} b_{11} & b_{12} \\ b_{21} & b_{22} \end{bmatrix} \begin{bmatrix} a_{11} & a_{12} \\ a_{21} & a_{22} \end{bmatrix} = \begin{bmatrix} a_{11}b_{11}+a_{21}b_{12} & a_{12}b_{11}+a_{22}b_{12} \\ a_{11}b_{21}+a_{21}b_{22} & a_{12}b_{21}+a_{22}b_{22} \end{bmatrix}$$

It is necessary to point out that multiplication of a matrix by a constant is one case where the order does not matter. The effect is to multiply *every element* of the matrix by the constant. It is important not to confuse this with the multiplication of a determinant by a constant.

Addition of matrices is comparatively straightforward. Each element of the sum is composed of the sum of the elements in the corresponding position in the matrices to be added. Similarly, each element of the difference of two matrices is the difference of their elements in the same position.

It is also necessary to state specifically what has already been assumed, namely that two matrices are equal only if each element of one is equal to the corresponding element of the other.

Application of Matrices to the Solution of Simultaneous Linear Equations—Inversion

If the matrix equation $\mathbf{y} = \mathbf{Ax}$ expresses a number of variables, y_1, y_2, y_3 in terms of other variables x_1, x_2, x_3, and we require to express x_1, x_2, x_3 in terms of y_1, y_2, y_3, it would appear that we merely have to divide both sides of the equation by $\mathbf{A}$ to obtain the required solution. No method of dividing matrices has been devised, however, and we are forced to use the expedient of multiplying both sides of the equation by "the inverse" of **A**. This may sound like hair-splitting, but the rub lies in the fact that we have first to find the inverse of **A**. This is a point of complete disillusionment, since basically the inverse can be found only with the aid of determinants in exactly the same way as would be used had matrices not been introduced.

The development of this procedure is as follows.

To reduce the expression **Ax** on the right of the equation to **x** it must be "pre-multiplied" by an inverse matrix $\mathbf{A}^{-1}$ such that

$$\mathbf{A}^{-1}(\mathbf{Ax}) = (\mathbf{A}^{-1}\mathbf{A})\mathbf{x} = \mathbf{Ux} = \mathbf{x}$$

where the product **U** of $\mathbf{A}^{-1}$ and **A** is a matrix which produces no change in a matrix multiplied by it. A little consideration shows that U, which is called the "unit" or "identity" matrix, is of the form

$$\begin{bmatrix} 1 & 0 & 0 & 0 & \cdots\cdots \\ 0 & 1 & 0 & 0 & \cdots\cdots \\ 0 & 0 & 1 & 0 & \cdots\cdots \\ 0 & 0 & 0 & 1 & \cdots\cdots \\ \cdots & \cdots & \cdots & \cdots & \cdots\cdots \\ \cdots & \cdots & \cdots & \cdots & \cdots\cdots \end{bmatrix}$$

with the appropriate number of rows and columns to make it conformable. It may be noted that here again is a particular case where the order of the terms of a product does not matter: $\mathbf{UA} = \mathbf{AU} = \mathbf{A}$.

Since any matrix, and therefore $[\mathbf{A}^{-1}]$, pre-multiplied by its own inverse gives the unit matrix **U**, $[\mathbf{A}^{-1}]^{-1}[\mathbf{A}^{-1}] = \mathbf{U}$. But the inverse of an inverse must be the original matrix, hence $[\mathbf{A}^{-1}]^{-1} = \mathbf{A}$, which leads to $\mathbf{AA}^{-1} = \mathbf{U}$. Since $\mathbf{A}^{-1}\mathbf{A}$ has already been defined as **U**, $\mathbf{AA}^{-1} = \mathbf{A}^{-1}\mathbf{A}$, here once again, the order does not matter.

However, in the derivation of **x** from **Ax**, the $\mathbf{A}^{-1}$ necessarily pre-multiplied **Ax** so that the $\mathbf{A}^{-1}$ and **A** might be multiplied together. It being essential to do exactly the same to the two sides of an equation, the $\mathbf{y}$ of the left-hand side must be pre-multiplied by $\mathbf{A}^{-1}$ to give the solution in the form

$$\mathbf{A}^{-1}\mathbf{y} = \mathbf{x} \quad \text{or} \quad \mathbf{x} = \mathbf{A}^{-1}\mathbf{y}$$

If the expanded equations represented by $\mathbf{y} = \mathbf{A}\mathbf{x}$ are solved by determinants and the results written in matrix arrays to correspond with the equation $\mathbf{x} = \mathbf{A}^{-1}\mathbf{y}$, it can be deduced that $\mathbf{A}^{-1}$ can be obtained by the following routine:

(1) Interchange the rows and columns of $\mathbf{A}$ to obtain the so-called "transpose" which is denoted by $\mathbf{A}_t$. Thus if

$$\mathbf{A} = \begin{bmatrix} a_{11} & a_{12} & a_{13} \\ a_{21} & a_{22} & a_{23} \\ a_{31} & a_{32} & a_{33} \end{bmatrix} \qquad \mathbf{A}_t = \begin{bmatrix} a_{11} & a_{21} & a_{31} \\ a_{12} & a_{22} & a_{32} \\ a_{13} & a_{23} & a_{33} \end{bmatrix}$$

(2) Replace each element of $\mathbf{A}_t$ by its own co-factor (i.e. minor with the appropriate sign) in the determinant formed by the array of $\mathbf{A}_t$. This gives

$$\begin{bmatrix} (a_{22}a_{33}-a_{32}a_{23}) & -(a_{12}a_{33}-a_{32}a_{13}) & (a_{12}a_{23}-a_{22}a_{13}) \\ -(a_{21}a_{33}-a_{31}a_{23}) & (a_{11}a_{33}-a_{31}a_{13}) & -(a_{11}a_{23}-a_{21}a_{13}) \\ (a_{21}a_{32}-a_{31}a_{22}) & -(a_{11}a_{32}-a_{31}a_{12}) & (a_{11}a_{22}-a_{21}a_{12}) \end{bmatrix}$$

(3) Divide this matrix, that is every element of it, by Δ, the determinant formed by the array of the whole matrix $\mathbf{A}$ or of its transpose $\mathbf{A}_t$. The inverse is thus obtained as

$$\mathbf{A}^{-1} = \frac{1}{\Delta}\begin{bmatrix} (a_{22}a_{33}-a_{32}a_{23}) & -(a_{12}a_{33}-a_{32}a_{13}) & (a_{12}a_{23}-a_{22}a_{13}) \\ -(a_{21}a_{33}-a_{31}a_{23}) & (a_{11}a_{33}-a_{31}a_{13}) & -(a_{11}a_{23}-a_{21}a_{13}) \\ (a_{21}a_{32}-a_{31}a_{22}) & -(a_{11}a_{32}-a_{31}a_{12}) & (a_{11}a_{22}-a_{21}a_{12}) \end{bmatrix}$$

$$= \begin{bmatrix} (a_{22}a_{33}-a_{32}a_{23})/\Delta & -(a_{12}a_{33}-a_{32}a_{13})/\Delta & (a_{12}a_{23}-a_{22}a_{13})/\Delta \\ -(a_{21}a_{33}-a_{31}a_{23})/\Delta & (a_{11}a_{33}-a_{31}a_{13})/\Delta & -(a_{11}a_{23}-a_{21}a_{13})/\Delta \\ (a_{21}a_{32}-a_{31}a_{22})/\Delta & -(a_{11}a_{32}-a_{31}a_{12})/\Delta & (a_{11}a_{22}-a_{21}a_{12})/\Delta \end{bmatrix}$$

Singular Matrices

It is apparent that if the determinant Δ is zero, there is a difficulty since some of the elements of the inverse are infinite and others indeterminate. In fact the inverse does not exist if $\Delta = 0$, and the implication is that either

(a) we are trying to solve for n variables with less than n independent equations, or

(b) we are trying to solve for n variables with more than n equations, in which case either

(i) some equations are redundant and can be discarded, leaving a non-zero determinant and enabling an inverse to be found, or

(ii) the equations are inconsistent so that it is not possible to decide which to discard, and no certain solution exists.

The fact that the inverse does not exist when $\Delta = 0$ does not represent a limitation of the matrix method, but merely indicates when a problem has been incorrectly formulated.

Matrices having zero-valued determinants are called "singular".

It is obviously a necessary, but not a sufficient, condition for a matrix to be non-singular that it be square, i.e. that it should have as many rows as it has columns.

The Transpose and Inverse of a Product

If $\mathbf{A} = \mathbf{PQ}$ and $\mathbf{B} = \mathbf{Q}_t\mathbf{P}_t$, comparison of a_{mn}, the nth element of the mth row of $\mathbf{A}$, with b_{nm}, the mth element of the nth row of $\mathbf{B}$ will show that the transpose of the product of two matrices is the product of the transposes of those two matrices *in the reverse order*:

$$a_{mn} = p_{m1}q_{1n} + p_{m2}q_{2n} + \dots .$$

If $(q_t)_{n1}$ is the first element of the nth row of $\mathbf{Q}_t$, it is equal to q_{1n}, the nth element of the first row of $\mathbf{Q}$, and so on. Hence

$$\begin{aligned} b_{nm} &= (q_t)_{n1}(p_t)_{1m} + (q_t)_{n2}(p_t)_{2m} + \dots \\ &= q_{1n}p_{m1} + q_{2n}p_{m2} + \dots \\ &= p_{m1}q_{1n} + p_{m2}q_{2n} + \dots \\ &= a_{mn} \end{aligned}$$

If $a_{mn} = b_{nm}$, then $\mathbf{B} = \mathbf{Q}_t\mathbf{P}_t$ is the transpose $\mathbf{A}_t$of $\mathbf{A} = \mathbf{PQ}$.

This can be extended to three and hence to any number of factors:

If $\mathbf{A} = \mathbf{PQR} = \mathbf{P(QR)}$, then $\mathbf{A}_t = (\mathbf{QR})_t\mathbf{P}_t = (\mathbf{R}_t\mathbf{Q}_t)\mathbf{P}_t = \mathbf{R}_t\mathbf{Q}_t\mathbf{P}_t$.

Similarly, the inverse of a product of any number of factors is the product of the inverses of the factors *in the reverse order*, i.e.

if $\mathbf{A} = \mathbf{PQR}$, then $\mathbf{A}^{-1} = \mathbf{R}^{-1}\mathbf{Q}^{-1}\mathbf{P}^{-1}$.

This may be proved by pre-multiplying both sides of the original equation $\mathbf{A} = \mathbf{PQR}$ by $\mathbf{A}^{-1}$ and post-multiplying both sides by $\mathbf{R}^{-1}\mathbf{Q}^{-1}\mathbf{P}^{-1}$:

$$
\begin{aligned}
\mathbf{A}^{-1}\mathbf{A}\mathbf{R}^{-1}\mathbf{Q}^{-1}\mathbf{P}^{-1} &= \mathbf{A}^{-1}\mathbf{PQR}\mathbf{R}^{-1}\mathbf{Q}^{-1}\mathbf{P}^{-1} \\
\mathbf{R}^{-1}\mathbf{Q}^{-1}\mathbf{P}^{-1} &= \mathbf{A}^{-1}\mathbf{PQ}\mathbf{Q}^{-1}\mathbf{P}^{-1} \\
&= \mathbf{A}^{-1}\mathbf{P}\mathbf{P}^{-1} \\
&= \mathbf{A}^{-1}
\end{aligned}
$$

Alternative Methods of Inversion

The labour of inverting a matrix increases rapidly as the number of rows and columns increases, so that five by five represents about the limit which can be tackled algebraically, or even numerically without the help of a digital computer. There are, however, a number of devices which help in the process in certain cases, and two of these will be considered here. The first may be referred to variously as "compound matrices", "sub-matrices", or "partitioning". The second is "reduction to diagonal form".

Compound Matrices

It is possible to have a matrix the elements of which are themselves matrices. Thus

$$
\mathbf{A} = \begin{bmatrix} a_{11} & a_{12} & a_{13} & a_{14} \\ a_{21} & a_{22} & a_{23} & a_{24} \\ a_{31} & a_{32} & a_{33} & a_{34} \\ a_{41} & a_{42} & a_{43} & a_{44} \end{bmatrix}
$$

can be represented as a "compound" matrix

$$
\begin{bmatrix} \mathbf{A}_{11} & \mathbf{A}_{12} \\ \mathbf{A}_{21} & \mathbf{A}_{22} \end{bmatrix}
$$

where $\mathbf{A}_{11}$, $\mathbf{A}_{12}$, $\mathbf{A}_{21}$, and $\mathbf{A}_{22}$ are "sub-matrices" which can be defined variously as

either $\mathbf{A}_{11} = [a_{11}]$ $\mathbf{A}_{12} = [a_{12} \quad a_{13} \quad a_{14}]$

$$\mathbf{A}_{21} = \begin{bmatrix} a_{21} \\ a_{31} \\ a_{41} \end{bmatrix} \qquad \mathbf{A}_{22} = \begin{bmatrix} a_{22} & a_{23} & a_{24} \\ a_{32} & a_{33} & a_{34} \\ a_{42} & a_{43} & a_{44} \end{bmatrix}$$

or

$$\mathbf{A}_{11} = \begin{bmatrix} a_{11} & a_{12} \\ a_{21} & a_{22} \end{bmatrix} \qquad \mathbf{A}_{12} = \begin{bmatrix} a_{13} & a_{14} \\ a_{23} & a_{24} \end{bmatrix}$$

$$\mathbf{A}_{21} = \begin{bmatrix} a_{31} & a_{32} \\ a_{41} & a_{42} \end{bmatrix} \qquad \mathbf{A}_{22} = \begin{bmatrix} a_{33} & a_{34} \\ a_{43} & a_{44} \end{bmatrix}$$

or

$$\mathbf{A}_{11} = \begin{bmatrix} a_{11} & a_{12} & a_{13} \\ a_{21} & a_{22} & a_{23} \\ a_{31} & a_{32} & a_{33} \end{bmatrix} \qquad \mathbf{A}_{12} = \begin{bmatrix} a_{14} \\ a_{24} \\ a_{34} \end{bmatrix}$$

$$\mathbf{A}_{21} = [a_{41} \quad a_{42} \quad a_{43}] \qquad \mathbf{A}_{22} = [a_{44}]$$

The choice between these possibilities is made according to the requirements of the particular problem, as will be seen later.

The variables of an equation may also be represented by compound matrices provided that the partitioning corresponds to that of the matrix of coefficients. If $\mathbf{y} = \mathbf{Ax}$, the three possibilities for $\mathbf{y}$ corresponding to those of $\mathbf{A}$ are therefore

either

$$\mathbf{y}_1 = [y_1] \qquad \mathbf{y}_2 = \begin{bmatrix} y_2 \\ y_3 \\ y_4 \end{bmatrix}$$

or

$$\mathbf{y}_1 = \begin{bmatrix} y_1 \\ y_2 \end{bmatrix} \qquad \mathbf{y}_2 = \begin{bmatrix} y_3 \\ y_4 \end{bmatrix}$$

or

$$\mathbf{y}_1 = \begin{bmatrix} y_1 \\ y_2 \\ y_3 \end{bmatrix} \qquad \mathbf{y}_2 = [y_4]$$

There are corresponding possibilities for the x's.

The equation $\mathbf{y} = \mathbf{A}\mathbf{x}$, which written in full is

$$\begin{bmatrix} y_1 \\ y_2 \\ y_3 \\ y_4 \end{bmatrix} = \begin{bmatrix} a_{11} & a_{12} & a_{13} & a_{14} \\ a_{21} & a_{22} & a_{23} & a_{24} \\ a_{31} & a_{32} & a_{33} & a_{34} \\ a_{41} & a_{42} & a_{43} & a_{44} \end{bmatrix} \begin{bmatrix} x_1 \\ x_2 \\ x_3 \\ x_4 \end{bmatrix}$$

may then be expressed in terms of the sub-matrices as

$$\begin{bmatrix} \mathbf{y}_1 \\ \mathbf{y}_2 \end{bmatrix} = \begin{bmatrix} \mathbf{A}_{11} & \mathbf{A}_{12} \\ \mathbf{A}_{21} & \mathbf{A}_{22} \end{bmatrix} \begin{bmatrix} \mathbf{x}_1 \\ \mathbf{x}_2 \end{bmatrix}$$

which form may be used to facilitate inversion.

If the solution of the equation $\mathbf{y} = \mathbf{A}\mathbf{x}$ is $\mathbf{x} = \mathbf{B}\mathbf{y}$, then $\mathbf{B} = \mathbf{A}^{-1}$ and $\mathbf{BA} = \mathbf{A}^{-1}\mathbf{A} = \mathbf{U}$. Expanded in terms of sub-matrices, this is

$$\begin{bmatrix} \mathbf{B}_{11} & \mathbf{B}_{12} \\ \mathbf{B}_{21} & \mathbf{B}_{22} \end{bmatrix} \begin{bmatrix} \mathbf{A}_{11} & \mathbf{A}_{12} \\ \mathbf{A}_{21} & \mathbf{A}_{22} \end{bmatrix} = \begin{bmatrix} \mathbf{U} & \mathbf{O} \\ \mathbf{O} & \mathbf{U} \end{bmatrix}$$

or

$$\begin{bmatrix} \mathbf{B}_{11}\mathbf{A}_{11} + \mathbf{B}_{12}\mathbf{A}_{21} & \mathbf{B}_{11}\mathbf{A}_{12} + \mathbf{B}_{12}\mathbf{A}_{22} \\ \mathbf{B}_{21}\mathbf{A}_{11} + \mathbf{B}_{22}\mathbf{A}_{21} & \mathbf{B}_{21}\mathbf{A}_{12} + \mathbf{B}_{22}\mathbf{A}_{22} \end{bmatrix} = \begin{bmatrix} \mathbf{U} & \mathbf{O} \\ \mathbf{O} & \mathbf{U} \end{bmatrix}$$

where $\mathbf{O}$ is a "null" matrix, i.e. one in which all the elements are zero. Thus $\begin{bmatrix} \mathbf{U} & \mathbf{O} \\ \mathbf{O} & \mathbf{U} \end{bmatrix}$ is itself a unit matrix partitioned to conform with the partitioning of $\mathbf{A}$ and $\mathbf{B}$.

From the first rows of the compound matrices in the above equation

$$\mathbf{B}_{11}\mathbf{A}_{11} + \mathbf{B}_{12}\mathbf{A}_{21} = \mathbf{U}$$

$$\mathbf{B}_{11}\mathbf{A}_{12} + \mathbf{B}_{12}\mathbf{A}_{22} = \mathbf{O}$$

Post-multiplying the second of these equations by $\mathbf{A}_{22}^{-1}$ gives

$$\mathbf{B}_{11}\mathbf{A}_{12}\mathbf{A}_{22}^{-1} + \mathbf{B}_{12}\mathbf{A}_{22}\mathbf{A}_{22}^{-1} = \mathbf{O}$$

from which $$\mathbf{B}_{12} = -\mathbf{B}_{11}\mathbf{A}_{12}\mathbf{A}_{22}^{-1}$$

Substituting this value of $\mathbf{B}_{12}$ in the first equation gives

$$\mathbf{B}_{11}\mathbf{A}_{11}-\mathbf{B}_{11}\mathbf{A}_{12}\mathbf{A}_{22}^{-1}\mathbf{A}_{21} = \mathbf{U}$$

from which
$$\mathbf{B}_{11}[\mathbf{A}_{11}-\mathbf{A}_{12}\mathbf{A}_{22}^{-1}\mathbf{A}_{21}] = \mathbf{U}$$

so that
$$\mathbf{B}_{11} = \mathbf{U}[\mathbf{A}_{11}-\mathbf{A}_{12}\mathbf{A}_{22}^{-1}\mathbf{A}_{21}]^{-1} = [\mathbf{A}_{11}-\mathbf{A}_{12}\mathbf{A}_{22}^{-1}\mathbf{A}_{21}]^{-1}$$

If the elements of $\mathbf{x}_2$ are of no interest and those of $\mathbf{y}_2$ are zero, we need not proceed further, since $\mathbf{x}_1$ is then equal to $\mathbf{B}_{11}\mathbf{y}_1$. (This situation arises when we do not require to know the currents in short-circuited windings, the terminal voltages of which are necessarily zero.)

Substituting this value of $\mathbf{B}_{11}$ in the expression for $\mathbf{B}_{12}$ leads to $\mathbf{B}_{12} = -[\mathbf{A}_{11}-\mathbf{A}_{12}\mathbf{A}_{22}^{-1}\mathbf{A}_{21}]^{-1}\mathbf{A}_{12}\mathbf{A}_{22}^{-1}$. This is sufficient to determine the elements of $\mathbf{x}_1$ even if the elements of $\mathbf{y}_2$ are not all zero, since $\mathbf{x}_1$ is equal to $\mathbf{B}_{11}\mathbf{y}_1+\mathbf{B}_{12}\mathbf{y}_2 = [\mathbf{A}_{11}-\mathbf{A}_{12}\mathbf{A}_{22}^{-1}\mathbf{A}_{21}]^{-1}[\mathbf{y}_1-\mathbf{A}_{12}\mathbf{A}_{22}^{-1}\mathbf{y}_2]$. (This arises when some of the currents we do not need to know flow in windings which are not short-circuited.)

If, however, a complete solution is required, we must consider a second pair of equations derived from the second rows of the compound matrices:

$$\mathbf{B}_{21}\mathbf{A}_{11}+\mathbf{B}_{22}\mathbf{A}_{21} = \mathbf{O}$$
$$\mathbf{B}_{21}\mathbf{A}_{12}+\mathbf{B}_{22}\mathbf{A}_{22} = \mathbf{U}$$

From the second of these, post-multiplying by $\mathbf{A}_{22}^{-1}$ gives

$$\mathbf{B}_{21}\mathbf{A}_{12}\mathbf{A}_{22}^{-1}+\mathbf{B}_{22} = \mathbf{A}_{22}^{-1}$$

whence
$$\mathbf{B}_{22} = \mathbf{A}_{22}^{-1}-\mathbf{B}_{21}\mathbf{A}_{12}\mathbf{A}_{22}^{-1}.$$

Substituting this value of $\mathbf{B}_{22}$ in the first equation of the second pair gives

$$\mathbf{B}_{21}\mathbf{A}_{11}+[\mathbf{A}_{22}^{-1}-\mathbf{B}_{21}\mathbf{A}_{12}\mathbf{A}_{22}^{-1}]\mathbf{A}_{21} = \mathbf{O}$$

from which
$$\mathbf{B}_{21}[\mathbf{A}_{11}-\mathbf{A}_{12}\mathbf{A}_{22}^{-1}\mathbf{A}_{21}] = -\mathbf{A}_{22}^{-1}\mathbf{A}_{21}$$

so that
$$\mathbf{B}_{21} = -\mathbf{A}_{22}^{-1}\mathbf{A}_{21}[\mathbf{A}_{11}-\mathbf{A}_{12}\mathbf{A}_{22}^{-1}\mathbf{A}_{21}]^{-1}$$

Hence
$$\mathbf{B}_{22} = \mathbf{A}_{22}^{-1}+\mathbf{A}_{22}^{-1}\mathbf{A}_{21}[\mathbf{A}_{11}-\mathbf{A}_{12}\mathbf{A}_{22}^{-1}\mathbf{A}_{21}]^{-1}\mathbf{A}_{12}\mathbf{A}_{22}^{-1}.$$

Since $\mathbf{B}_{11}$, $\mathbf{B}_{12}$, $\mathbf{B}_{21}$ and $\mathbf{B}_{22}$ are now all known, the solution is complete. It will be noted that only two inversions have been required, namely $\mathbf{A}_{22}^{-1}$ and $[\mathbf{A}_{11}-\mathbf{A}_{12}\mathbf{A}_{22}^{-1}\mathbf{A}_{21}]^{-1}$. The latter has the same number of rows and columns as $\mathbf{A}_{11}$. Thus the method requires the inversion of two matrices which together have as many rows and columns as the given matrix **A**, and in addition a considerable amount of multiplication, addition, and subtraction. Nevertheless, this process may involve much less work than the direct inversion of **A**. Thus, in the example, the routine inversion of a four-by-four matrix would require the calculation of 16 three-by-three determinants and 1 four-by-four determinant, whereas if done by partitioning into 4 two-by-two sub-matrices, the inversion of only 2 two-by-two matrices is required, and this is trivial.

In practice it is not necessary to write out the sub-matrices. It is sufficient to partition the ordinary matrices by dotted lines as in

$$\begin{bmatrix} y_1 \\ y_2 \\ \hdashline y_3 \\ y_4 \end{bmatrix} = \left[\begin{array}{cc:cc} a_{11} & a_{12} & a_{13} & a_{14} \\ a_{21} & a_{22} & a_{23} & a_{24} \\ \hdashline a_{31} & a_{32} & a_{33} & a_{34} \\ a_{41} & a_{42} & a_{43} & a_{44} \end{array}\right] \begin{bmatrix} x_1 \\ x_2 \\ \hdashline x_3 \\ x_4 \end{bmatrix}$$

Linear Transformation

Linear transformation means no more than a change of the variables of the equations to another set of variables related to the first set by a set of linear equations. We have thus already encountered this idea on p. 7 in the consideration of matrix multiplication.

Suppose that we have an equation $\mathbf{y} = \mathbf{Ax}$ and that we replace the variable $\mathbf{x}$ by $\mathbf{x}'$ where $\mathbf{x} = \mathbf{Bx}'$. We may also replace the variable $\mathbf{y}$ by $\mathbf{y}'$ where $\mathbf{y}' = \mathbf{Cy}$.† Then

$$\mathbf{y}' = \mathbf{Cy} = \mathbf{CAx} = \mathbf{CABx}' = \mathbf{A}'\mathbf{x}'$$

where $\mathbf{A}' = \mathbf{CAB}$.

† Note the difference between the ways in which $\mathbf{x}'$ and $\mathbf{y}'$ have been defined.

The "transformed" equation is thus $\mathbf{y}' = \mathbf{A}'\mathbf{x}'$, which is identical *in form* with the original equation $\mathbf{y} = \mathbf{A}\mathbf{x}$.

The purpose of making such a linear transformation is often to obtain a simpler solution for $\mathbf{x}$ in terms of $\mathbf{y}$, and this depends on making a suitable choice of transformation so that $\mathbf{A}'$ is much more easily inverted than $\mathbf{A}$. The solution is then obtained in the form

$$\mathbf{x} = \mathbf{B}\mathbf{x}' = \mathbf{B}[\mathbf{A}']^{-1}\mathbf{y}' = \mathbf{B}[\mathbf{A}']^{-1}\mathbf{C}\mathbf{y}$$

It is usual to use the same symbols for both the original and the transformed quantities and to indicate the latter by means of primes as above. Sometimes two or more transformations are performed in succession, in which case double or even triple primes may be used. For example, after replacing $\mathbf{x}$ by $\mathbf{x}'$ according to the equation $\mathbf{x} = \mathbf{B}\mathbf{x}'$, we might replace $\mathbf{x}'$ by $\mathbf{x}''$ according to an equation $\mathbf{x}' = \mathbf{D}\mathbf{x}''$, so that $\mathbf{x} = \mathbf{B}\mathbf{D}\mathbf{x}''$. The double transformation may be performed either as two separate operations, or as a single operation after having first calculated the product $\mathbf{BD}$.

Reduction to Diagonal Form

If a linear transformation applied to a matrix $\mathbf{A}$ produces a matrix $\mathbf{A}'$ of the form

$$\mathbf{A}' = \begin{bmatrix} a'_{11} & 0 & 0 & 0 & \cdots\cdots \\ 0 & a'_{22} & 0 & 0 & \cdots\cdots \\ 0 & 0 & a'_{33} & 0 & \cdots\cdots \\ 0 & 0 & 0 & a'_{44} & \cdots\cdots \\ \cdots & \cdots & \cdots & \cdots & \cdots\cdots \\ \cdots & \cdots & \cdots & \cdots & \cdots\cdots \end{bmatrix}$$

the transformation is said to reduce $\mathbf{A}$ to diagonal form, the characteristic of which is that all the elements are zero except for those on the diagonal from top left to bottom right, which is called the "principal diagonal".†

† Some of the elements on this diagonal could, of course, be zero, but then the determinant would be zero and the matrix would be singular and would have no inverse.

The advantage of this form is that the inverse is obtained by simply inverting each non-zero element:

$$[\mathbf{A}']^{-1} = \begin{bmatrix} 1/a'_{11} & 0 & 0 & 0 & \cdots\cdots \\ 0 & 1/a'_{22} & 0 & 0 & \cdots\cdots \\ 0 & 0 & 1/a'_{33} & 0 & \cdots\cdots \\ 0 & 0 & 0 & 1/a'_{44} & \cdots\cdots \\ \cdots & \cdots & \cdots & \cdots & \cdots\cdots \\ \cdots & \cdots & \cdots & \cdots & \cdots\cdots \end{bmatrix}$$

This can be checked by applying the routine inversion technique of p. 10.

There are a number of ways of finding suitable transformations for reducing a matrix to diagonal form, although only one such transformation is of practical importance for the present work.†

The Advantages of Matrices

The advantages of matrices for our present purpose are that in a set of linear differential equations, or rather in the Laplace transform equations, all the variables of one kind can be represented by a single symbol and so also all the parameters of each kind. The set of equations is thus represented by a single equation. The solution can be expressed in the same symbols and matrix algebra gives us a routine procedure for finding the solution. When this task becomes heavy, there are a number of devices which ease the burden and these also are most conveniently employed in matrix form. Of these, linear transformation and partitioning are the most important. Since matrices can be multiplied together according to simple rules, either manually or, in numerical cases, also by computer, they form an ideal tool for effecting linear transformations and also for indicating advantageous transformations.

† See the transformation from three-phase to symmetrical component quantities, pp. 28 and 106.

Types of Matrix

Certain forms of matrix are important because of the frequency with which they occur and/or because of their special properties. Some such have already been mentioned, but are included here for completeness.

(1) A "square" matrix has as many rows as it has columns. There is no necessity for a matrix to be square, as there is with a determinant, although its use may be restricted if it is not square. A matrix may be made square by adding zeros to the bottom or the right, but it is still "singular" and cannot be inverted.

(2) A "symmetric" matrix is a square matrix in which the elements are symmetrical about the "principal" diagonal, i.e. that from top left to bottom right. In such a matrix the element a_{ij} is equal to the element a_{ji}, where the first index shows the row, and the second the column, in which the element occurs.

(3) A "skew-symmetric" matrix is a square matrix in which the elements on the principal diagonal are all zero and the other elements are symmetrical about the principal diagonal except that they are of opposite sign, i.e. $a_{ij} = -a_{ji}$, and $a_{ii} = 0$.

(4) A "hermitian" matrix is a square matrix with only real elements on the principal diagonal and the other elements symmetrical except that their imaginary parts are of opposite sign, i.e. are complex conjugates. A hermitian matrix is thus the sum of a real symmetric matrix and an imaginary skew-symmetric matrix.

(5) A "diagonal" matrix is a square matrix in which all elements other than those on the principal diagonal are zero.

(6) A "scalar" matrix is a diagonal matrix in which all the elements on the principal diagonal are equal. In multiplication it has the effect of multiplying every element of the other matrix by the value of its non-zero elements. It is equal to the product of the "unit" matrix and a single quantity, or "scalar", of this value.

(7) A "unit" or "identity" matrix is a scalar matrix in which all the non-zero elements are unity. In multiplication it has the effect of multiplying every element of the other matrix by unity, thus leaving

it unchanged. It is represented by **U** or **I**. In the present work **U** is used to avoid confusion with the current symbol. It is the matrix equivalent to the unity of ordinary algebra.

(8) A "null" matrix is one in which every element is zero. It is represented by **O**. It is the matrix equivalent to the zero of ordinary algebra.

(9) A "singular" matrix is one for which the determinant formed by the same array is of zero value. A non-square matrix is necessarily singular; a square matrix may or may not be singular according to the values of its elements.

(10) An "orthogonal" matrix is one in which the transpose and inverse are identical, $\mathbf{A}_t = \mathbf{A}^{-1}$. It is so called because the linear transformation defined by an orthogonal matrix, when represented geometrically, leaves orthogonal axes orthogonal. The value of the determinant of an orthogonal matrix is necessarily ± 1, but the converse is not true. An orthogonal matrix is necessarily square.

The product of orthogonal matrices is orthogonal:

$$(\mathbf{AB})_t = \mathbf{B}_t\mathbf{A}_t = \mathbf{B}^{-1}\mathbf{A}^{-1} = (\mathbf{AB})^{-1}$$

(11) A "unitary" (not to be confused with "unit") matrix is one of which the inverse is equal to the complex conjugate of the transpose, the complex conjugate of a matrix being obtained by replacing every element by its conjugate. The complex conjugate is denoted by an asterisk, hence, for a unitary matrix, $\mathbf{A}^{-1} = \mathbf{A}_t^*$. The *magnitude* of the determinant of a unitary matrix is unity.

Differentiation and Integration of a Matrix

A matrix is differentiated with respect to a single variable, i.e. a scalar, by differentiating every element of the matrix with respect to that variable. Similarly, a matrix is integrated with respect to a scalar by integrating every element with respect to that scalar.

CHAPTER 3

Application of Matrix Algebra to Static Electrical Networks

Laplace Transform Equations

Consider the three circuits shown in Fig. 1 in which there are mutual couplings between all three coils.

The voltage equations of these circuits may be written

$$v_1 = R_1 i_1 + L_1 \frac{di_1}{dt} + \frac{1}{C_1}\int_0^t i_1 . dt + M_{12}\frac{di_2}{dt} + M_{13}\frac{di_3}{dt}$$

$$v_2 = M_{21}\frac{di_1}{dt} + R_2 i_2 + L_2\frac{di_2}{dt} + \frac{1}{C_2}\int_0^t i_2 . dt + M_{23}\frac{di_3}{dt}$$

$$v_3 = M_{31}\frac{di_1}{dt} + M_{32}\frac{di_2}{dt} + R_3 i_3 + L_3\frac{di_3}{dt} + \frac{1}{C_3}\int_0^t i_3 . dt$$

The mutual inductances of this system are all positive† with the conventional directions shown in Fig. 1.

The first of these equations can be written in operational form as

$$v_1 = (R_1 + L_1 p + 1/C_1 p) i_1 + M_{12} p i_2 + M_{13} p i_3$$

or in terms of the Laplace transforms for zero initial conditions as

$$\bar{v}_1 = (R_1 + L_1 s + 1/C_1 s)\bar{i}_1 + M_{12} s \bar{i}_2 + M_{13} s \bar{i}_3$$

The other two equations can be similarly expressed.

† See footnote on p. 37.

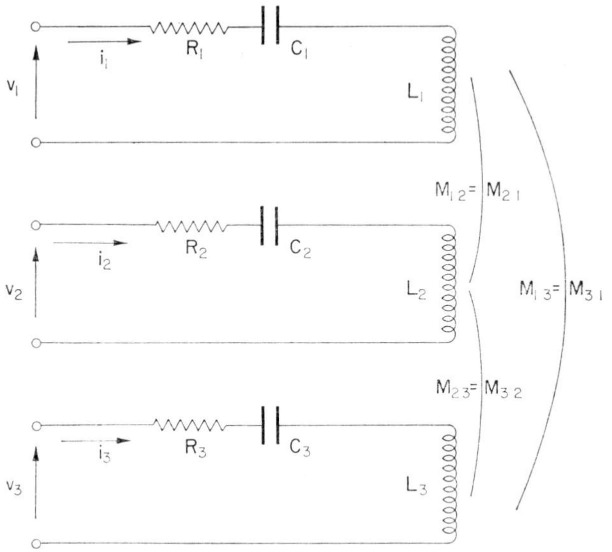

FIG. 1. Coupled circuits.

The operational equations can be abbreviated to

$$v_1 = Z_1 i_1 + X_{12} i_2 + X_{13} i_3$$
$$v_2 = X_{21} i_1 + Z_2 i_2 + X_{23} i_3$$
$$v_3 = X_{31} i_1 + X_{32} i_2 + Z_3 i_3$$

where $Z = R+Lp+1/Cp$, and $X = Mp$ are called "transient impedances".†

The corresponding Laplace transform equations

$$\bar{v}_1 = Z_1 \bar{i}_1 + X_{12} \bar{i}_2 + X_{13} \bar{i}_3$$
$$\bar{v}_2 = X_{21} \bar{i}_1 + Z_2 \bar{i}_2 + X_{23} \bar{i}_3$$
$$\bar{v}_3 = X_{31} \bar{i}_1 + X_{32} \bar{i}_2 + Z_3 \bar{i}_3$$

where $Z = R+Ls+1/Cs$, and $X = Ms$, form a set of linear algebraic equations with constant coefficients. To determine the currents, given

† This general use of the term "transient impedance" must be distinguished from the specific use in synchronous machine theory. See "direct-axis transient reactance", p. 235.

the terminal voltages v, it is necessary to solve these equations for the current transforms $\bar{i}$ and then to find the inverse transforms i.

The inversion of the "impendance" matrix

$$\begin{bmatrix} Z_1 & X_{12} & X_{13} \\ X_{21} & Z_2 & X_{23} \\ X_{31} & X_{32} & Z_3 \end{bmatrix}$$

is therefore one way of performing the principal labour of finding the solution.

Notation

Before we proceed three minor changes of notation will be made. Firstly Z_1, Z_2, Z_3 will be written Z_{11}, Z_{22}, Z_{33} by analogy with the X_{12}, X_{13}, etc. We can then say that in all the impedance terms the first subscript indicates the circuit in which the voltage is experienced and the second subscript denotes the circuit in which the current flows to produce this voltage. For complete generality the X's could be written as Z's, being impedances in which the resistive or "in-phase" component is zero. It is usually more convenient, however, to retain the distinction by using both X's and Z's.

The second change of notation is to write the indices of the currents as supercripts rather than as subscripts, that is i^1, i^2, i^3 in place of i_1, i_2, i_3. Confusion with i squared or i cubed will be avoided by writing these as $(i)^2$ and $(i)^3$. Since the indices of the impedance terms are still written subscript, there will be no need to use brackets for the inverse impedance elements such as Z_{11}^{-1}. There are two reasons for this change. Firstly, it permits a simple check on the accuracy of the terms. When the equations are written

$$\begin{aligned} v_1 &= Z_{11}i^1 + X_{12}i^2 + X_{13}i^3 \\ v_2 &= X_{21}i^1 + Z_{22}i^2 + X_{23}i^3 \\ v_3 &= X_{31}i^1 + X_{32}i^2 + Z_{33}i^3 \end{aligned}$$

it can be seen that the "net" index of every term in any one equation is identical, if it is assumed that the same index appearing superscript and subscript in a product "cancels". Thus in the first equation the voltage has a subscript 1; the first term on the right-hand side is $Z_{11}i^1$

which has a net index subscript 1; so also have the second term $X_{12}i^2$ and the third term $X_{13}i^3$.

The second reason for using this superscript index† for current is as an introduction to such usage in tensor notation.

A further change will be to write expanded matrices in Kron's egg-box form, instead of using the square brackets of the normal mathematicians' form. The reason for this is that it is essential to be able to identify an element as relating to a particular part of the system, without relying on its indices, as, for example, in a numerical case. The identifying index is written outside the egg-box for both row (the first index) and column (the second index), so that the voltage equation $\mathbf{v}=\mathbf{Zi}$ will henceforth be written in full as

$$
\begin{array}{c|c|}
 & \\
\hline
1 & v_1 \\
\hline
2 & v_2 \\
\hline
3 & v_3 \\
\hline
\end{array}
\; = \;
\begin{array}{c|c|c|c|}
 & 1 & 2 & 3 \\
\hline
1 & Z_{11} & X_{12} & X_{13} \\
\hline
2 & X_{21} & Z_{22} & X_{23} \\
\hline
3 & X_{31} & X_{32} & Z_{33} \\
\hline
\end{array}
\;
\begin{array}{c|c|}
 & \\
\hline
1 & i^1 \\
\hline
2 & i^2 \\
\hline
3 & i^3 \\
\hline
\end{array}
$$

When an element is zero it is customary to leave that space of the egg-box empty. This causes no difficulty in completed work but it is advisable to write in the zeros during calculation, since this confirms that that element has been checked and found to be zero, whereas if the space were left blank it may well be that a non-zero element had been inadvertently omitted and re-checking may be necessary.

Linear Transformation in Electrical Circuit Analysis

Linear transformations are common in electric circuit analysis, among them being:

the change from phase currents and voltages to line currents and voltages in polyphase circuits,

† The use of superscript indices in matrix analysis is not usual, but the advantages make it desirable. Since the indices in the present work relate to specific circuits, they are "closed" indices in tensor notation, but since "open" indices are not used here, the usual brackets can be omitted.

the change from phase quantities to symmetrical components,
the change from three-phase quantities to equivalent two-phase quantities,
the change from branch currents to mesh currents.

It will be noted that the first of these differs from the others in that phase and line currents and voltages all exist physically in the circuit whereas mesh currents, symmetrical components, etc., are merely mathematical devices to ease computation and have no separate physical existence. There are, of course, many such linear transformations used in circuitry which are appropriate only to particular cases and not widely applicable, as are most of those quoted above. The change from branch currents to mesh currents has, of course, to be performed according to the connections of the particular circuit. The

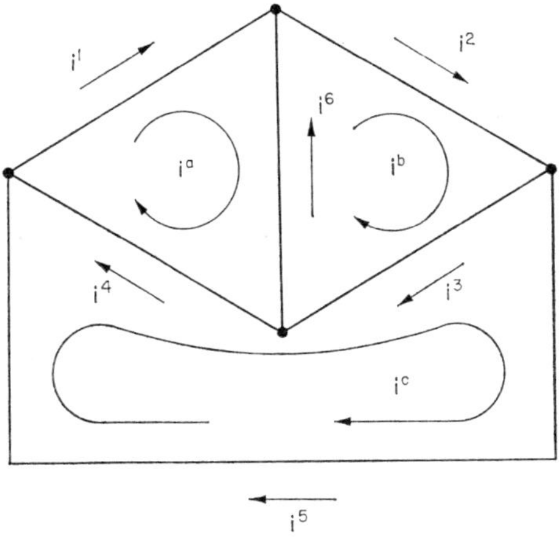

FIG. 2. Bridge network.

matrix required for this transformation is obtained from the equation for the branch or "old" currents in terms of the mesh or "new" currents. This particular type of transformation matrix, being associated with the connection of the branches, is called a "connection" matrix and is usually represented by **C**.

As an example consider the bridge network shown in Fig. 2.

With the conventional directions shown

$$\begin{aligned}
i^1 &= i^a \\
i^2 &= i^b \\
i^3 &= i^b - i^c \\
i^4 &= i^a - i^c \\
i^5 &= i^c \\
i^6 &= -i^a + i^b
\end{aligned}$$

In matrix form this is

$$\begin{array}{c|c|} 1 & i^1 \\ 2 & i^2 \\ 3 & i^3 \\ 4 & i^4 \\ 5 & i^5 \\ 6 & i^6 \end{array} = \begin{array}{c|c|c|c|} & a & b & c \\ 1 & 1 & 0 & 0 \\ 2 & 0 & 1 & 0 \\ 3 & 0 & 1 & -1 \\ 4 & 1 & 0 & -1 \\ 5 & 0 & 0 & 1 \\ 6 & -1 & 1 & 0 \end{array} \begin{array}{c|c|} a & i^a \\ b & i^b \\ c & i^c \end{array}$$

or

$$\mathbf{i} = \mathbf{C}\mathbf{i}'$$

where

$$\mathbf{C} = \begin{array}{c|c|c|c|} & a & b & c \\ 1 & 1 & 0 & 0 \\ 2 & 0 & 1 & 0 \\ 3 & 0 & 1 & -1 \\ 4 & 1 & 0 & -1 \\ 5 & 0 & 0 & 1 \\ 6 & -1 & 1 & 0 \end{array}$$

and where the current matrix **i** represents the old (branch) currents and **i**′ the new (mesh) currents.

The connection matrix is thus formed from the array of the coefficients of the new currents in the equations for the old currents expressed in terms of the new currents.

It will be noted that this matrix is not square and is therefore singular. The implication is that since this matrix cannot be inverted, it is not possible to reverse the process and to find the mesh currents from the branch currents. This is patently false. There are several ways of regarding this inconsistency, but for the present purpose we can regard the branch currents as not being independent† and hence they are not all required when the mesh currents are expressed in terms of the branch currents. Inspection of the circuit confirms this and also enables the difficulty of inverting **C** to be avoided.

As a second example we will consider the transformation from three-phase to symmetrical component quantities.‡

Suppose that we have an impedance matrix

	a	b	c
a	Z_{aa}	X_{ab}	X_{ac}
b	X_{ba}	Z_{bb}	X_{bc}
c	X_{ca}	X_{cb}	Z_{cc}

in which $Z_{aa} = Z_{bb} = Z_{cc} = Z$

$$X_{ab} = X_{bc} = X_{ca} = X_f$$

$$X_{ac} = X_{cb} = X_{ba} = X_r \quad \text{say, but that } X_f \neq X_r\text{§}$$

as can happen in a rotating machine, but not, of course, in a static network, where X_{ab} and X_{ba} are necessarily equal.

† An alternative interpretation, given in ref. 4, is beyond the scope of this book.

‡ See p. 106.

§ Such a matrix is called a circulant.

The voltage equation can therefore be written

$$\begin{array}{c|c|} & \\ \hline \text{a} & v_\text{a} \\ \hline \text{b} & v_\text{b} \\ \hline \text{c} & v_\text{c} \\ \hline \end{array} = \begin{array}{c|c|c|c|} & \text{a} & \text{b} & \text{c} \\ \hline \text{a} & Z & X_\text{f} & X_\text{r} \\ \hline \text{b} & X_\text{r} & Z & X_\text{f} \\ \hline \text{c} & X_\text{f} & X_\text{r} & Z \\ \hline \end{array} \begin{array}{c|c|} & \\ \hline \text{a} & i^\text{a} \\ \hline \text{b} & i^\text{b} \\ \hline \text{c} & i^\text{c} \\ \hline \end{array}$$

where the v's are the terminal phase voltages and the i's are the phase currents.

If the linear transformation defined by†

$$\mathbf{C} = \begin{array}{c|c|c|c|} & \text{p} & \text{n} & \text{o} \\ \hline \text{a} & 1 & 1 & 1 \\ \hline \text{b} & h^2 & h & 1 \\ \hline \text{c} & h & h^2 & 1 \\ \hline \end{array}$$

where h is the 120° operator‡ represented by the complex number $(-\frac{1}{2}+j\sqrt{3}/2)$, is applied to the current, and the same transformation

† The standard symbols for positive and negative sequence are 1 and 2 respectively. These can be confusing in some circumstances in the present work and p and n are used here instead. Note also that the use of the symbol **C** has been extended from representing a connection matrix to representing any transformation matrix.

‡ Since h is the 120° operator, its application twice in succession should produce rotation through 240°:

$$\begin{aligned} h\times h &= (-\tfrac{1}{2}+j\sqrt{3}/2)(-\tfrac{1}{2}+j\sqrt{\ })3/2 \\ &= (-\tfrac{1}{2}-j\sqrt{3}/2) \end{aligned}$$

which is identifiable as the 240° operator. It is logical to express $h\times h$ as h^2.

Similarly,

$$h^3 = h\times h\times h = h\times h^2 = (-\tfrac{1}{2}+j\sqrt{3}/2)(-\tfrac{1}{2}-j\sqrt{3}/2) = 1,$$

is a 360° operator, or identity operator. Furthermore,

$$h^4 = h\times h^3 = h\times 1 = h, \qquad h^{-1} = h^{-1}\times h^3 = h^2, \qquad \text{etc.}$$

h^2, h, 1 are recognizable as the three cube roots of unity.

Lastly, it is worth noting that $h^2+h+1 = 0$.

is also applied to the voltage, we get

$$\begin{array}{c|c|} & \\ \text{a} & v_{\text{a}} \\ \text{b} & v_{\text{b}} \\ \text{c} & v_{\text{c}} \end{array} = \begin{array}{c|c|c|c|} & \text{p} & \text{n} & \text{o} \\ \text{a} & 1 & 1 & 1 \\ \text{b} & h^2 & h & 1 \\ \text{c} & h & h^2 & 1 \end{array} \begin{array}{c|c|} & \\ \text{p} & v_{\text{p}} \\ \text{n} & v_{\text{n}} \\ \text{o} & v_{\text{o}} \end{array}$$

and

$$\begin{array}{c|c|} & \\ \text{a} & i^{\text{a}} \\ \text{b} & i^{\text{b}} \\ \text{c} & i^{\text{c}} \end{array} = \begin{array}{c|c|c|c|} & \text{p} & \text{n} & \text{o} \\ \text{a} & 1 & 1 & 1 \\ \text{b} & h^2 & h & 1 \\ \text{c} & h & h^2 & 1 \end{array} \begin{array}{c|c|} & \\ \text{p} & i^{\text{p}} \\ \text{n} & i^{\text{n}} \\ \text{o} & i^{\text{o}} \end{array}$$

Substituting these in the voltage equation gives

$$\begin{array}{c|c|c|c|} & \text{p} & \text{n} & \text{o} \\ \text{a} & 1 & 1 & 1 \\ \text{b} & h^2 & h & 1 \\ \text{c} & h & h^2 & 1 \end{array} \begin{array}{c|c|} & \\ \text{p} & v_{\text{p}} \\ \text{n} & v_{\text{n}} \\ \text{o} & v_{\text{o}} \end{array}$$

$$= \begin{array}{c|c|c|c|} & \text{a} & \text{b} & \text{c} \\ \text{a} & Z & X_{\text{f}} & X_{\text{r}} \\ \text{b} & X_{\text{r}} & Z & X_{\text{f}} \\ \text{c} & X_{\text{f}} & X_{\text{r}} & Z \end{array} \begin{array}{c|c|c|c|} & \text{p} & \text{n} & \text{o} \\ \text{a} & 1 & 1 & 1 \\ \text{b} & h^2 & h & 1 \\ \text{c} & h & h^2 & 1 \end{array} \begin{array}{c|c|} & \\ \text{p} & i^{\text{p}} \\ \text{n} & i^{\text{n}} \\ \text{o} & i^{\text{o}} \end{array}$$

$$= \begin{array}{c|c|c|c|} & \text{p} & \text{n} & \text{o} \\ \text{a} & (Z+h^2X_{\text{f}}+hX_{\text{r}}) & (Z+hX_{\text{f}}+h^2X_{\text{r}}) & (Z+X_{\text{f}}+X_{\text{r}}) \\ \text{b} & (h^2Z+hX_{\text{f}}+X_{\text{r}}) & (hZ+h^2X_{\text{f}}+X_{\text{r}}) & (Z+X_{\text{f}}+X_{\text{r}}) \\ \text{c} & (hZ+X_{\text{f}}+h^2X_{\text{r}}) & (h^2Z+X_{\text{f}}+hX_{\text{r}}) & (Z+X_{\text{f}}+X_{\text{r}}) \end{array} \begin{array}{c|c|} & \\ \text{p} & i^{\text{p}} \\ \text{n} & i^{\text{n}} \\ \text{o} & i^{\text{o}} \end{array}$$

To proceed we must pre-multiply both sides of this equation by the inverse $\mathbf{C}^{-1}$ of **C**, which entails first finding the inverse. Transposing **C**, gives

$\mathbf{C}_t =$

	a	b	c
p	1	h^2	h
n	1	h	h^2
o	1	1	1

The determinant of the matrix is

$$(h-h^2)-(h^2-h)+(h^4-h^2)$$
$$= 3(h-h^2)$$
$$= 3h(1-h)$$

The inverse of **C** is therefore

$\mathbf{C}^{-1} = \dfrac{1}{3h(1-h)}$

	a	b	c
p	$(h-h^2)$	(h^2-1)	$(1-h)$
n	$(h-h^2)$	$(1-h)$	(h^2-1)
o	$(h-h^2)$	$(h-h^2)$	$(h-h^2)$

$= \dfrac{1}{3}$

	a	b	c
p	1	$-(h+1)/h$	$1/h$
n	1	$1/h$	$-(h+1)/h$
o	1	1	1

$= \dfrac{1}{3}$

	a	b	c
p	1	h	h^2
n	1	h^2	h
o	1	1	1

This may be recognized as the matrix of the coefficients in the equations for the symmetrical components in terms of the phase quantities. It must, of course, be so, since, being the inverse of **C**, it represents the solution of the original equations which expressed the phase quantities in terms of the components.

Pre-multiplying both sides of the equation on p. 29 by this inverse gives

$$\begin{array}{c|c|} \mathrm{p} & v_{\mathrm{p}} \\ \mathrm{n} & v_{\mathrm{n}} \\ \mathrm{o} & v_{\mathrm{o}} \end{array} = \frac{1}{3}\; \begin{array}{c|c|c|c|} & \mathrm{a} & \mathrm{b} & \mathrm{c} \\ \hline \mathrm{p} & 1 & h & h^2 \\ \mathrm{n} & 1 & h^2 & h \\ \mathrm{o} & 1 & 1 & 1 \\ \hline \end{array} \times$$

$$\times \begin{array}{c|c|c|c|} & \mathrm{p} & \mathrm{n} & \mathrm{o} \\ \hline \mathrm{a} & (Z+h^2X_{\mathrm{f}}+hX_{\mathrm{r}}) & (Z+hX_{\mathrm{f}}+h^2X_{\mathrm{r}}) & (Z+X_{\mathrm{f}}+X_{\mathrm{r}}) \\ \mathrm{b} & (h^2Z+hX_{\mathrm{f}}+X_{\mathrm{r}}) & (hZ+h^2X_{\mathrm{f}}+X_{\mathrm{r}}) & (Z+X_{\mathrm{f}}+X_{\mathrm{r}}) \\ \mathrm{c} & (hZ+X_{\mathrm{f}}+h^2X_{\mathrm{r}}) & (h^2Z+X_{\mathrm{f}}+hX_{\mathrm{r}}) & (Z+X_{\mathrm{f}}+X_{\mathrm{r}}) \\ \hline \end{array} \; \begin{array}{c|c|} \mathrm{p} & i^{\mathrm{p}} \\ \mathrm{n} & i^{\mathrm{n}} \\ \mathrm{o} & i^{\mathrm{o}} \end{array}$$

$$= \frac{1}{3}\; \begin{array}{c|c|c|c|} & \mathrm{p} & \mathrm{n} & \mathrm{o} \\ \hline \mathrm{p} & 3(Z+h^2X_{\mathrm{f}}+hX_{\mathrm{r}}) & 0 & 0 \\ \mathrm{n} & 0 & 3(Z+hX_{\mathrm{f}}+h^2X_{\mathrm{r}}) & 0 \\ \mathrm{o} & 0 & 0 & 3(Z+X_{\mathrm{f}}+X_{\mathrm{r}}) \\ \hline \end{array} \; \begin{array}{c|c|} \mathrm{p} & i^{\mathrm{p}} \\ \mathrm{n} & i^{\mathrm{n}} \\ \mathrm{o} & i^{\mathrm{o}} \end{array}$$

or

$$\begin{array}{c|c|} \mathrm{p} & v_{\mathrm{p}} \\ \mathrm{n} & v_{\mathrm{n}} \\ \mathrm{o} & v_{\mathrm{o}} \end{array} = \begin{array}{c|c|c|c|} & \mathrm{p} & \mathrm{n} & \mathrm{o} \\ \hline \mathrm{p} & Z_{\mathrm{pp}} & 0 & 0 \\ \mathrm{n} & 0 & Z_{\mathrm{nn}} & 0 \\ \mathrm{o} & 0 & 0 & Z_{\mathrm{oo}} \\ \hline \end{array} \; \begin{array}{c|c|} \mathrm{p} & i^{\mathrm{p}} \\ \mathrm{n} & i^{\mathrm{n}} \\ \mathrm{o} & i^{\mathrm{o}} \end{array}$$

where $Z_{\mathrm{pp}} = (Z+h^2X_{\mathrm{f}}+hX_{\mathrm{r}})$ is called the impedance to positive-sequence current, $Z_{\mathrm{nn}} = (Z+hX_{\mathrm{f}}+h^2X_{\mathrm{r}})$ is called the impedance to negative-sequence current, and $Z_{oo} = (Z+X_f+X_r)$ is called the impedance to zero-sequence current.

This example of a linear transformation shows how a new set of linear equations in terms of new variables is obtained in a routine manner from the old set. It is possible in some cases to solve the new equations much more easily than the old and, after doing so, to transform the solution back to the old variables. This is the case in the present example. Since the new impedance matrix is diagonal, its inverse can be written down and the solution in terms of the new variables is

$$\begin{array}{c|c|} & \\ \text{p} & i^{\text{p}} \\ \text{n} & i^{\text{n}} \\ \text{o} & i^{\text{o}} \end{array} = \begin{array}{c|c|c|c|} & \text{p} & \text{n} & \text{o} \\ \text{p} & 1/Z_{\text{pp}} & 0 & 0 \\ \text{n} & 0 & 1/Z_{\text{nn}} & 0 \\ \text{o} & 0 & 0 & 1/Z_{\text{oo}} \end{array} \begin{array}{c|c|} & \\ \text{p} & v_{\text{p}} \\ \text{n} & v_{\text{n}} \\ \text{o} & v_{\text{o}} \end{array}$$

For the old currents in terms of the new voltages it is

$$\begin{array}{c|c|} & \\ \text{a} & i^{\text{a}} \\ \text{b} & i^{\text{b}} \\ \text{c} & i^{\text{c}} \end{array} = \begin{array}{c|c|c|c|} & \text{p} & \text{n} & \text{o} \\ \text{a} & 1 & 1 & 1 \\ \text{b} & h^2 & h & 1 \\ \text{c} & h & h^2 & 1 \end{array} \begin{array}{c|c|} & \\ \text{p} & i^{\text{p}} \\ \text{n} & i^{\text{n}} \\ \text{o} & i^{\text{o}} \end{array}$$

$$= \begin{array}{c|c|c|c|} & \text{p} & \text{n} & \text{o} \\ \text{a} & 1 & 1 & 1 \\ \text{b} & h^2 & h & 1 \\ \text{c} & h & h^2 & 1 \end{array} \begin{array}{c|c|c|c|} & \text{p} & \text{n} & \text{o} \\ \text{p} & 1/Z_{\text{pp}} & 0 & 0 \\ \text{n} & 0 & 1/Z_{\text{nn}} & 0 \\ \text{o} & 0 & 0 & 1/Z_{\text{oo}} \end{array} \begin{array}{c|c|} & \\ \text{p} & v_{\text{p}} \\ \text{n} & v_{\text{n}} \\ \text{o} & v_{\text{o}} \end{array}$$

The new voltages of this equation are defined in terms of the old voltages by the equation $\mathbf{v}' = \mathbf{C}^{-1}\mathbf{v}$, obtained by pre-multiplying

$\mathbf{v} = \mathbf{C}\mathbf{v}'$ by $\mathbf{C}^{-1}$,

$$\begin{array}{c|c|} & \\ \mathrm{p} & v_{\mathrm{p}} \\ \mathrm{n} & v_{\mathrm{n}} \\ \mathrm{o} & v_{\mathrm{o}} \end{array} = \frac{1}{3} \begin{array}{c|c|c|c|} & \mathrm{a} & \mathrm{b} & \mathrm{c} \\ \mathrm{p} & 1 & h & h^2 \\ \mathrm{n} & 1 & h^2 & h \\ \mathrm{o} & 1 & 1 & 1 \end{array} \begin{array}{c|c|} & \\ \mathrm{a} & v_{\mathrm{a}} \\ \mathrm{b} & v_{\mathrm{b}} \\ \mathrm{c} & v_{\mathrm{c}} \end{array}$$

Choice of Transformations—Invariance of Power

As has already been shown, the choice of transformations is made so as to yield an easier solution to the problem. There are, however, two transformations to consider: that of current and that of voltage. From considerations of matrix algebra alone, there is no reason why these two should not be selected entirely arbitrarily, i.e. the voltage transformation may be selected without any reference to that for the current. This was in fact done in the example of symmetrical components on p. 29. The same transformation was applied to both current and voltage without any consideration of the consequences. The three currents i^{a}, i^{b}, i^{c} were replaced by three others i^{p}, i^{n}, i^{o} which together make up only the current i^{a}, since $i^{\mathrm{p}}+i^{\mathrm{n}}+i^{\mathrm{o}} = i^{\mathrm{a}}$. Similarly, the three voltages v_{p}, v_{n}, v_{o} together make up only v_{a}. As a result the power in terms of the new variables, namely $v_{\mathrm{p}}i^{\mathrm{p}}+v_{\mathrm{n}}i^{\mathrm{n}}+v_{\mathrm{o}}i^{\mathrm{o}}$, represents only the power in one phase and not the total power in the original system. In many ways this is undesirable and where electrodynamic phenomena are involved it is almost invariably safer to ensure that the power is the same in terms of both new and old variables, taking the system as a whole in both cases. This principle is known in tensor analysis as the "invariance of power". It is, however, only a generalization of what electrical engineers are accustomed to do in "referring" secondary quantities to the primary in the analysis of transformers and induction motors.

Transformation of Voltage and Impedance for Invariant Power with a Given Current Transformation

The total power input to a number of circuits is given by the sum of the products of terminal voltage and current of each input. This is given in matrix form by the product of the voltage matrix written as a row, i.e. transposed, and the current matrix written as a column thus:

$$\mathbf{v}_t\mathbf{i} = \begin{array}{c} \begin{array}{ccc} 1 & 2 & 3 \end{array} \\ \begin{array}{|c|c|c|} \hline v_1 & v_2 & v_3 \\ \hline \end{array} \end{array} \quad \begin{array}{c|c|} \cline{2-2} 1 & i^1 \\ \cline{2-2} 2 & i^2 \\ \cline{2-2} 3 & i^3 \\ \cline{2-2} \end{array} = \begin{array}{|c|} \hline v_1 i^1 + v_2 i^2 + v_3 i^3 \\ \hline \end{array}$$

in which all the indices "cancel" and the result is a one-by-one matrix, i.e. a single quantity or "scalar", having no indices.

Alternatively, the total terminal power can be expressed as $\mathbf{i}_t\mathbf{v}$. This can be proved either by performing the multiplication, as for $\mathbf{v}_t\mathbf{i}$, or as follows:

Since the product $\mathbf{v}_t\mathbf{i}$ is a one-by-one matrix, it is equal to its own transpose:

$$\mathbf{v}_t\mathbf{i} = [\mathbf{v}_t\mathbf{i}]_t = \mathbf{i}_t[\mathbf{v}_t]_t$$

But the transpose of a transpose is obviously the original matrix, $[\mathbf{v}_t]_t = \mathbf{v}$, and hence

$$\mathbf{v}_t\mathbf{i} = \mathbf{i}_t\mathbf{v}$$

The above applies to instantaneous or d.c. values. In steady-state a.c. conditions the active power input to a pair of terminals is defined in terms of complex current and voltage as the real part of the product of one and the conjugate of the other. The total active power input to a number of circuits is correspondingly given by the real part of one of the matrix products $\mathbf{v}_t\mathbf{i}^*$ or $\mathbf{i}_t^*\mathbf{v}$, where the asterisk denotes the complex conjugate, obtained by taking the conjugate of every element of the matrix. The imaginary part of $\mathbf{v}_t\mathbf{i}^*$ or $\mathbf{i}_t^*\mathbf{v}$ is the reactive power.†

For generality the remainder of this analysis will be expressed in complex terms, since the real case can then be regarded as a particular one.

If $\mathbf{i}'$ and $\mathbf{v}'$ are the transformed current and voltage, $\mathbf{i}$ and $\mathbf{v}$ being the old current and voltage, the active power is $\mathrm{Re}\,(\mathbf{i}_t'^*\mathbf{v}')$ in terms of the new variables and $\mathrm{Re}\,(\mathbf{i}_t^*\mathbf{v})$ in terms of the old ones. If the power is to be invariant under the transformation, these expressions must be equal. If the reactive power is also made invariant, the imaginary parts are also equal, so that

$$\mathbf{i}_t'^*\mathbf{v}' = \mathbf{i}_t^*\mathbf{v}$$

Since $\mathbf{i} = \mathbf{C}\mathbf{i}'$, $\quad \mathbf{i}_t = \mathbf{i}_t'\mathbf{C}_t$ and $\mathbf{i}_t^* = \mathbf{i}_t'^*\mathbf{C}_t^*$.

Hence
$$\mathbf{i}_t'^*\mathbf{v}' = \mathbf{i}_t'^*\mathbf{C}_t^*\mathbf{v}$$

It is tempting to "cancel" the $\mathbf{i}_t'^*$ here to obtain $\mathbf{v}' = \mathbf{C}_t^*\mathbf{v}$, but it must be remembered that there is no matrix division and that this cancelling would have to be done by pre-multiplying both sides of the equation by the inverse of $\mathbf{i}_t'^*$. Since $\mathbf{i}$ is a column matrix, $\mathbf{i}_t'^*$ is a row matrix and is therefore singular, so that no inverse exists. In fact it is very simple to set up a numerical example in the form $\mathbf{ax} = \mathbf{ay}$ where $\mathbf{x}$ and $\mathbf{y}$ are most obviously not equal. If, however, this equation is valid for all values of the elements of $\mathbf{a}$ and these are independent, we can argue as follows:

Suppose that the equation written in full is

$$\begin{bmatrix} a^1 & a^2 & a^3 \end{bmatrix} \begin{bmatrix} x_1 \\ x_2 \\ x_3 \end{bmatrix} = \begin{bmatrix} a^1 & a^2 & a^3 \end{bmatrix} \begin{bmatrix} y_1 \\ y_2 \\ y_3 \end{bmatrix}$$

† The active power is also given by the real part of $\mathbf{v}_t^*\mathbf{i}$ or $\mathbf{i}_t\mathbf{v}^*$. The imaginary parts of these two products are, however, opposite in sign to those of $\mathbf{v}_t\mathbf{i}^*$ and $\mathbf{i}_t^*\mathbf{v}$. The standard convention associates a positive sign with the reactive power flowing into an inductive reactance. $\mathbf{v}_t^*\mathbf{i}$ and $\mathbf{i}_t\mathbf{v}^*$ do not conform to this, whereas $\mathbf{v}_t\mathbf{i}^*$ and $\mathbf{i}_t^*\mathbf{v}$ do, and are accordingly to be preferred.

which leads to

$$a^1x_1+a^2x_2+a^3x_3 = a^1y_1+a^2y_2+a^3y_3.$$

If a change is made to a^1, a^2 and a^3 being unchanged, both sides of the equation must change by the same amount. This requires that $x_1 = y_1$. Similarly, changes to a^2 or a^3 require that $x_2 = y_2$ and $x_3 = y_3$ respectively. Since all the corresponding elements of the two matrices are equal, $\mathbf{x} = \mathbf{y}$.

In the present case, the equation $\mathbf{i}_t'^*\mathbf{v}' = \mathbf{i}_t'^*\mathbf{C}_t^*\mathbf{v}$ is valid for all the values of the independent elements of $\mathbf{i}_t'^*$. It follows that every element of $\mathbf{v}'$ must be equal to the corresponding element of the product $\mathbf{C}_t^*\mathbf{v}$ and that

$$\mathbf{v}' = \mathbf{C}_t^*\mathbf{v}$$

is the required transformation for voltage.

The voltage equation $\mathbf{v} = \mathbf{Zi}$ thus transforms† to

$$\begin{aligned}\mathbf{v}' = \mathbf{C}_t^*\mathbf{v} = \mathbf{C}_t^*(\mathbf{Zi}) &= \mathbf{C}_t^*\mathbf{Zi} = \mathbf{C}_t^*\mathbf{Z}(\mathbf{Ci}')\\ &= \mathbf{C}_t^*\mathbf{ZCi}' = (\mathbf{C}_t^*\mathbf{ZC})\mathbf{i}' = \mathbf{Z}'\mathbf{i}'\end{aligned}$$

When the transformations of both current and voltage are specified, the transformation of impedance is automatically determined and may not be arbitrarily chosen. For invariant power it is obviously $\mathbf{Z}' = \mathbf{C}_t^*\mathbf{ZC}$.

It will be noted that in all this it is not necessary to invert the matrix $\mathbf{C}$, since the inverse is not required in deriving the equation $\mathbf{v}' = \mathbf{Z}'\mathbf{i}'$, nor will it be required to obtain $\mathbf{i}$ from $\mathbf{i}'$ subsequently. This is indeed fortunate, since $\mathbf{C}$ is frequently singular, as, for example, when it represents an interconnection.

† Note the *invariance of form* of the voltage equation under transformation.

CHAPTER 4

Transformers

The Two-winding Transformer

The two-winding transformer consists of two coils, and if the positive directions of current are assumed to be as shown in Fig. 3, the voltage equation is

$$\begin{array}{c|c|} & \\ 1 & v_1 \\ \hline 2 & v_2 \end{array} = \begin{array}{c|c|c|} & 1 & 2 \\ \hline 1 & R_{11}+L_{11}p & M_{12}p \\ \hline 2 & M_{21}p & R_{22}+L_{22}p \end{array} \begin{array}{c|c|} & \\ 1 & i^1 \\ \hline 2 & i^2 \end{array}\,\dagger$$

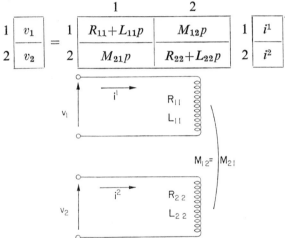

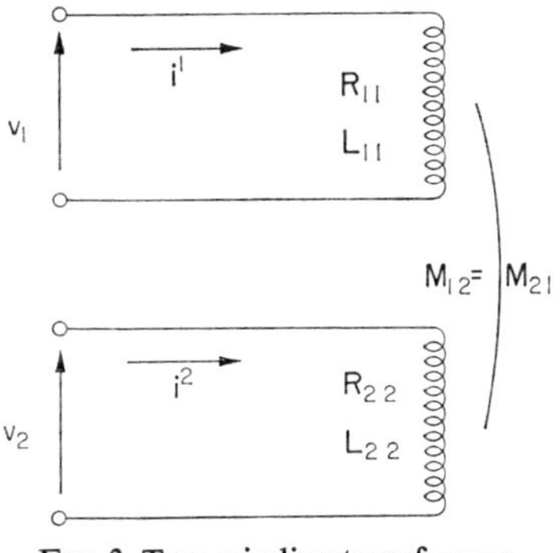

FIG. 3. Two-winding transformer.

† *The sign of mutual inductance.* A mutual inductance M_{ab}, or reactance $\omega M_{ab} = X_{ab}$, is positive when a positive rate of change of current in circuit a produces an induced e.m.f. in circuit b of the same polarity as does a positive rate of change of current in circuit b itself. In a.c. terms this means that if the currents in circuits a and b are in phase with each other, then so also are the e.m.f.s which they induce in circuit a, or those which they induce in circuit b. If the e.m.f.s are in opposition, the mutual inductance is negative.

Alternatively we may say that a mutual inductance is positive if the flux set up by positive current in one circuit links the other circuit in the same direction as the flux set up by positive current in that circuit itself.

Consider the operation of referring the secondary quantities to the primary.

If the turns ratio $k = N_2/N_1$, where N_1 and N_2 are the numbers of turns of windings 1 and 2 respectively, the secondary current referred to the primary will be ki^2 and the currents in the old system in terms of the currents in the new system, i.e. the referred system, are expressed by

$$i^1 = i^{1'}$$
$$i^2 = i^{2'}/k$$

The process of referring to the primary is thus equivalent to a linear transformation, $\mathbf{i} = \mathbf{C}\mathbf{i}'$, where

$$\mathbf{C} = \begin{array}{c|c|c|} & 1' & 2' \\ \hline 1 & 1 & \\ \hline 2 & & 1/k \\ \hline \end{array}$$

The voltages in the referred system are therefore $\mathbf{v}' = \mathbf{C}_t\mathbf{v}$, or

$$\begin{array}{c|c|} 1' & v_1' \\ \hline 2' & v_2' \\ \hline \end{array} = \begin{array}{c|c|c|} & 1 & 2 \\ \hline 1' & 1 & \\ \hline 2' & & 1/k \\ \hline \end{array} \begin{array}{c|c|} 1 & v_1 \\ \hline 2 & v_2 \\ \hline \end{array} = \begin{array}{c|c|} 1' & v_1 \\ \hline 2' & v_2/k \\ \hline \end{array}$$

and the impedance matrix becomes

$$\mathbf{Z}' = \mathbf{C}_t\mathbf{Z}\mathbf{C} = \begin{array}{c|c|c|} & 1 & 2 \\ \hline 1' & 1 & \\ \hline 2' & & 1/k \\ \hline \end{array} \begin{array}{c|c|c|} & 1 & 2 \\ \hline 1 & R_{11}+L_{11}p & M_{12}p \\ \hline 2 & M_{21}p & R_{22}+L_{22}p \\ \hline \end{array} \begin{array}{c|c|c|} & 1' & 2' \\ \hline 1 & 1 & \\ \hline 2 & & 1/k \\ \hline \end{array}$$

$$= \begin{array}{c|c|c|} & 1' & 2' \\ \hline 1' & R_{11}+L_{11}p & (1/k)M_{12}p \\ \hline 2' & (1/k)M_{21}p & (1/k)^2(R_{22}+L_{22}p) \\ \hline \end{array}$$

$$= \begin{array}{c|c|c|} & 1' & 2' \\ \hline 1' & R_{11}'+L_{11}'p & M_{12}'p \\ \hline 2' & M_{21}'p & R_{22}'+L_{22}'p \\ \hline \end{array} \quad \text{say.}$$

The conventional equivalent circuit of the two-winding transformer, neglecting any core loss, is shown in Fig. 4.

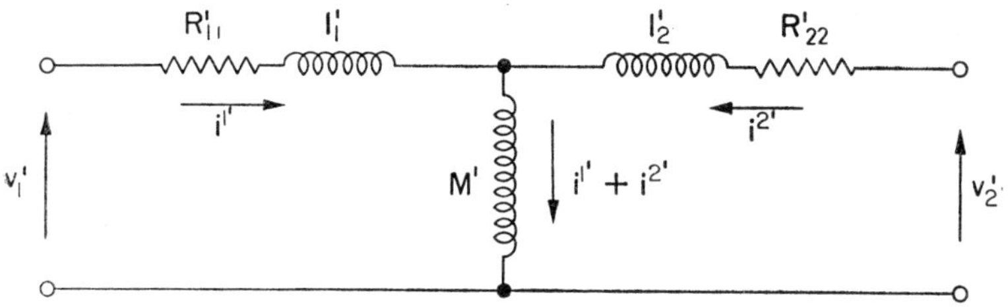

FIG. 4. Simple equivalent circuit.

The voltage equation of this circuit can be written down by inspection as

$$\begin{array}{c|c|} & \\ 1' & v_1' \\ \hline 2' & v_2' \end{array} = \begin{array}{c|c|c|} & 1' & 2' \\ 1' & R_{11}'+(l_1'+M')p & M'p \\ \hline 2' & M'p & R_{22}'+(l_2'+M')p \end{array} \begin{array}{c|c|} & \\ 1' & i^{1'} \\ \hline 2' & i^{2'} \end{array}$$

These equations are obviously identical with those of the transformer in referred terms if

$$l_1'+M' = L_{11}', \quad l_2'+M' = L_{22}' \quad \text{and} \quad M' = M_{12}' = M_{21}'$$

This circuit is therefore a satisfactory representation of a transformer, except for saturation and core loss in the case of a transformer with a ferro-magnetic core. Now it is apparent that the flux of which the path lies wholly in the core must link all the turns of both windings and hence be represented by the product of M' and the currents. This product will also include the effect of such flux as links some turns of both coils, although part or all of its path lies in the air. The fluxes represented by the products of l_1' and l_2' and the currents link only the winding producing them and must, in most cases, necessarily lie for a large part of their paths in air. It is customarily assumed therefore that the saturation and iron loss effects are associated only with M' and that l_1' and l_2' are substantially constant.

The iron loss is usually represented in the circuit by a resistance in parallel with M', so that the equivalent circuit is then as shown in Fig. 5.†

Since the forms of the equations of the transformer are identical, irrespective of whether they are expressed in actual or referred terms, it is pertinent to ask why it was necessary to use the referred values for

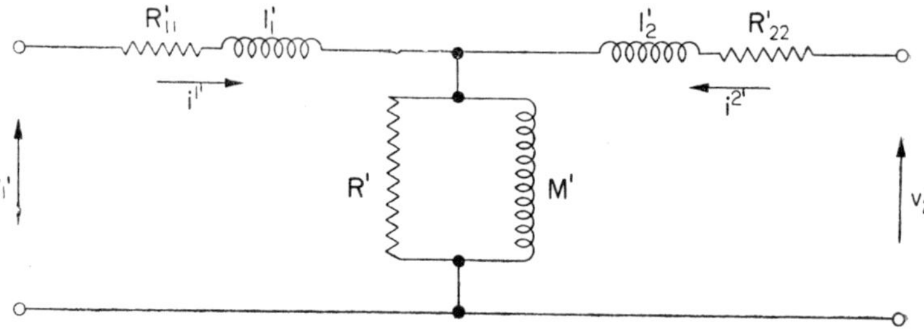

FIG. 5. Conventional equivalent circuit.

the equivalent circuit. For both l_1' and l_2' to be positive it is essential for both L_{11}' and L_{22}' to be greater than M'. This can be ensured only by treating the transformer as if both windings had the same number of turns, i.e. with both windings referred to the same number of turns. For the non-referred case we would have

$$l_1 = L_{11} - M_{12} = L_{11}' - kM_{12}'$$

and

$$l_2 = L_{22} - M_{21} = k^2 L_{22}' - kM_{21}'$$

It is apparent that when k is sufficiently larger than unity l_1 is negative and when k is sufficiently smaller than unity l_2 is negative.

An equivalent circuit in terms of non-referred values is perfectly valid in theory, irrespective of the signs of l_1 and l_2. It has the practical disadvantage, however, that l_1 and l_2 vary considerably with saturation and are also more difficult to visualize in terms of the fluxes.

When all the inductances are referred to the same number of turns, the differences $L_{11}' - M' = l_1'$ and $L_{22}' - M' = l_2'$ are the "leakage"

† See ref. 3, p. 146.

inductances, which, when multiplied by the angular frequency, give the "leakage reactances" of classical theory. It is both more precise and more convenient here to define the leakage inductances as above, and thus to regard them as merely arithmetical differences, than to relate them directly to the magnetic flux patterns which gave rise to the concept† of leakage.

Parameters

The steady-state d.c. voltage equation of the two-winding transformer is obtained by substituting zero for p in the transient equation on p. 37 and is

$$\begin{array}{c|c|}\hline 1 & V_1 \\ \hline 2 & V_2 \\ \hline\end{array} = \begin{array}{c|c|c|}\multicolumn{1}{c}{} & \multicolumn{1}{c}{1} & \multicolumn{1}{c}{2} \\ \hline 1 & R_{11} & \\ \hline 2 & & R_{22} \\ \hline\end{array} \begin{array}{c|c|}\hline 1 & I^1 \\ \hline 2 & I^2 \\ \hline\end{array}$$

The steady-state a.c. voltage equation is obtained by substituting $j\omega$ for p, and if at the same time we write X_{11}, X_{22}, X_m for ωL_{11}, ωL_{22}, ωM, we get

$$\begin{array}{c|c|}\hline 1 & V_1 \\ \hline 2 & V_2 \\ \hline\end{array} = \begin{array}{c|c|c|}\multicolumn{1}{c}{} & \multicolumn{1}{c}{1} & \multicolumn{1}{c}{2} \\ \hline 1 & R_{11}+jX_{11} & jX_m \\ \hline 2 & jX_m & R_{22}+jX_{22} \\ \hline\end{array} \begin{array}{c|c|}\hline 1 & I^1 \\ \hline 2 & I^2 \\ \hline\end{array}$$

If we require to measure the parameters of the transformer, it is clear that steady-state d.c. tests will give R_{11} as V_1/I^1 and R_{22} as V_2/I^2.

To measure the inductances, however, we must perform either transient tests or steady-state a.c. tests.‡ Since the normal operation is usually a steady-state a.c. condition, the latter method is normally preferable.

It is evident that with both voltages and both currents present we shall have only two equations and at least three unknowns: X_{11}, X_{22},

† See ref. 3, p. 52.
‡ See ref. 3, p. 194.

and X_m. If, however, one of the currents is zero, we shall have two equations and only two of these particular unknowns. For example, if winding 2 is open-circuit and winding 1 excited, the voltage equations become

$$V_1 = (R_{11}+jX_{11})I^1$$
$$V_2 = jX_m I^1$$

If the relative phase, as well as the magnitudes, of V_1, V_2, and I^1 are measured, the first of these equations will give both R_{11} and X_{11}, whilst the magnitudes of V_2 and I^1 will be sufficient to determine X_m. The value of R_{11} thus obtained will, however, differ from the value obtained from a d.c. test because it will include the effect of eddy currents in the conductors. This difference may be trivial, but if the transformer has an iron core, the purported value of R_{11} obtained by the a.c. test will also include the effect of the iron loss and will differ considerably from the value obtained by a d.c. test.

There is a further complication with an iron-cored transformer in that both X_{11} and X_m vary widely with saturation, whereas the transformer performance depends primarily on the differences $x_1' = \omega l_1'$ and $x_2' = \omega l_2'$, which, at normal currents are substantially independent of saturation. These differences are small compared to X_{11}', X_{22}', and X_m' themselves, and the error in determining x_1' as $X_{11}'-X_m'$ would be considerable. The "open-circuit test" is, therefore, suitable only for finding X_m' which is most conveniently found in the form $X_{11}'-x_1'$.

An alternative test would be to apply a voltage V_1 to the primary winding with the secondary winding short-circuited, so that $V_2 = 0$. By inverting the impedance matrix we get the current equation for this condition as

$$\begin{array}{c|c|} & \\ 1 & I^1 \\ \hline 2 & I^2 \end{array} = \frac{1}{\Delta} \begin{array}{c|c|c|} & 1 & 2 \\ 1 & R_{22}+jX_{22} & -jX_m \\ \hline 2 & -jX_m & R_{11}+jX_{11} \end{array} \begin{array}{c|c|} & \\ 1 & V_1 \\ \hline 2 & 0 \end{array}$$

where $$\Delta = (R_{11}+jX_{11})\,(R_{22}+jX_{22})+X_m^2$$

Therefore

$$I^1 = V_1 \frac{R_{22}+jX_{22}}{(R_{11}+jX_{11})(R_{22}+jX_{22})+X_m^2}$$
$$= \frac{V_1}{R_{11}+jX_{11}+X_m^2/(R_{22}+jX_{22})}$$

The measured effective impedance is therefore

$$V_1/I^1 = R_{11}+jX_{11}+X_m^2/(R_{22}+jX_{22})$$
$$= R_{11}+jX_{11}+X_m^2(R_{22}-jX_{22})/(R_{22}^2+X_{22}^2)$$
$$= \{R_{11}+R_{22}X_m^2/(R_{22}^2+X_{22}^2)\}+j\{X_{11}-X_{22}X_m^2/(R_{22}^2+X_{22}^2)\}$$

If, as is usual, R_{22}^2 is wholly negligible compared to X_{22}^2, this impedance reduces to

$$\{R_{11}+R_{22}(X_m/X_{22})^2\}+j\{X_{11}-X_m^2/X_{22}\}.$$

(X_m/X_{22}) is approximately equal to the turns ratio.

Consequently $R_{11}+R_{22}(X_m/X_{22})^2$ is very nearly equal to the sum of the primary resistance and the secondary resistance referred to the primary number of turns.

The reactance $\{X_{11}-X_m^2/X_{22}\}$ is of the form characteristic of the reactance measured, as in this case, at the terminals of a winding coupled magnetically to a short-circuited winding of negligible resistance.

If we put $X_{11} = X_m+x_1$ and $X_{22} = X_m+x_2$, the measured reactance is

$$X_m+x_1-X_m^2/(X_m+x_2) = X_m+x_1-X_m/(1+x_2/X_m)$$
$$= X_m+x_1-X_m(1+x_2/X_m)^{-1}$$
$$= X_m+x_1-X_m(1-x_2/X_m+x_2^2/X_m^2- \ldots)$$

If x_2 is sufficiently small compared to X_m for squares and higher powers of x_2/X_m to be neglected, this reduces to

$$X_m+x_1-X_m+x_2 = x_1+x_2.$$

This condition is usually fulfilled if all values have been referred to a common number of turns, in which case (x_1+x_2) is the sum of the "leakage reactances" of classical theory, referred to that number of turns. Since the test was performed with all measurements made on the primary side, it automatically yields the sum of the leakage reactances referred to the primary number of turns.

It has already been explained that it was not possible to obtain a value of x_1 with sufficient accuracy from the open-circuit test. This applies equally to x_2. In fact it is not possible by any test conducted solely at the winding terminals to get accurate values for x_1 and x_2 separately. In the absence of other information, it is usual to assume that the primary and secondary leakage reactances are equal. All parameters can then be determined from the open-circuit, short-circuit, and resistance tests.

The Three-winding Transformer

Taking the conventional directions for the three-winding transformer as shown in Fig. 6,† where all windings are treated as sinks, all voltage equations are of the same form, giving the matrix equation

$$\begin{array}{c|c|}
 & \\
\hline
1 & v_1 \\
\hline
2 & v_2 \\
\hline
3 & v_3 \\
\hline
\end{array}
=
\begin{array}{c|c|c|c|}
 & 1 & 2 & 3 \\
\hline
1 & R_{11}+L_{11}p & M_{12}p & M_{13}p \\
\hline
2 & M_{21}p & R_{22}+L_{22}p & M_{23}p \\
\hline
3 & M_{31}p & M_{32}p & R_{33}+L_{33}p \\
\hline
\end{array}
\;
\begin{array}{c|c|}
 & \\
\hline
1 & i^1 \\
\hline
2 & i^2 \\
\hline
3 & i^3 \\
\hline
\end{array}$$

This can be referred to a common base, say the winding 1, by a transformation matrix

$$\mathbf{C} = \begin{array}{c|c|c|c|}
 & 1' & 2' & 3' \\
\hline
1 & 1 & & \\
\hline
2 & & 1/k_2 & \\
\hline
3 & & & 1/k_3 \\
\hline
\end{array}$$

where $k_2 = N_2/N_1$ and $k_3 = N_3/N_1$, N_1, N_2, N_3 being the numbers of turns of windings 1, 2, 3 respectively.

† The relative positions of the three windings of Fig. 6 are intended to imply that M_{23} is the smallest of the three mutual inductances when all are referred to a common base.

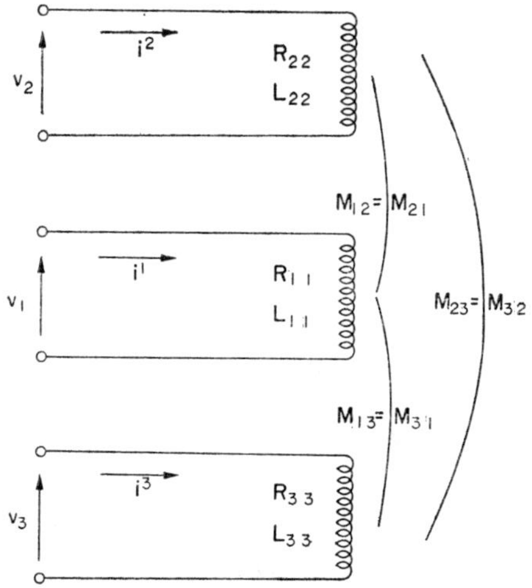

FIG. 6. Three-winding transformer.

The referred impedance matrix is

$$\mathbf{Z}' = \mathbf{C}_t\mathbf{Z}\mathbf{C} =$$

$$\begin{array}{c|c|c|c|}
 & 1 & 2 & 3 \\ \hline
1' & 1 & & \\ \hline
2' & & 1/k_2 & \\ \hline
3' & & & 1/k_3 \\ \hline
\end{array}
\quad
\begin{array}{c|c|c|c|}
 & 1 & 2 & 3 \\ \hline
1 & R_{11}+L_{11}p & M_{12}p & M_{13}p \\ \hline
2 & M_{21}p & R_{22}+L_{22}p & M_{23}p \\ \hline
3 & M_{31}p & M_{32}p & R_{33}+L_{33}p \\ \hline
\end{array}
\quad
\begin{array}{c|c|c|c|}
 & 1' & 2' & 3' \\ \hline
1 & 1 & & \\ \hline
2 & & 1/k_2 & \\ \hline
3 & & & 1/k_3 \\ \hline
\end{array}$$

$$= \begin{array}{c|c|c|c|}
 & 1' & 2' & 3' \\ \hline
1' & R_{11}+L_{11}p & (1/k_2)M_{12}p & (1/k_3)M_{13}p \\ \hline
2' & (1/k_2)M_{21}p & (1/k_2)^2(R_{22}+L_{22}p) & (1/k_2k_3)M_{23}p \\ \hline
3' & (1/k_3)M_{31}p & (1/k_2k_3)M_{32}p & (1/k_3)^2(R_{33}+L_{33}p) \\ \hline
\end{array}$$

$$= \begin{array}{c|c|c|c|}
 & 1' & 2' & 3' \\ \hline
1' & R'_{11}+L'_{11}p & M'_{12}p & M'_{13}p \\ \hline
2' & M'_{21}p & R'_{22}+L'_{22}p & M'_{23}p \\ \hline
3' & M'_{31}p & M'_{32}p & R'_{33}+L'_{33}p \\ \hline
\end{array} \quad \text{say.}$$

Consider now the equations of the circuit shown in Fig. 7.

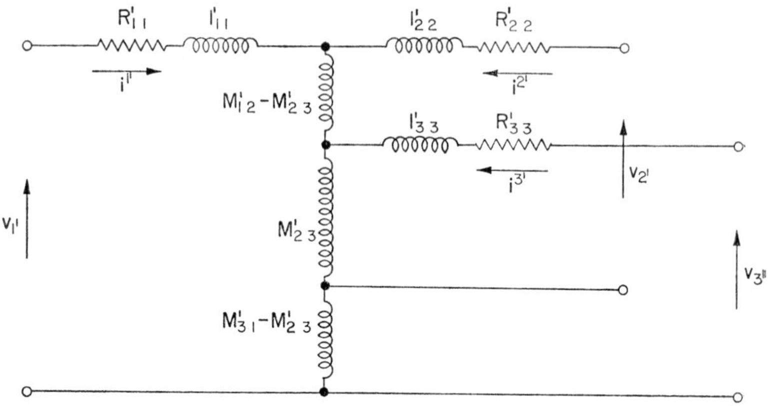

FIG. 7. Equivalent circuit.

The impedance matrix of this circuit is

	1′	2′	3′
1′	$R'_{11}+(l'_{11}+M'_{12}+M'_{31}-M'_{23})p$	$M'_{12}p$	$M'_{31}p$
2′	$M'_{12}p$	$R'_{22}+(l'_{22}+M'_{12})p$	$M'_{23}p$
3′	$M'_{31}p$	$M'_{23}p$	$R'_{33}+(l'_{33}+M'_{31})p$

which is identical with that of the transformer in referred terms if

$$l'_{11}+M'_{12}+M'_{31}-M'_{23} = L'_{11}$$
$$l'_{22}+M'_{12} = L'_{22}$$
$$l'_{33}+M'_{31} = L'_{33}$$

that is, provided that

$$l'_{11} = L'_{11}-M'_{12}-M'_{31}+M'_{23}$$
$$l'_{22} = L'_{22}-M'_{12}$$
$$l'_{33} = L'_{33}-M'_{31}$$

It should be noted that whilst l'_{22} and l'_{33} are leakage inductances, l'_{11} is not so simple. l'_{22} is the leakage of winding 2 with respect to winding 1, and l'_{33} the leakage of winding 3 with respect to winding 1. l'_{11}, however, is the leakage inductance of winding 1 with respect to winding 2 (i.e. $L'_{11}-M'_{12}$) *minus* the amount by which M'_{31} exceeds M'_{23}, the last being necessarily the smallest mutual inductance when the branches are selected as in Fig. 7. If, therefore, $(M'_{31}-M'_{23})$ is greater than $(L'_{11}-M'_{12})$, which it well might be, l'_{11} is a negative inductance. This cannot, of course, be regarded as a capacitance unless all the currents and voltages are steady-state single-frequency sinusoidal functions of time.

The differences of mutual inductances, like the leakage inductances, must be due to fluxes which have a considerable part of their path in air and are thus substantially constant. The saturation and core loss must therefore be largely associated with the smallest mutual inductance (here taken as M'_{23}) and the core loss may be approximately represented by a resistance in parallel with M'_{23}.

If a further approximation is made by moving the branch M'_{23} to the other ends of the branches representing the differences of the mutual inductances, the conventional equivalent circuit of the three-winding transformer, Fig. 8, is obtained. In this circuit one of the inductances, but only one, may be negative.

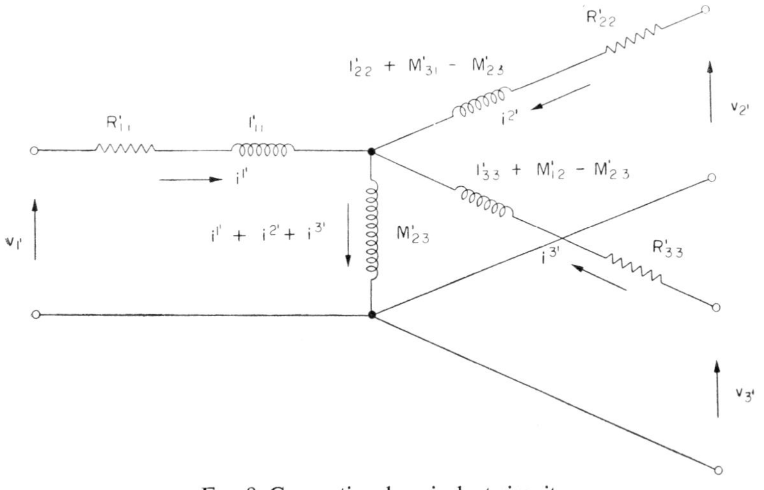

FIG. 8. Conventional equivalent circuit.

The three "leakage" inductances of this circuit are

$$
\begin{aligned}
l'_{11} = L'_{11} - M'_{12} - M'_{31} + M'_{23} &= (L'_{11} - M'_{12}) - (M'_{31} - M'_{23}) \\
l'_{22} + M'_{31} - M'_{23} &= L'_{22} - M'_{12} + M'_{31} - M'_{23} \\
&= (L'_{22} - M'_{12}) + (M'_{31} - M'_{23}) \\
l'_{33} + M'_{12} - M'_{23} &= L'_{33} - M'_{31} + M'_{12} - M'_{23} \\
&= (L'_{33} - M'_{31}) + (M'_{12} - M'_{23})
\end{aligned}
$$

Now $(L'_{11} - M'_{12})$, $(L'_{22} - M'_{12})$, and $(L'_{33} - M'_{31})$ are all genuine leakage inductances and are all necessarily positive. By definition, M'_{23} is the smallest mutual inductance in this case, therefore $(M'_{31} - M'_{23})$ and $(M'_{12} - M'_{23})$ are positive and l'_{11} is the only inductance in the circuit which can be, but is not necessarily, negative. In general, therefore, the inductance in the branch of the circuit *not* associated with the smallest mutual inductance is the one which may be negative.

Parameters

Steady-state a.c. tests on a three-winding transformer, analogous to the open-circuit test on a two-winding transformer, can be made by exciting one winding with both of the others open-circuit. Again, tests analogous to the two-winding transformer short-circuit test can be made by exciting one winding, with one other short-circuited and one open-circuited. Consideration of the equations or of the equivalent circuits shows that three such tests could lead to approximate values for the sums, two at a time, of the reactances corresponding to l'_{11}, $(l'_{22} + M'_{31} - M'_{23})$, $(l'_{33} + M'_{12} - M'_{23})$. From these sums, the inductances of the three branches of the circuit Fig. 8 can be found separately. Data in this form are sufficient in practice to enable the performance under given conditions to be determined, although, as already shown, these branch inductances are not simple leakage inductances. It is not possible to determine values for the separate true leakage inductances in the two-winding sense of the term, any more than it is in the two-winding transformer itself.

More Complicated Magnetic Circuits

A useful approach[5] to systems with more complicated magnetic circuits is based on the ability to write down a matrix of m.m.f.s $\mathbf{f}$, a matrix of reluctances $\mathbf{S}$, and a matrix of fluxes ϕ from consideration of the physical arrangement of the system. The matrix equation $\mathbf{f} = \mathbf{S}\phi$ then relates the three matrices. There are also two other matrix relationships $\mathbf{f} = \mathbf{Ni}$ and $\mathbf{v} = \mathbf{Ri}+\mathbf{N}_t\,\mathrm{d}\phi/\mathrm{d}t$, where $\mathbf{N}$, the matrix of the numbers of turns, will, in general, be a rectangular matrix with the number of elements of $\mathbf{f}$ greater than the number of elements of $\mathbf{i}$.

We have therefore $\quad \mathbf{S}\phi = \mathbf{f} = \mathbf{Ni}$

or $\quad \phi = \mathbf{S}^{-1}\mathbf{Ni}$

and $\quad \mathbf{v} = \mathbf{Ri}+\mathbf{N}_t\,\mathrm{d}(\mathbf{S}^{-1}\mathbf{Ni})/\mathrm{d}t$

Since $\mathbf{N}_t$ is constant, this may be written

$$\mathbf{v} = \mathbf{Ri}+\mathrm{d}(\mathbf{N}_t\mathbf{S}^{-1}\mathbf{N})\mathbf{i}/\mathrm{d}t$$

so that $\mathbf{N}_t\mathbf{S}^{-1}\mathbf{N} = \mathbf{N}_t\mathbf{\Lambda}\mathbf{N} = \mathbf{L}$, where $\mathbf{\Lambda} = \mathbf{S}^{-1}$ is the permeance matrix.

Thus the $L = N^2\Lambda$ of a single electric circuit and single magnetic circuit is replaced by the matrix equation $\mathbf{L} = \mathbf{N}_t\mathbf{\Lambda}\mathbf{N}$ for a system with a number of electric and magnetic circuits.

Whether it is possible to write down $\mathbf{\Lambda}$ directly, or whether it is necessary to write down $\mathbf{S}$ and to invert it, will depend upon the configuration of the system. In general, some parts of the magnetic circuit will be common to several flux paths and it will be necessary to work in terms of reluctance. It may be helpful to consider the electric circuit analogue of the magnetic circuit in determining the reluctance matrix.

This method is useful in considering three-phase transformers. The six windings of a simple three-phase transformer require a six-by-six impedance matrix. It is not possible to learn anything about the performance until the relative magnitudes of the 36 inductances have been determined. This method offers a means of doing this and thereby showing the difference between the three-limb core type and other types of three-phase transformer, particularly in respect of third harmonic and fundamental zero-sequence impedance. Three-phase transformers are, however, too complicated for an introduction to the method and as an illustration a two-winding single-phase transformer with the windings on opposite limbs will be considered.

There are three flux paths as shown diagrammatically in Fig. 9.

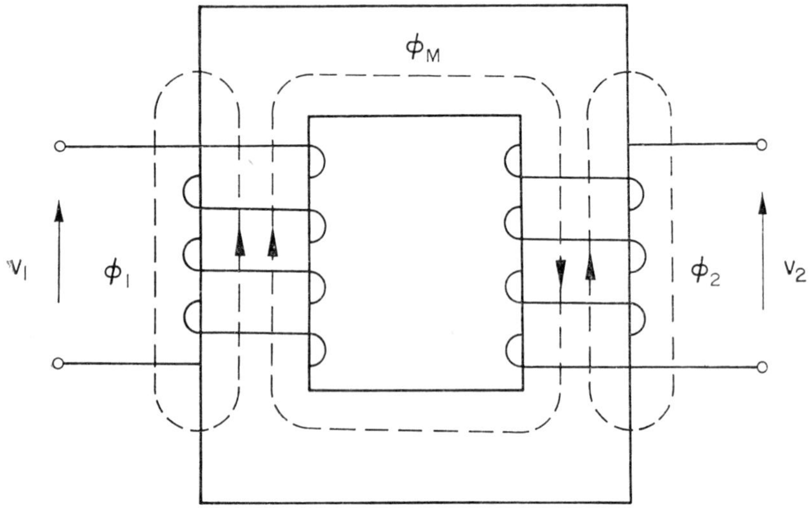

FIG. 9. Flux paths.

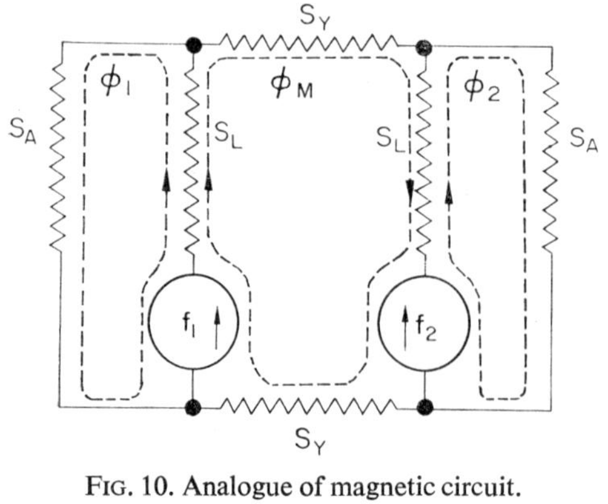

FIG. 10. Analogue of magnetic circuit.

The electric circuit analogue[†] of the magnetic circuit is shown in Fig. 10 in which S_Y is the reluctance of one yoke, S_L is the reluctance of one limb, S_A is the reluctance of an air-leakage path which is assumed to be the same for both windings, and f_1 and f_2 are the m.m.f.s of windings 1 and 2 respectively.

The "mesh" equations of the magnetic circuits are

$$\begin{aligned} f_1 &= (S_L+S_A)\phi_1+S_L\phi_M \\ f_1-f_2 &= S_L\phi_1+2(S_L+S_Y)\phi_M-S_L\phi_2 \\ f_2 &= -S_L\phi_M+(S_L+S_A)\phi_2 \end{aligned}$$

which in matrix form is

$$\begin{array}{c|c|} 1 & f_1 \\ M & f_1-f_2 \\ 2 & f_2 \end{array} = \begin{array}{c|c|c|c|} & 1 & M & 2 \\ \hline 1 & S_L+S_A & S_L & \\ M & S_L & 2(S_L+S_Y) & -S_L \\ 2 & & -S_L & S_L+S_A \end{array} \quad \begin{array}{c|c|} 1 & \phi_1 \\ M & \phi_M \\ 2 & \phi_2 \end{array}$$

This reluctance matrix can be inverted to give a permeance matrix of the form

$$\mathbf{S}^{-1} = \mathbf{\Lambda} = \begin{array}{c|c|c|c|} & 1 & M & 2 \\ \hline 1 & \Lambda_{11} & \Lambda_{1M} & \Lambda_{12} \\ M & \Lambda_{1M} & \Lambda_{MM} & \Lambda_{2M} \\ 2 & \Lambda_{12} & \Lambda_{2M} & \Lambda_{22} \end{array}$$

and

$$\begin{array}{c|c|} 1 & \phi_1 \\ M & \phi_M \\ 2 & \phi_2 \end{array} = \begin{array}{c|c|c|c|} & 1 & M & 2 \\ \hline 1 & \Lambda_{11} & \Lambda_{1M} & \Lambda_{12} \\ M & \Lambda_{1M} & \Lambda_{MM} & \Lambda_{2M} \\ 2 & \Lambda_{12} & \Lambda_{2M} & \Lambda_{22} \end{array} \quad \begin{array}{c|c|} 1 & f_1 \\ M & f_M \\ 2 & f_2 \end{array}$$

† See ref. 3, p. 41.

but

$$
\begin{array}{c|c|}
 & \\ \hline
1 & f_1 \\ \hline
M & f_M \\ \hline
2 & f_2 \\ \hline
\end{array}
=
\begin{array}{c|c|}
 & \\ \hline
1 & f_1 \\ \hline
M & f_1 - f_2 \\ \hline
2 & f_2 \\ \hline
\end{array}
=
\begin{array}{c|c|c|}
 & 1 & 2 \\ \hline
1 & N_1 & \\ \hline
M & N_1 & -N_2 \\ \hline
2 & & N_2 \\ \hline
\end{array}
\begin{array}{c|c|}
 & \\ \hline
1 & i^1 \\ \hline
2 & i^2 \\ \hline
\end{array}
$$

Hence

$$
\mathbf{\Lambda N} =
\begin{array}{c|c|c|}
 & 1 & 2 \\ \hline
1 & N_1(\Lambda_{11}+\Lambda_{1M}) & N_2(\Lambda_{12}-\Lambda_{1M}) \\ \hline
M & N_1(\Lambda_{1M}+\Lambda_{MM}) & N_2(\Lambda_{2M}-\Lambda_{MM}) \\ \hline
2 & N_1(\Lambda_{12}+\Lambda_{2M}) & N_2(\Lambda_{22}-\Lambda_{2M}) \\ \hline
\end{array}
$$

and

$$
\mathbf{N_t \Lambda N} =
\begin{array}{c|c|c|}
 & 1 & 2 \\ \hline
1 & N_1^2(\Lambda_{11}+2\Lambda_{1M}+\Lambda_{MM}) & N_1N_2(\Lambda_{12}+\Lambda_{2M}-\Lambda_{1M}-\Lambda_{MM}) \\ \hline
2 & N_1N_2(\Lambda_{12}+\Lambda_{2M}-\Lambda_{1M}-\Lambda_{MM}) & N_2^2(\Lambda_{22}-2\Lambda_{2M}+\Lambda_{MM}) \\ \hline
\end{array}
$$

The inductance matrix can thus be deduced from the magnetic circuit and winding data, and the impedance matrix follows.

CHAPTER 5

The Matrix Equations of the Basic Rotating Machines

The Matrix Equation of the Basic Commutator Machine

The most convenient commutator machine for initial consideration has the two-pole configuration shown diagrammatically in Fig. 11. There is a pair of brushes q in the normal d.c. machine position†

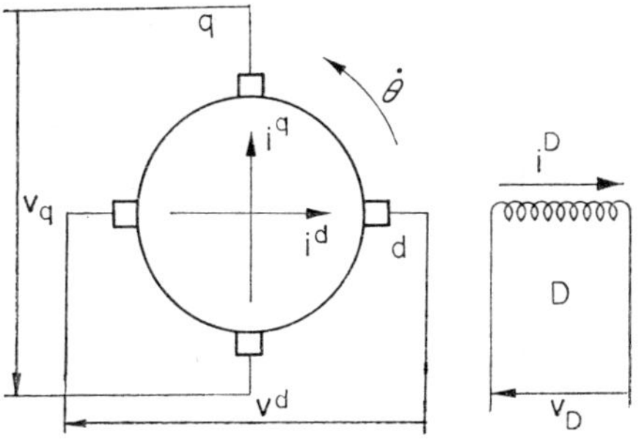

FIG. 11. Basic commutator machine.

relative to the "field" winding D, and also a pair of brushes d at right angles electrically. This is the metadyne‡ arrangement.

The voltage equations can be written down on the assumption that the components to be included for each winding are:

† Brushes are shown diagrammatically on the axis along which the m.m.f. due to their current acts. This is not the physical position of the brushes of an actual machine, which is dependent on the configuration of the armature coil end-windings, but is as if the brushes made direct contact with the conductors of the coils undergoing commutation. See ref. 3, p. 266.

‡ See ref. 3, p. 284.

(i) the terminal voltage;
(ii) the resistance drop;
(iii) the voltage due to the changing current in the winding itself, i.e. the self-induced voltage, treated as a voltage drop;
(iv) the voltages due to the changing currents in all the other windings which produce a flux linking the winding under consideration, i.e. the mutually induced voltages;
(v) in the armature circuits only, the voltages generated by the rotation of the windings in the fluxes set up by all currents in coils having an axis which is not parallel to the axis of the armature circuit under consideration, irrespective of whether these currents flow in stationary windings or in the armature itself.

These equations are sometimes called the Maxwell–Lorentz equations.

The voltage equations of the simple machine shown are then

$$\begin{aligned} v_d &= R_a i^d + L_d p i^d + G_{dq}\dot{\theta} i^q + M_{dD} p i^D \\ v_q &= -G_{qd}\dot{\theta} i^d + R_a i^q + L_q p i^q - G_{qD}\dot{\theta} i^D \\ v_D &= M_{Dd} p i^d + R_D i^D + L_D p i^D \end{aligned}$$

where $\dot{\theta} = d\theta/dt$ is the angular velocity of the armature relative to the field (stator) in electrical radians per second. It, therefore, has the dimension $(\text{Time})^{-1}$. G_{dq}, G_{qd}, and G_{qD} are constants depending on the permeance of the magnetic circuits and the winding arrangement of the armature and field. Since $G_{dq}\dot{\theta} i^q$ is a voltage, G_{dq}, and similarly G_{qd} and G_{qD}, must have the dimensions of inductance.

The minus signs attached to $G_{qd}\dot{\theta} i^d$ and $G_{qD}\dot{\theta} i^D$ are justified as follows.

The positive currents i^d, i^D are assumed to magnetize from left to right. Armature conductors in the vicinity of the right-hand d brush, therefore, have generated e.m.f.s perpendicularly into the plane of the paper, i.e. aiding the positive terminal voltage v_q in magnetizing vertically upwards. These e.m.f.s would, therefore, be positive on the left-hand side of the equation, but have negative signs as voltage drops on the right-hand side.

A similar reasoning will show that $G_{dq}\dot{\theta} i^q$ term is positive.

If the flux waves are sinusoidally distributed in space, the e.m.f.s generated by rotation at a speed $\dot{\theta} = \omega$ will be exactly equal in magnitude to those produced in a stationary winding by the same flux alternating at an angular frequency ω, although of course displaced in time and space by $\pi/2$. Under these conditions therefore, $G_{dq} = L_q$, $G_{qd} = L_d$, and $G_{qD} = M_{dD}$. These equalities do not normally occur in a salient-pole commutator machine since the flux waves of such machines are not even approximately sinusoidal.

The voltage equations can be written in matrix form as $\mathbf{v} = \mathbf{Zi}$:

$$\begin{array}{c|c|} & \\ \mathrm{d} & v_d \\ \mathrm{q} & v_q \\ \mathrm{D} & v_D \end{array} = \begin{array}{c|c|c|c|} & \mathrm{d} & \mathrm{q} & \mathrm{D} \\ \hline \mathrm{d} & R_a + L_d p & G_{dq}\dot{\theta} & M_{dD}p \\ \mathrm{q} & -G_{qd}\dot{\theta} & R_a + L_q p & -G_{qD}\dot{\theta} \\ \mathrm{D} & M_{Dd}p & 0 & R_D + L_D p \end{array} \begin{array}{c|c|} & \\ \mathrm{d} & i^d \\ \mathrm{q} & i^q \\ \mathrm{D} & i^D \end{array}$$

Matrix Equations of Slip-ring and Squirrel-cage Machines

The basis of the analysis of machines in which connections are made not through a commutator, but through slip-rings to specific points on the rotor windings, and of machines in which the rotor windings are short-circuited internally is the equation

$$v = Ri + \frac{\mathrm{d}}{\mathrm{d}t}(\text{flux-linkage})$$

where v is the terminal voltage, i the current and R the resistance of the circuit. This equation is sometimes called the Maxwell circuit equation.

Assuming linearity, i.e. neglecting changes of saturation, the flux-linkage of each winding may be expressed as the sum of the flux-linkages due to each individual current, that is as the sum of the products of inductance and current. The equations for a machine with a number of windings are then of the form

$$v_1 = R_{11}i^1 + \frac{\mathrm{d}}{\mathrm{d}t}(L_{11}i^1 + M_{12}i^2 + M_{13}i^3 + \ldots)$$

$$v_2 = R_{22}i^2 + \frac{\mathrm{d}}{\mathrm{d}t}(M_{21}i^1 + L_{22}i^2 + M_{23}i^3 + \ldots)$$

$$\ldots\ldots\ldots\ldots\ldots\ldots\ldots\ldots\ldots\ldots\ldots\ldots$$

Writing p for d/dt we can express these as

$$\begin{aligned} v_1 &= R_{11}i^1+p(L_{11}i^1)+p(M_{12}i^2)+p(M_{13}i^3)+ \dots \\ v_2 &= p(M_{21}i^1)+R_{22}i^2+p(L_{22}i^2)+p(M_{23}i^3)+ \dots \\ &\dots\dots\dots\dots\dots\dots\dots\dots\dots\dots\dots\dots \end{aligned}$$

and then as

$$\begin{aligned} v_1 &= (R_{11}+pL_{11})i^1+(pM_{12})i^2+(pM_{13})i^3+ \dots \\ v_2 &= (pM_{21})i^1+(R_{22}+pL_{22})i^2+(pM_{23})i^3+ \dots \\ &\dots\dots\dots\dots\dots\dots\dots\dots\dots\dots\dots\dots \end{aligned}$$

provided it is remembered that the p applies to the product of the inductance within the bracket and the current outside the bracket in every case.

With this same proviso, the equations may be written in matrix form as

$$\begin{array}{|c|} \hline v_1 \\ \hline v_2 \\ \hline - \\ \hline \\ \end{array} = \begin{array}{|c|c|c|c|c|c|} \hline R_{11}+pL_{11} & pM_{12} & pM_{13} & - & - & \\ \hline pM_{21} & R_{22}+pL_{22} & pM_{23} & - & - & \\ \hline - & - & - & - & - & \\ \hline & & & & & \\ \end{array} \begin{array}{|c|} \hline i^1 \\ \hline i^2 \\ \hline - \\ \hline \\ \end{array}$$

$$= \begin{array}{|c|c|c|c|} \hline R_{11} & & & \\ \hline & R_{22} & & \\ \hline & & - & \\ \hline & & & \\ \end{array} \begin{array}{|c|} \hline i^1 \\ \hline i^2 \\ \hline - \\ \hline \\ \end{array}$$

$$+p \begin{array}{|c|c|c|c|c|} \hline L_{11} & M_{12} & M_{13} & - & \\ \hline M_{21} & L_{22} & M_{23} & - & \\ \hline - & - & - & - & \\ \hline & & & & \\ \end{array} \begin{array}{|c|} \hline i^1 \\ \hline i^2 \\ \hline - \\ \hline \\ \end{array}$$

or $$\mathbf{v} = \mathbf{Ri}+p(\mathbf{Li})$$

where the **L** matrix includes both the self-inductances and the mutual inductances.

It must be noted that in many cases the inductances are functions of the angular position of the rotor and hence are time functions. The differential equations, although linear, do not, in general, have constant coefficients and cannot be integrated in a simple routine manner. The first objective must therefore be to apply a transformation leading to equations which have constant coefficients. This will be treated by considering first the simplest case, namely a balanced two-phase machine with a uniform air-gap, i.e. not having salient poles. This machine is structurally a two-phase induction motor.

The Matrix Equation of the Balanced Two-phase Machine with a Uniform Air-gap (Induction Machine)

Consider a two-pole machine of induction motor type with balanced two-phase windings on both stator and rotor and a uniform air-gap. In Fig. 12 the stator windings are designated D and Q, and the rotor windings α and β. The axis of the α-phase winding and that of the D winding are at an angle θ which increases as the rotor revolves

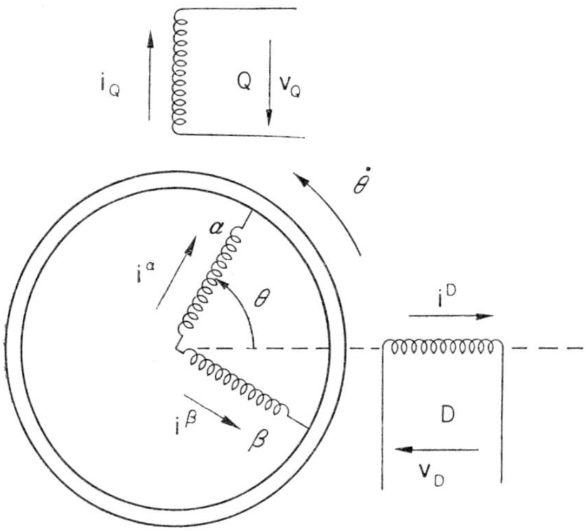

FIG. 12. Balanced two-phase machine with uniform air-gap.

in a counter-clockwise direction. θ is thus a function of time and $d\theta/dt$, which will again be written $\dot{\theta}$, is the angular velocity of the rotor relative to the stator.

The terminal voltages of the four windings can be expressed as resistance voltage drops and rates of change of flux-linkage in terms of currents and inductances as

$$
\begin{aligned}
v_\alpha &= R_{\alpha\alpha}i^\alpha + p(L_{\alpha\alpha}i^\alpha) + p(M_{\alpha\beta}i^\beta) + p(M_{\alpha D}i^D) + p(M_{\alpha Q}i^Q)\\
v_\beta &= p(M_{\beta\alpha}i^\alpha) + R_{\beta\beta}i^\beta + p(L_{\beta\beta}i^\beta) + p(M_{\beta D}i^D) + p(M_{\beta Q}i^Q)\\
v_D &= p(M_{D\alpha}i^\alpha) + p(M_{D\beta}i^\beta) + R_{DD}i^D + p(L_{DD}i^D) + p(M_{DQ}i^Q)\\
v_Q &= p(M_{Q\alpha}i^\alpha) + p(M_{Q\beta}i^\beta) + p(M_{QD}i^D) + R_{QQ}i^Q + p(L_{QQ}i^Q)
\end{aligned}
$$

Before proceeding further we must inquire into the nature of these inductances. Since the air-gap is uniform, all the self-inductances will be independent of the angular position of the rotor and may be regarded as constant if saturation changes are ignored. On this assumption, the windings being balanced,

$$L_{\alpha\alpha} = L_{\beta\beta} = L_r \text{ say,} \quad \text{and} \quad L_{DD} = L_{QQ} = L_S \text{ say,}$$

where L_r and L_S are constant.

Again, the symmetry of the arrangement shows that there will be no linkage with any winding by flux set up by a current in the winding at right angles to it, and consequently the mutual inductances $M_{\alpha\beta} = M_{\beta\alpha}$ and $M_{DQ} = M_{QD}$ are all zero.

The remaining mutual inductances $M_{\alpha D} = M_{D\alpha}$, $M_{\alpha Q} = M_{Q\alpha}$, $M_{\beta D} = M_{D\beta}$ and $M_{\beta Q} = M_{Q\beta}$ are all functions of θ and are therefore functions of time. It is apparent that their variation with θ is cyclic with a period corresponding to one revolution of the rotor. For simplicity it is desirable to assume that they vary sinusoidally, although this may be only approximately true in practice. Since the maximum will occur in each case when the axes of the two windings are coincident, and since all pairs of windings involved are similar, and since, moreover, the air-gap is uniform, the maximum value will be the same in all cases.

These inductances can therefore be written

$$M_{\alpha D} = M\cos\theta \qquad M_{\beta D} = M\sin\theta$$
$$M_{\alpha Q} = M\sin\theta \qquad M_{\beta Q} = -M\cos\theta$$

$M_{\beta Q}$ is negative since, for values of θ between $-\pi/2$ and $\pi/2$, i.e. positive values of cos θ, with the chosen positive directions of Fig. 12, positive current in β produces a flux component linking Q in the opposite direction to the flux set up by a positive current in Q itself.

Since the rotor windings are balanced, $R_{\alpha\alpha} = R_{\beta\beta} = R_r$ say, and similarly for balanced stator windings $R_{DD} = R_{QQ} = R_S$ say.

M being constant, the voltage equations are of the form

$$\begin{aligned} v_\alpha &= R_r i^\alpha + L_r p i^\alpha + Mp(\cos\theta i^D) + Mp(\sin\theta i^Q) \\ v_\beta &= R_r i^\beta + L_r p i^\beta + Mp(\sin\theta i^D) - Mp(\cos\theta i^Q) \\ v_D &= Mp(\cos\theta i^\alpha) + Mp(\sin\theta i^\beta) + R_S i^D + L_S p i^D \\ v_Q &= Mp(\sin\theta i^\alpha) - Mp(\cos\theta i^\beta) + R_S i^Q + L_S p i^Q \end{aligned}$$

which can be expressed in the matrix form $\mathbf{v} = \mathbf{Zi}$ as

$$\begin{array}{c|c|} \alpha & v_\alpha \\ \beta & v_\beta \\ D & v_D \\ Q & v_Q \end{array} = \begin{array}{c|c|c|c|c|} & \alpha & \beta & D & Q \\ \alpha & R_r + L_r p & & Mp\cos\theta & Mp\sin\theta \\ \beta & & R_r + L_r p & Mp\sin\theta & -Mp\cos\theta \\ D & Mp\cos\theta & Mp\sin\theta & R_S + L_S p & \\ Q & Mp\sin\theta & -Mp\cos\theta & & R_S + L_S p \end{array} \begin{array}{c|c|} \alpha & i^\alpha \\ \beta & i^\beta \\ D & i^D \\ Q & i^Q \end{array}$$

in which it is understood† that the "p"s of **Z** operate on the products of the functions of θ in **Z** itself and the currents in **i**.

If the differentiations are performed, the terms such as $Mp(\cos\theta i^D)$ result in two terms, in this case $M\cos\theta p i^D - M\sin\theta\dot{\theta} i^D$. The presence

† In all this work, p is presumed to operate on all the functions *following it* in the particular term of the full algebraic equation.

of terms like these shows that although we have a set of linear differential equations, a number of the coefficients are functions of time unless the rotor is stationary, which is, of course, an unacceptable restriction. We lack a simple routine process for solving such equations as these and must therefore look for some means of changing the variables so that we obtain equations which we can integrate in a simple routine manner.

The third and fourth equations give us a clue. They can be written

$$v_D = Mp(\cos\theta i^\alpha + \sin\theta i^\beta) + R_S i^D + L_S p i^D$$
$$v_Q = Mp(\sin\theta i^\alpha - \cos\theta i^\beta) + R_S i^Q + L_S p i^Q$$

If we introduce two new variables i^d, i^q defined by

$$i^d = \cos\theta i^\alpha + \sin\theta i^\beta$$
$$i^q = \sin\theta i^\alpha - \cos\theta i^\beta$$

hese two equations become

$$v_D = Mpi^d + R_S i^D + L_S p i^D$$
$$v_Q = Mpi^q + R_S i^Q + L_S p i^Q$$

Our object has thus been achieved for these equations by means of the transformation

$$\begin{array}{c|c|} & \\ \hline d & i^d \\ \hline q & i^q \\ \hline \end{array} = \begin{array}{c|c|c|} & \alpha & \beta \\ \hline d & \cos\theta & \sin\theta \\ \hline q & \sin\theta & -\cos\theta \\ \hline \end{array} \begin{array}{c|c|} & \\ \hline \alpha & i^\alpha \\ \hline \beta & i^\beta \\ \hline \end{array}$$

It remains to consider the effect of this transformation on the first two equations.

Before we can substitute in them for i^α, i^β in terms of i^d, i^q, we must solve the transformation equation which is of the form $\mathbf{i}' = \mathbf{C}^{-1}\mathbf{i}$.

The determinant of the matrix is -1. Since it is symmetric, its transpose is the same as the matrix itself. The inverse can therefore be written down as

$$\mathbf{C} = -\begin{array}{c|c|c|} & \text{d} & \text{q} \\ \hline \alpha & -\cos\theta & -\sin\theta \\ \hline \beta & -\sin\theta & \cos\theta \\ \hline \end{array}$$

$$= \begin{array}{c|c|c|} & \text{d} & \text{q} \\ \hline \alpha & \cos\theta & \sin\theta \\ \hline \beta & \sin\theta & -\cos\theta \\ \hline \end{array}$$

which is again identical with the original matrix, which is therefore orthogonal as well as symmetric. Hence

$$i^{\alpha} = \cos\theta i^{\text{d}} + \sin\theta i^{\text{q}}$$
$$i^{\beta} = \sin\theta i^{\text{d}} - \cos\theta i^{\text{q}}$$

and substituting for i^{α} and i^{β} in the first two equations,

$$\begin{aligned} v_{\alpha} &= R_{\text{r}}(\cos\theta i^{\text{d}} + \sin\theta i^{\text{q}}) + L_{\text{r}}p(\cos\theta i^{\text{d}} + \sin\theta i^{\text{q}}) \\ &\quad + Mp(\cos\theta i^{\text{D}}) + Mp(\sin\theta i^{\text{Q}}) \\ v_{\beta} &= R_{\text{r}}(\sin\theta i^{\text{d}} - \cos\theta i^{\text{q}}) + L_{\text{r}}p(\sin\theta i^{\text{d}} - \cos\theta i^{\text{q}}) \\ &\quad + Mp(\sin\theta i^{\text{D}}) - Mp(\cos\theta i^{\text{Q}}) \end{aligned}$$

Performing the differentiations,

$$\begin{aligned} v_{\alpha} = R_{\text{r}}(\cos\theta i^{\text{d}} + \sin\theta i^{\text{q}}) &+ L_{\text{r}}(\cos\theta p i^{\text{d}} + \sin\theta p i^{\text{q}} - \sin\theta\dot{\theta} i^{\text{d}} + \cos\theta\dot{\theta} i^{\text{q}}) \\ &+ M(\cos\theta p i^{\text{D}} + \sin\theta p i^{\text{Q}} - \sin\theta\dot{\theta} i^{\text{D}} + \cos\theta\dot{\theta} i^{\text{Q}}) \\ v_{\beta} = R_{\text{r}}(\sin\theta i^{\text{d}} - \cos\theta i^{\text{q}}) &+ L_{\text{r}}(\sin\theta p i^{\text{d}} - \cos\theta p i^{\text{q}} + \cos\theta\dot{\theta} i^{\text{d}} + \sin\theta\dot{\theta} i^{\text{q}}) \\ &+ M(\sin\theta p i^{\text{D}} - \cos\theta p i^{\text{Q}} + \cos\theta\dot{\theta} i^{\text{D}} + \sin\theta\dot{\theta} i^{\text{Q}}) \end{aligned}$$

These obviously have not got constant coefficients, but they are hybrid equations in that they express v_{α}, v_{β} in terms of i^{d}, i^{q}, i^{D}, i^{Q}. We must therefore complete the transformation by expressing the voltages in terms of v_{d}, v_{q}. The third and fourth voltage equations are in terms of v_{D}, v_{Q} and do not require any futher change, since we have retained i^{D} and i^{Q} in the transformation. It is important to note that we could not have conveniently made the change to v_d, v_q before performing the differentiations, since the only functions of θ to be differentiated are those associated with the change from i^α, i^β to i^d, i^q. The functions of θ associated with the change from v_α, v_β to v_d, v_q must not be differentiated.

The current transformation matrix was

$$\mathbf{C} = \begin{array}{c|c|c|} & \text{d} & \text{q} \\ \hline \alpha & \cos\theta & \sin\theta \\ \hline \beta & \sin\theta & -\cos\theta \\ \hline \end{array}$$

For invariant power, the voltage transformation matrix is the transpose, which is

$$\mathbf{C}_t = \begin{array}{c|c|c|} & \alpha & \beta \\ \hline \text{d} & \cos\theta & \sin\theta \\ \hline \text{q} & \sin\theta & -\cos\theta \\ \hline \end{array}$$

Hence

$$\begin{aligned}
v_d &= \cos\theta v_\alpha + \sin\theta v_\beta \\
&= R_r(\cos^2\theta i^d + \sin\theta\cos\theta i^q + \sin^2\theta i^d - \sin\theta\cos\theta i^q) \\
&\quad + L_r(\cos^2\theta p i^d + \sin\theta\cos\theta p i^q + \sin^2\theta p i^d - \sin\theta\cos\theta p i^q \\
&\quad - \sin\theta\cos\theta\dot{\theta} i^d + \cos^2\theta\dot{\theta} i^q + \sin\theta\cos\theta\dot{\theta} i^d + \sin^2\theta\dot{\theta} i^q) \\
&\quad + M(\cos^2\theta p i^D + \sin\theta\cos\theta p i^Q + \sin^2\theta p i^D - \sin\theta\cos\theta p i^Q \\
&\quad - \sin\theta\cos\theta\dot{\theta} i^D + \cos^2\theta\dot{\theta} i^Q + \sin\theta\cos\theta\dot{\theta} i^D + \sin^2\theta\dot{\theta} i^Q) \\
&= R_r i^d + L_r p i^d + L_r\dot{\theta} i^q + M p i^D + M\dot{\theta} i^Q
\end{aligned}$$

and

$$\begin{aligned}
v_q &= \sin\theta v_\alpha - \cos\theta v_\beta \\
&= R_r(\sin\theta\cos\theta i^d + \sin^2\theta i^q - \sin\theta\cos\theta i^d + \cos^2\theta i^q) \\
&\quad + L_r(\sin\theta\cos\theta p i^d + \sin^2\theta p i^q - \sin\theta\cos\theta p i^d + \cos^2\theta p i^q \\
&\quad - \sin^2\theta\dot{\theta} i^d + \sin\theta\cos\theta\dot{\theta} i^q - \cos^2\theta\dot{\theta} i^d - \sin\theta\cos\theta\dot{\theta} i^q) \\
&\quad + M(\sin\theta\cos\theta p i^D + \sin^2\theta p i^Q - \sin\theta\cos\theta p i^D + \cos^2\theta p i^Q \\
&\quad - \sin^2\theta\dot{\theta} i^D + \sin\theta\cos\theta\dot{\theta} i^Q - \cos^2\theta\dot{\theta} i^D - \sin\theta\cos\theta\dot{\theta} i^Q) \\
&= R_r i^q + L_r p i^q - L_r\dot{\theta} i^d + M p i^Q - M\dot{\theta} i^D
\end{aligned}$$

The four equations are now

$$
\begin{aligned}
v_d &= R_r i^d + L_r p i^d + L_r\dot{\theta} i^q + M p i^D + M\dot{\theta} i^Q \\
v_q &= -L_r\dot{\theta} i^d + R_r i^q + L_r p i^q - M\dot{\theta} i^D + M p i^Q \\
v_D &= M p i^d + R_S i^D + L_S p i^D \\
v_Q &= M p i^q + R_S i^Q + L_S p i^Q
\end{aligned}
$$

Thus we now have all four equations with constant coefficients provided that $L_r\dot{\theta}$ and $M\dot{\theta}$ are constant. Since L_r and M are constants, this limitation is that $\dot{\theta}$ is constant. Solution by routine application of Laplace transforms is therefore restricted to *constant angular velocity.*

These equations can be written in matrix form as

$$
\begin{array}{c|c|}
 & \\
\hline
d & v_d \\
q & v_q \\
D & v_D \\
Q & v_Q \\
\hline
\end{array}
=
\begin{array}{c|c|c|c|c|}
 & d & q & D & Q \\
\hline
d & R_r + L_r p & L_r\dot{\theta} & Mp & M\dot{\theta} \\
q & -L_r\dot{\theta} & R_r + L_r p & -M\dot{\theta} & Mp \\
D & Mp & & R_S + L_S p & \\
Q & & Mp & & R_S + L_S p \\
\hline
\end{array}
\begin{array}{c|c|}
 & \\
\hline
d & i^d \\
q & i^q \\
D & i^D \\
Q & i^Q \\
\hline
\end{array}
$$

The transformation

$$
\mathbf{C} = \begin{array}{c|c|c|}
 & d & q \\
\hline
\alpha & \cos\theta & \sin\theta \\
\beta & \sin\theta & -\cos\theta \\
\hline
\end{array}
$$

is thus of paramount importance, since it enables us to determine the performance under steady-state and transient conditions of machines with polyphase windings on the rotor without the necessity of seeking a special method for integrating the differential equations—assuming that a method could be found. The solution of the differential equations is now a routine process according to "classical", operational or Laplace transform method, as we prefer.

In effecting the transformation to equations with constant coefficients we have, however, introduced some restrictions:

Saturation changes have had to be ignored.
The variation of inductance with position has been assumed to be sinusoidal.
The angular velocity has had to be taken as constant.
The windings of both members have been taken as balanced.
The air-gap has been assumed to be uniform.

It has not been shown so far that it is essential to assume that the inductances vary sinusoidally with position, but it is easy to do so. If it had been assumed that $M_{\alpha D} = M_1 \cos \theta + M_3 \cos 3\theta$ say, the transformation would obviously have led to functions of multiples of θ and the resulting transformed equations would have been complicated and would not have had constant coefficients.

The assumption that the windings were balanced is only partly necessary. It is essential that the rotor windings be balanced,† but it is not necessary for the stator windings to be balanced. This distinction arises because we have not transformed the stator winding currents but only the rotor currents. Inspection of the original equations on p. 59 shows that at that stage there was no difference of form between the four equations. We could therefore have selected the first two rather than the second two in order to deduce a suitable transformation. We would then have changed the currents i^D, i^Q and not i^α, i^β. It would then have been necessary for the stator windings, but not the rotor windings, to be balanced.

Furthermore we have assumed that the air-gap is uniform. This is not necessary. The stator may in fact be salient, but not the rotor. This distinction again arises from the fact that we transformed the rotor currents not the stator currents.

Had we transformed the stator currents instead of the rotor currents, the rotor, but not the stator, could have been salient.

† See Appendix A.

That it is not necessary for the stator windings to be balanced and that the stator can be salient will be apparent when the transformation is applied to such a system in the next section.

That the rotor must be "round" or non-salient follows directly from the requirements for balanced rotor windings and for constant stator winding self-inductance.

In all the preceding work the requirements for rotor and stator would have been interchanged had we chosen to transform i^D and i^Q in preference to i^α and i^β. Other things being equal, it is, however, more convenient to consider ourselves at rest relative to the stator rather than relative to the rotor, and to apply the transformation to the currents in the windings which are then moving relative to us, i.e. the rotor currents.

The Matrix Equation of the Balanced Two-phase Revolving-armature Salient-pole Machine (Synchronous Machine)

The transformation deduced in the previous section will now be applied to the machine with a salient-pole stator. Figure 13 shows diagrammatically the two-pole salient-pole structure with the field

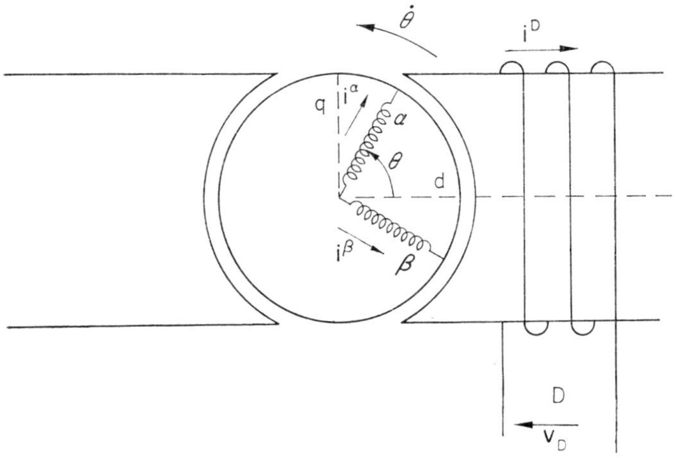

FIG. 13. Balanced two-phase salient-pole machine.

winding designated D, the axes of the two armature phase-windings designated α and β, and also two hypothetical stationary winding axes, d and q, the axis of the d winding being coincident with that of the D winding and that of the q at an angle $\pi/2$ with that of the D winding. The instantaneous position of the armature is again described by the angle θ measured from the D axis to the α axis. Positive rotation is thus counterclockwise.

The voltage equations in these terms† are

$$\begin{aligned}
v_\alpha &= R_{\alpha\alpha}i^\alpha + p(L_{\alpha\alpha}i^\alpha) + p(M_{\alpha\beta}i^\beta) + p(M_{\alpha D}i^D)\\
v_\beta &= p(M_{\beta\alpha}i^\alpha) + R_{\beta\beta}i^\beta + p(L_{\beta\beta}i^\beta) + p(M_{\beta D}i^D)\\
v_D &= p(M_{D\alpha}i^\alpha) + p(M_{D\beta}i^\beta) + R_{DD}i^D + p(L_{DD}i^D)
\end{aligned}$$

It is now necessary to consider how the inductance terms of these equations depend upon the angle θ. Given an existing machine, one could measure the inductances for a number of values of θ. For general purposes however, we will assume that the inductances, the values of which are obviously periodic functions of θ, can be represented with sufficient accuracy by taking only the lowest order non-zero harmonic term. We can then resolve and combine m.m.f.s and fluxes as space vectors. In particular we can resolve winding m.m.f.s. into their components along the axes of symmetry and resolve fluxes on the axes of symmetry into their components along the winding axes.

$L_{\alpha\alpha}$

The current i^α produces a m.m.f. $N_\alpha i^\alpha$ on its own axis, where N_α is the effective number of turns of the α-phase winding. This m.m.f. has a component $N_\alpha i^\alpha \cos\theta$ in the d direction and a component $N_\alpha i^\alpha \sin\theta$ in the q direction.

If Λ_d and Λ_q are the permeances of the main magnetic circuits on the d and q axes respectively, the component fluxes along these two axes are $\Lambda_d N_\alpha i^\alpha \cos\theta$ and $\Lambda_q N_\alpha i^\alpha \sin\theta$ respectively. The components of these fluxes along the α axis are $\Lambda_d N_\alpha i^\alpha \cos^2\theta$ and $\Lambda_q N_\alpha i^\alpha \sin^2\theta$

† It is not necessary for our present purpose to complicate matters by including a "quadrature-axis" Q winding on the stator, in addition to the "direct-axis" D winding.

and hence the flux-linkage of the α-phase winding due to the fluxes in the main paths is

$$N_\alpha(\Lambda_d N_\alpha i^\alpha \cos^2\theta + \Lambda_q N_\alpha i^\alpha \sin^2\theta)$$
$$= N_\alpha^2(\Lambda_d \cos^2\theta + \Lambda_q \sin^2\theta)i^\alpha .$$

If the armature flux leakage paths have an additional permeance λ, assumed to be independent of θ, the leakage flux is $\lambda N_\alpha i^\alpha$ and the flux-linkage due to it is $N_\alpha^2\lambda i^\alpha = N_\alpha^2(\lambda\cos^2\theta + \lambda\sin^2\theta)i^\alpha$.

The total flux-linkage is thus

$$N_\alpha^2(\Lambda_d \cos^2\theta + \Lambda_q \sin^2\theta)i^\alpha + N_\alpha^2(\lambda\cos^2\theta + \lambda\sin^2\theta)i^\alpha$$
$$= N_\alpha^2\{(\Lambda_d + \lambda)\cos^2\theta + (\Lambda_q + \lambda)\sin^2\theta\}i^\alpha .$$

It follows that the self-inductance of the α-phase winding, being equal to the flux-linkage for unit current, is $N_\alpha^2\{(\Lambda_d + \lambda)\cos^2\theta + (\Lambda_q + \lambda)\sin^2\theta\}$. If changes of saturation are neglected, N_α, Λ_d, Λ_q, and λ are all constant, so that $L_{\alpha\alpha}$ may be written $(L_d\cos^2\theta + L_q\sin^2\theta)$, where $L_d = N_\alpha^2(\Lambda_d + \lambda)$ and $L_q = N_\alpha^2(\Lambda_q + \lambda)$ are constants.

Alternatively,

$$L_{\alpha\alpha} = L_d(1+\cos 2\theta)/2 + L_q(1-\cos 2\theta)/2$$
$$= (L_d + L_q)/2 + \{(L_d - L_q)/2\}\cos 2\theta.$$

The inductance represented by $N_\alpha^2\lambda$ is the armature leakage inductance l_a.

$L_{\beta\beta}$

Because the rotor windings are balanced, the self-inductance of the β phase will be identical with that of the α phase except that the angle θ will be replaced by $(\theta - \pi/2)$. Hence

$$L_{\beta\beta} = L_d\cos^2(\theta-\pi/2) + L_q\sin^2(\theta-\pi/2)$$
$$= (L_d\sin^2\theta + L_q\cos^2\theta)$$
$$= (L_d + L_q)/2 - \{(L_d - L_q)/2\}\cos 2\theta$$

$M_{\beta\alpha}$

The main flux components due to a current i^α have already been found to be $\Lambda_d N_\alpha i^\alpha \cos\theta$ and $\Lambda_q N_\alpha i^\alpha \sin\theta$ in the d and q directions respectively. The components in the direction of the axis of the β-phase winding are therefore $\Lambda_d N_\alpha i^\alpha \cos\theta \sin\theta$ and $-\Lambda_q N_\alpha i^\alpha \sin\theta \cos\theta$ and the flux-linkage of the β-phase winding is $N_\beta(\Lambda_d N_\alpha i^\alpha - \Lambda_q N_\alpha i^\alpha)$ $\sin\theta\cos\theta$. But since the windings are balanced, $N_\beta = N_\alpha$ and the total linkage is

$$N_\alpha^2(\Lambda_d - \Lambda_q)\sin\theta\cos\theta i^\alpha$$

and the mutual inductance is

$$\{N_\alpha^2(\Lambda_d - \Lambda_q)/2\}\sin 2\theta = N_\alpha^2[\{(\Lambda_d+\lambda)-(\Lambda_q+\lambda)\}/2]\sin 2\theta$$
$$= \{(L_d - L_q)/2\}\sin 2\theta$$

$M_{\alpha\beta}$

The inductance $M_{\alpha\beta}$ is necessarily equal to $M_{\beta\alpha}$, but its value can be derived in a similar manner to that of $M_{\beta\alpha}$ if desired.

$M_{D\alpha}$

The only component of flux linking the D winding due to a current i^α is $\Lambda_d N_\alpha i^\alpha \cos\theta$ and the linkage is therefore $N_D N_\alpha \Lambda_d i^\alpha \cos\theta$ and the mutual inductance is $N_D N_\alpha \Lambda_d \cos\theta$, which may be written $L_{Dd}\cos\theta$ where $L_{Dd} = N_D N_\alpha \Lambda_d$ and is constant.

If L_{Dd} were referred to the armature number of turns, it would become $N_D N_\alpha \Lambda_d (N_\alpha/N_D) = N_\alpha^2 \Lambda_d$. It can be seen, therefore, that the armature self-inductance term L_d can be regarded as the arithmetic sum of the armature leakage inductance l_a and the maximum value L_{Dd} of the mutual inductance $M_{D\alpha}$ referred to the armature.

$M_{D\beta}$

The derivation of $M_{D\beta}$ is similar to that of $M_{D\alpha}$ except that the angle θ is replaced by $(\theta-\pi/2)$, hence $M_{D\beta} = L_{Dd}\cos(\theta-\pi/2) = L_{Dd}\sin\theta$.

L_{DD}

The self-inductance of the field winding is independent of θ since the armature, being non-salient, always presents the same permeance to the field. This inductance is therefore simply L_{DD}, a constant.

Whilst the deduction of the form of the inductances above is plausible, it conceals inherent restrictions. For it to be justified to resolve m.m.f.s and fluxes as has been done in the preceding page, it is necessary for the system to be linear, i.e. to ignore changes of saturation, and also for the said m.m.f.s and fluxes to be sinusoidally distributed around the periphery of the armature. These do not, of course, represent additional limitations, since, as we have already seen, it is also necessary to make the same assumptions in order that the transformation shall lead to equations with constant coefficients. In actual practice the m.m.f.s and fluxes will not be of purely sinusoidal distribution but will contain space harmonics.† Within the scope of the present work it is not possible to take any account of these except in connection with the impedance to zero sequence, for which it is essential to do so.‡

Since the armature winding is balanced, $R_{\alpha\alpha} = R_{\beta\beta}$ and both may be represented by R_a.

The voltage equations can now be written

$$\begin{aligned} v_\alpha &= R_a i^\alpha + p(L_d \cos^2\theta + L_q \sin^2\theta) i^\alpha \\ &\quad + p(L_d - L_q)\sin\theta\cos\theta\, i^\beta + pL_{Dd}\cos\theta\, i^D \\ v_\beta &= p(L_d - L_q)\sin\theta\cos\theta\, i^\alpha + R_a i^\beta + p(L_d \sin^2\theta \\ &\quad + L_q \cos^2\theta) i^\beta + pL_{Dd}\sin\theta\, i^D \\ v_D &= pL_{Dd}\cos\theta\, i^\alpha + pL_{Dd}\sin\theta\, i^\beta + R_{DD} i^D + pL_{DD} i^D \end{aligned}$$

in which the p operates on both the functions of θ and on the currents.

These equations can be written in matrix form as

$$\begin{array}{c|c|} & \\ \alpha & v_\alpha \\ \hline \beta & v_\beta \\ \hline D & v_D \end{array} = \begin{array}{c|c|c|c|} & \alpha & \beta & D \\ \hline \alpha & R_a + p(L_d\cos^2\theta + L_q\sin^2\theta) & p(L_d - L_q)\sin\theta\cos\theta & pL_{Dd}\cos\theta \\ \hline \beta & p(L_d - L_q)\sin\theta\cos\theta & R_a + p(L_d\sin^2\theta + L_q\cos^2\theta) & pL_{Dd}\sin\theta \\ \hline D & pL_{Dd}\cos\theta & pL_{Dd}\sin\theta & R_{DD} + pL_{DD} \\ \hline \end{array} \begin{array}{c|c|} & \\ \alpha & i^\alpha \\ \hline \beta & i^\beta \\ \hline D & i^D \end{array}$$

† See refs. 6 and 7.

‡ See p. 133.

where again it must be remembered that the p operates both on the functions of θ in the impedance matrix and on the elements of the current matrix.

That some specific meaning can be given to L_d and L_q is apparent if the machine is examined when the axis of the α-phase winding coincides with that of the D winding, i.e. when $\theta = 0$. The equations then become

$$\begin{aligned} v_\alpha &= R_a i^\alpha + pL_d i^\alpha + pL_{Dd} i^D \\ v_\beta &= R_a i^\beta + pL_q i^\beta \\ v_D &= pL_{Dd} i^\alpha + R_{DD} i^D + pL_{DD} i^D \end{aligned}$$

Since L_d, L_q and L_{DD} are constant, these equations may be written

$$\begin{aligned} v_\alpha &= R_a i^\alpha + L_d p i^\alpha + L_{Dd} p i^D \\ v_\beta &= R_a i^\beta + L_q p i^\beta \\ v_D &= L_{Dd} p i^\alpha + R_{DD} i^D + L_{DD} p i^D \end{aligned}$$

From these it can be seen that L_d is the self-inductance of one phase of the armature winding when its axis coincides with the axis of the field, and L_q is the self-inductance when it is at right angles to the field axis.† L_d will normally be the maximum value of the armature winding inductance and L_q the minimum.

We will now apply the same transformation as for the uniform air-gap machine but in matrix form.

It is not necessary to transform the current i^D, consequently the complete transformation matrix is

$$\mathbf{C} = \begin{array}{c|c|c|c|} & \text{d} & \text{q} & \text{D} \\ \hline \alpha & \cos\theta & \sin\theta & 0 \\ \hline \beta & \sin\theta & -\cos\theta & 0 \\ \hline \text{D} & 0 & 0 & 1 \\ \hline \end{array}$$

† This is true only of the two-phase winding and not of the three-phase winding. See p. 138.

The application of this transformation is not, however, a matter for simple $\mathbf{C}_t\mathbf{ZC}$. In the equation $\mathbf{v} = \mathbf{Zi}$ in α, β, D axes, the p's of $\mathbf{Z}$ operate on the products of the functions of θ in $\mathbf{Z}$ and the currents $\mathbf{i}$. In the intermediate equation $\mathbf{v} = \mathbf{ZCi}'$, in which $\mathbf{Ci}'$, has replaced $\mathbf{i}$, the p's of $\mathbf{Z}$ operate on the triple products of the functions of θ in $\mathbf{Z}$, those of $\mathbf{C}$, and the currents of $\mathbf{i}'$, which is in d, q, D axes. It is therefore necessary to form the product $\mathbf{ZCi}'$ and then to perform the differentiations before premultiplying by $\mathbf{C}_t$ to obtain $\mathbf{v}'$, as has already been noted on p. 54.

It is for this reason that the operator p has been retained in distinction from the Laplace transform complex number represented in this work by s.

$$\mathbf{C} = \begin{array}{c|c|c|c|} & \alpha & \beta & \mathrm{D} \\ \hline \alpha & \begin{array}{c} R_a + L_d p \cos^2\theta \\ + L_q p \sin^2\theta \end{array} & \begin{array}{c} (L_d - L_q) \\ \times p \sin\theta\cos\theta \end{array} & L_{Dd} p \cos\theta \\ \hline \beta & \begin{array}{c} (L_d - L_q) \\ \times p \sin\theta\cos\theta \end{array} & \begin{array}{c} R_a + L_d p \sin^2\theta \\ + L_q\, p \cos^2\theta \end{array} & L_{Dd} p \sin\theta \\ \hline \mathrm{D} & L_{Dd} p \cos\theta & L_{Dd} p \sin\theta & R_{DD} + L_{DD} p \\ \hline \end{array} \quad \begin{array}{c|c|c|c|} & \mathrm{d} & \mathrm{q} & \mathrm{D} \\ \hline \alpha & \cos\theta & \sin\theta & 0 \\ \hline \beta & \sin\theta & -\cos\theta & 0 \\ \hline \mathrm{D} & 0 & 0 & 1 \\ \hline \end{array}$$

$$\mathbf{Ci}' = \begin{array}{c|c|c|c|} & \mathrm{d} & \mathrm{q} & \mathrm{D} \\ \hline \alpha & R_a\cos\theta + L_d p \cos\theta & R_a \sin\theta + L_q p \sin\theta & L_{Dd} p \cos\theta \\ \hline \beta & R_a \sin\theta + L_d p \sin\theta & -R_a\cos\theta - L_q p \cos\theta & L_{Dd} p \sin\theta \\ \hline \mathrm{D} & L_{Dd} p & 0 & R_{DD} + L_{DD} p \\ \hline \end{array} \quad \begin{array}{c|c|} \mathrm{d} & i^d \\ \hline \mathrm{q} & i^q \\ \hline \mathrm{D} & i^D \\ \hline \end{array}$$

$$= \begin{array}{c|c|} \alpha & R_a \cos\theta i^d + L_d p(\cos\theta i^d) + R_a \sin\theta i^q + L_q p(\sin\theta i^q) + L_{Dd} p(\cos\theta i^D) \\ \hline \beta & R_a \sin\theta i^d + L_d p(\sin\theta i^d) - R_a\cos\theta i^q - L_q p(\cos\theta i^q) + L_{Dd} p(\sin\theta i^D) \\ \hline \mathrm{D} & L_{Dd} p i^d \qquad\qquad + R_{DD} i^D + L_{DD} p i^D \\ \hline \end{array}$$

$$
= \begin{array}{c|l|}
\hline
\alpha & R_a \cos\theta i^d + L_d \cos\theta p i^d + R_a \sin\theta i^q + L_q \sin\theta p i^q + L_{Dd}\cos\theta p i^D \\
 & \quad -L_d \sin\theta\,.\,\dot{\theta} i^d \qquad\qquad\quad +L_q \cos\theta\,.\,\dot{\theta} i^q \qquad -L_{Dd}\sin\theta\,.\,\dot{\theta} i^D \\
\hline
\beta & R_a \sin\theta i^d + L_d \sin\theta p i^d - R_a \cos\theta i^q - L_q \cos\theta p i^q + L_{Dd}\sin\theta p i^D \\
 & \quad +L_d \cos\theta\,.\,\dot{\theta} i^d \qquad\qquad\quad +L_q \sin\theta\,.\,\dot{\theta} i^q \qquad +L_{Dd}\cos\theta\,.\,\dot{\theta} i^D \\
\hline
D & L_{Dd} p i^d \qquad\qquad\qquad\qquad + R_{DD} i^D + L_{DD} p i^D \\
\hline
\end{array}
$$

$$
= \begin{array}{c|c|c|c|}
 & d & q & D \\
\hline
\alpha & \begin{matrix} R_a \cos\theta + L_d \cos\theta p \\ -L_d \sin\theta\,.\,\dot{\theta} \end{matrix} & \begin{matrix} R_a \sin\theta + L_q \sin\theta p \\ +L_q \cos\theta\,.\,\dot{\theta} \end{matrix} & \begin{matrix} L_{Dd}\cos\theta p \\ -L_{Dd}\sin\theta\,.\,\dot{\theta} \end{matrix} \\
\hline
\beta & \begin{matrix} R_a \sin\theta + L_d \sin\theta p \\ +L_d \cos\theta\,.\,\dot{\theta} \end{matrix} & \begin{matrix} -R_a \cos\theta - L_q \cos\theta p \\ +L_q \sin\theta\,.\,\dot{\theta} \end{matrix} & \begin{matrix} L_{Dd}\sin\theta p \\ +L_{Dd}\cos\theta\,.\,\dot{\theta} \end{matrix} \\
\hline
D & L_{Dd} p & & R_{DD} + L_{DD} p \\
\hline
\end{array}
\quad
\begin{array}{c|c|}
\hline
d & i^d \\
\hline
q & i^q \\
\hline
D & i^D \\
\hline
\end{array}
$$

where the p now operates only on the current following it.

To complete the transformation this must be pre-multiplied by the transpose of **C**, which, however, is equal to **C** since it is symmetric. If this multiplication is performed and the resulting impedance matrix simplified, it is found to be

$$
\begin{array}{c|c|c|c|}
 & d & q & D \\
\hline
d & R_a + L_d p & L_q \dot{\theta} & L_{Dd} p \\
\hline
q & -L_d \dot{\theta} & R_a + L_q p & -L_{Dd} \dot{\theta} \\
\hline
D & L_{Dd} p & & R_{DD} + L_{DD} p \\
\hline
\end{array}
$$

In this case also, therefore, the transformation leads to linear differential equations with constant coefficients in the form

$$
\begin{array}{c|c|}
\hline
d & v_d \\
\hline
q & v_q \\
\hline
D & v_D \\
\hline
\end{array}
=
\begin{array}{c|c|c|c|}
 & d & q & D \\
\hline
d & R_a + L_d p & L_q \dot{\theta} & L_{Dd} p \\
\hline
q & -L_d \dot{\theta} & R_a + L_q p & -L_{Dd} \dot{\theta} \\
\hline
D & L_{Dd} p & & R_{DD} + L_{DD} p \\
\hline
\end{array}
\quad
\begin{array}{c|c|}
\hline
d & i^d \\
\hline
q & i^q \\
\hline
D & i^D \\
\hline
\end{array}
$$

provided that the speed is constant.

It will be noted that this impedance matrix is identical with that of the metadyne on p. 55, if the assumption is made that the fluxes of the metadyne are, like those of the two-phase machine, sinusoidally distributed around the air-gap. This result is particularly important in that it emphasizes the essential identity of the different types of electrical machine.

We have, incidentally, shown that balanced stator windings are not a necessary condition, since we have obtained constant coefficients with only one stator winding, namely D.

The Form of the Transformed Impedance Matrix

In terms of the d, q, D reference axes the impedance matrix **Z** consists of three parts:

1. A resistance matrix **R** =

$$\begin{array}{c|c|c|c|} & \text{d} & \text{q} & \text{D} \\ \hline \text{d} & R_a & & \\ \hline \text{q} & & R_a & \\ \hline \text{D} & & & R_{DD} \\ \hline \end{array}$$

which is diagonal.

2. An inductance matrix **L** given by the coefficients of p as

$$\mathbf{L} = \begin{array}{c|c|c|c|} & \text{d} & \text{q} & \text{D} \\ \hline \text{d} & L_d & & L_{Dd} \\ \hline \text{q} & & L_q & \\ \hline \text{D} & L_{Dd} & & L_{DD} \\ \hline \end{array}$$

which is symmetric.

3. A matrix formed by the coefficients of $\dot{\theta}$ to which the symbol **G** is given.

$$\mathbf{G} = \begin{array}{c|c|c|c|} & \text{d} & \text{q} & \text{D} \\ \hline \text{d} & & L_q & \\ \hline \text{q} & -L_d & & -L_{Dd} \\ \hline \text{D} & & & \\ \hline \end{array}$$

which has no obvious form.

Of these matrices the first two are the resistances and inductances which would exist in a stationary machine having the d and D axes coincident. The third, **G**, has elements only in the rows d, q which have replaced the moving windings α and β. These elements result in components of voltage proportional to speed, i.e. in the rotationally generated voltages.

The voltage equation can thus be written

$$\mathbf{v} = \mathbf{Zi} = (\mathbf{R}+\mathbf{L}p+\mathbf{G}\dot{\theta})\mathbf{i} = \mathbf{Ri}+\mathbf{L}p\mathbf{i}+\mathbf{G}\dot{\theta}\mathbf{i}$$

The voltage equation of the uniform air-gap machine with balanced two-phase windings given on p. 63 can be similarly expressed.

The recognition that **Z** consists of these three parts leads to a systematic way of writing down the impedance matrix in terms of "stationary" reference axes such as d, q, D as follows:

1. Write in the principal diagonal the terms representing the resistances of the windings. Since the d and q relate to the non-salient member, their resistance values are equal to those of the balanced windings of that member. Those relating to the other member are the actual values, because these axes are not transformed.

2. Also along the principal diagonal write the Lp terms corresponding to the self-inductances of the windings. Here L_d and L_q will not be equal if the other member is salient, even though the windings are, of course, identical.

3. Write in the mutual inductance terms Mp wherever windings have a common axis. Each term will appear twice, because mutual terms are always symmetric.

4. Where Lp or Mp appears in a row replacing that corresponding to a moving winding (i.e. in a d or q row) write $L\dot{\theta}$ or $M\dot{\theta}$ in the same column of the other such row (i.e. in the q or d row), prefixing those in a q row with a negative sign.

Applying this procedure to the machine of Fig. 13 (p. 65) the results of the various steps are as follows:

1.

	d	q	D
d	R_a		
q		R_a	
D			R_{DD}

2.

	d	q	D
d	$L_d p$		
q		$L_q p$	
D			$L_{DD} p$

3.

	d	q	D
d			$L_{dD} p$
q			
D	$L_{dD} p$		

4. The relevant terms of the above two matrices are

	d	q	D
d	$L_d p$		$L_{dD} p$
q		$L_q p$	
D			

which leads to

	d	q	D
d		$L_q \dot{\theta}$	
q	$-L_d \dot{\theta}$		$-L_{dD} \dot{\theta}$
D			

Adding this to the results of steps 1, 2, and 3 above gives the complete matrix

	d	q	D
d	$R_a+L_d p$	$L_q\dot{\theta}$	$L_{dD} p$
q	$-L_d\dot{\theta}$	$R_a+L_q p$	$-L_{dD}\dot{\theta}$
D	$L_{dD} p$		$R_{DD}+L_{DD} p$

It is important to note that this procedure leads to the correct results irrespective of how many windings there are on two members of the machine.

Thus the impedance matrix of the most complicated machine can be written down by a routine procedure in terms of axes such as d, q, D, Q, provided that the machine conforms to the limitations specified below.

Limitations of the Method

It is convenient to review the position at this stage.

The fundamental equations of an electro-dynamic system are inherently integro-differential equations. In most electrical machine analysis they reduce to differential equations without further differentiation since the effects of capacitance can be ignored.†

The routine solution of such equations by means of Laplace transforms is restricted to equations with constant coefficients and involves the solution of a set of simultaneous linear equations, which can be most conveniently performed in matrix terms. This particular method of solution is therefore restricted to devices for which the differential equations either have constant coefficients, or which can be transformed by a linear transformation into equations having constant coefficients. The machines which comply with this condition are

1. Commutator machines, for which the equations can be written down as for the metadyne on p. 53.

† This does not mean that connection of external capacitors to machines cannot be considered without further complication.

2. Polyphase machines which:

(a) Have the windings of one member uniformly distributed as a balanced polyphase winding—including squirrel-cage windings. The windings of the other member may be distributed or concentrated, balanced or unbalanced (pp. 65–73).

(b) Have the balanced windings on a non-salient member, i.e. one with a cylindrical surface. This is necessary in order that balanced windings shall be possible, but it also results in constant self-inductances for the windings of the other member, since the permeances of their magnetic circuits are independent of the angular position of the members (p. 68).
The other member may be salient or non-salient (pp. 57–73).

(c) Are such that space harmonics may be ignored. This is equivalent to requiring sinusoidal variation of inductances with position (p. 64 and 66).
With both commutator and polyphase machines it is necessary to assume that:

(a) Changes of saturation can be ignored. This is necessary so that the inductances shall not be functions of the currents (p. 58).

(b) The angular velocity is constant. This has the effect of making terms involving $\dot{\theta}$ constant (p. 63). Strictly speaking this is not essential, but if this is not assumed it is necessary to know the mechanical torque equation so that $\dot{\theta}$ (and $\ddot{\theta}$) can be eliminated. The form of the equations resulting from this elimination, and hence the possibility of routine solution, depend upon the particular case.

In spite of the theoretical restriction to constant speed, the routine method can be used for approximate solutions of acceleration and oscillation problems provided that the mechanical time-constants are very large compared with the electrical time-constants. This is a usual assumption and is justified in most cases of rotating machines.

The other restrictions mean that the method cannot be used for machines in which both members are salient, as in inductor alternators, or for machines with unbalanced windings on both sides of the air-gap, as in the single-phase, synchronous generator (alternator). In such cases it is possible to write down the equations in terms of the actual winding axes, but transformation does not lead to constant coefficients and thus offers no advantage. Whether such equations can be solved depends upon the particular case and on what, if any, simplifying assumptions may be justified.

The single-phase alternator is discussed in Chapter 13 (pp. 305–308).

To avoid repetitions in the analysis, with only minor differences, three further restrictions will be imposed without any loss of generality:

(a) Only two-pole machines will be shown diagrammatically and analysed, in order to avoid ambiguity with differing electrical and mechanical angles. To apply the results to multipolar machines it will be necessary to reduce the speed and to increase the torque according to the number of pole-pairs.

This is equivalent to taking $\dot{\theta}$ as the electrical angular velocity in all cases, so that for a multipolar machine the mechanical angular velocity will be $\dot{\theta}/p$ and the torque will be pT, where p is the number of pole-pairs and T is the mechanical power divided by the electrical angular velocity, i.e. the torque as for a two-pole machine.

(b) Because commutator machines are more conveniently constructed and analysed as revolving-armature machines, it will be assumed that in all cases the rotor is the non-salient member with a polyphase or commutator-type winding. The equations for the revolving-field machines are, of course, identical with those of the revolving-armature machines except for such details as the direction of rotation regarded as positive. They can, however, be derived directly if desired by exactly similar techniques.

(c) Because some circuits are of necessity "sinks", and in some cases all circuits are sinks, it is convenient so to treat all circuits in all machines. This means in effect that all machines are analysed as motors, or rather with the sign conventions normally used when machines are regarded as motors. Negative terminal power and negative mechanical output power thus indicate generation.

To obtain the equations of generators conforming to the usual generator conventions, it will be necessary to change the conventional directions of the armature currents. It will, therefore, be necessary to change the signs of the elements in the corresponding columns of the impedance matrix.

Alternatively, the problem can be solved using "motor" conventions and the sign of the armature currents changed as a final step. This latter method is less liable to error.

CHAPTER 6

The Torque Expressions

THE following analysis to determine the torque in terms of the currents is based on the principle of the conservation of energy. It is postulated that that part of the electrical input power which cannot be otherwise accounted for will appear as mechanical output power. Electrical input power may be stored as energy associated with the magnetic or electric fields, dissipated as electromagnetic radiation or heat, or converted into the desired mechanical form. It is assumed that the capacitances and voltages of the system are so small that the energy associated with the electric fields at any time is wholly negligible, so that the only energy to be considered is that in the magnetic fields. The power represented by electromagnetic radiation will be trivially small at the rates of change of currents normally encountered in electrical machines and can consequently also be ignored.

It is apparent that energy will be given out as heat as a result of the resistance of the conductors and magnetic hysteresis and eddy currents in the iron parts of the magnetic circuit. For the present purpose, however, as has already been shown, the inductances of the system in any given position have to be regarded as independent of the currents. This is equivalent to replacing the hysteresis loop by a straight line through the origin, i.e. equivalent to neglecting both changes of saturation and hysteresis. Since eddy currents also have to be neglected in a simple analysis, the consequence is that the electrical input power is assumed to be entirely accounted for by the sum of that which increases the energy of the magnetic fields, that dissipated as heat in the form of the I^2R losses and that given out in the form of mechanical work.

This last assumption may also be made if the hysteresis loop is replaced by a single *curved* line through the origin, i.e. if changes of saturation, but not hysteresis, are taken into account. The torque expressions which result cannot be applied directly and simply to electrical machines, but are derived in Appendix B.

When changes of saturation are neglected, the inductances are functions solely of θ, the angular position of the rotor.

The Energy stored in the Magnetic Fields

In order to deduce an expression for the torque of a machine, it is necessary to find first a value for the energy stored in the magnetic fields. When saturation, and therefore also hysteresis, are neglected, magnetic energy is dependent only on the inductances and currents at the instant considered. It is independent of the means by which that state is reached. Let us suppose, therefore, that the inductances are first adjusted and that the currents are then brought up from zero to the specified values. Because the inductances are constant during the charging process, the voltage equation $\mathbf{v} = \mathbf{Ri}+p(\mathbf{Li})$ reduces to $\mathbf{v} = \mathbf{Ri}+\mathbf{L}p\mathbf{i}$. For a system with three windings, taken as an example, the voltages equations are, therefore,

$$v_1 = R_{11}i^1+L_{11}\frac{\mathrm{d}i^1}{\mathrm{d}t}+M_{12}\frac{\mathrm{d}i^2}{\mathrm{d}t}+M_{13}\frac{\mathrm{d}i^3}{\mathrm{d}t}$$

$$v_2 = M_{21}\frac{\mathrm{d}i^1}{\mathrm{d}t}+R_{22}i^2+L_{22}\frac{\mathrm{d}i^2}{\mathrm{d}t}+M_{23}\frac{\mathrm{d}i^3}{\mathrm{d}t}$$

$$v_3 = M_{31}\frac{\mathrm{d}i^1}{\mathrm{d}t}+M_{32}\frac{\mathrm{d}i^2}{\mathrm{d}t}+R_{33}i^3+L_{33}\frac{\mathrm{d}i^3}{\mathrm{d}t}$$

Multiplying these equations by i^1, i^2, and i^3 respectively and adding gives the total input power at the terminals as

$$\begin{aligned} i^1v_1+i^2v_2+i^3v_3 = {} & R_{11}(i^1)^2+L_{11}i^1\frac{\mathrm{d}i^1}{\mathrm{d}t}+M_{12}i^1\frac{\mathrm{d}i^2}{\mathrm{d}t}+M_{13}i^1\frac{\mathrm{d}i^3}{\mathrm{d}t} \\ & +R_{22}(i^2)^2+L_{22}i^2\frac{\mathrm{d}i^2}{\mathrm{d}t}+M_{21}i^2\frac{\mathrm{d}i^1}{\mathrm{d}t}+M_{23}i^2\frac{\mathrm{d}i^3}{\mathrm{d}t} \\ & +R_{33}(i^3)^2+L_{33}i^3\frac{\mathrm{d}i^3}{\mathrm{d}t}+M_{31}i^3\frac{\mathrm{d}i^1}{\mathrm{d}t}+M_{32}i^3\frac{\mathrm{d}i^2}{\mathrm{d}t} \end{aligned}$$

Of this total, $R_{11}(i^1)^2+R_{22}(i^2)^2+R_{33}(i^3)^2$ is obviously the power dissipated in the resistances, and, because the machine is at rest, the remainder must be the power being stored in the magnetic fields. This power may be expressed in matrix form, if desired, as

$$\begin{array}{ccc} 1 & 2 & 3 \\ \hline i^1 & i^2 & i^3 \end{array} \quad \begin{array}{c|ccc} & 1 & 2 & 3 \\ \hline 1 & L_{11} & M_{12} & M_{13} \\ 2 & M_{21} & L_{22} & M_{23} \\ 3 & M_{31} & M_{32} & L_{33} \end{array} \quad \begin{array}{c|c} 1 & di^1/dt \\ 2 & di^2/dt \\ 3 & di^3/dt \end{array}$$

$$= \begin{array}{ccc} 1 & 2 & 3 \\ \hline i^1 & i^2 & i^3 \end{array} \quad \begin{array}{c|ccc} & 1 & 2 & 3 \\ \hline 1 & L_{11}p & M_{12}p & M_{13}p \\ 2 & M_{21}p & L_{22}p & M_{23}p \\ 3 & M_{31}p & M_{32}p & L_{33}p \end{array} \quad \begin{array}{c|c} 1 & i^1 \\ 2 & i^2 \\ 3 & i^3 \end{array}$$

Noting that $M_{12} = M_{21}$, etc., and integrating with respect to time from zero initial conditions, gives the total stored energy as

$$\int_0^t \left\{ L_{11}i^1 \frac{di^1}{dt} + L_{22}i^2 \frac{di^2}{dt} + L_{33}i^3 \frac{di^3}{dt} + M_{12}\left(i^1 \frac{di^2}{dt} + i^2 \frac{di^1}{dt}\right) \right.$$
$$\left. + M_{23}\left(i^2 \frac{di^3}{dt} + i^3 \frac{di^2}{dt}\right) + M_{31}\left(i^3 \frac{di^1}{dt} + i^1 \frac{di^3}{dt}\right) \right\} dt$$

$$= \int_0^{i^1} L_{11}i^1\, di^1 + \int_0^{i^2} L_{22}i^2\, di^2 + \int_0^{i^3} L_{33}i^3\, di^3$$
$$+ \int_{0,\,0}^{i^1,\,i^2} M_{12}(i^1\, di^2 + i^2\, di^1) + \int_{0,\,0}^{i^2,\,i^3} M_{23}(i^2\, di^3 + i^3\, di^2)$$
$$+ \int_{0,\,0}^{i^3,\,i^1} M_{31}(i^3\, di^1 + i^1\, di^3)$$

$$= \tfrac{1}{2}L_{11}(i^1)^2 + \tfrac{1}{2}L_{22}(i^2)^2 + \tfrac{1}{2}L_{33}(i^3)^2$$
$$+ M_{12}i^1i^2 + M_{23}i^2i^3 + M_{31}i^3i^1$$

$$= \frac{1}{2} \begin{array}{c} \begin{array}{ccc} 1 & 2 & 3 \end{array} \\ \begin{array}{|c|c|c|} \hline i^1 & i^2 & i^3 \\ \hline \end{array} \end{array} \begin{array}{c|c|c|c|} & 1 & 2 & 3 \\ \hline 1 & L_{11} & M_{12} & M_{31} \\ \hline 2 & M_{12} & L_{22} & M_{23} \\ \hline 3 & M_{31} & M_{23} & L_{33} \\ \hline \end{array} \begin{array}{c|c|} \hline 1 & i^1 \\ \hline 2 & i^2 \\ \hline 3 & i^3 \\ \hline \end{array} = \tfrac{1}{2}\mathbf{i}_t\mathbf{L}\mathbf{i}$$

For the general case, both sides of the voltage equation $\mathbf{v} = \mathbf{Ri}+\mathbf{L}p\mathbf{i}$ are premultiplied by $\mathbf{i}_t$ to obtain the power equation

$$\mathbf{i}_t\mathbf{v} = \mathbf{i}_t\mathbf{Ri}+\mathbf{i}_t\mathbf{L}p\mathbf{i}$$

Now $\mathbf{R}$ is a diagonal matrix, so that the scalar $\mathbf{i}_t\mathbf{Ri}$ is of the form $\sum_{k=1\ldots n} i^k R_{kk} i^k = \sum_{k=1\ldots n} R_{kk}(i^k)^2$, where n is the number of circuits. $\mathbf{i}_t\mathbf{Ri}$ is thus the total power dissipated as heat in the resistances of the system.

Since the system is stationary, there is no mechanical power output, and the input power $\mathbf{i}_t\mathbf{v}$ can only be dissipated as heat or stored. It follows that the power being stored is $\mathbf{i}_t\mathbf{L}p\mathbf{i} = \mathbf{i}_t\mathbf{L}(\mathrm{d}\mathbf{i}/\mathrm{d}t)$. The stored energy is therefore

$$\int_0^t \mathbf{i}_t\mathbf{L}\frac{\mathrm{d}\mathbf{i}}{\mathrm{d}t}\,\mathrm{d}t$$

Now $\mathbf{i}_t\mathbf{L}(\mathrm{d}\mathbf{i}/\mathrm{d}t)$ is a scalar and is, therefore, equal to its own transpose $(\mathrm{d}\mathbf{i}/\mathrm{d}t)_t\,\mathbf{L}_t\mathbf{i}$, which is equal to $(\mathrm{d}\mathbf{i}_t/\mathrm{d}t)\mathbf{Li}$, because $\mathbf{L}$ is symmetric and is therefore equal to its own transpose, whilst the differential of a transpose is equal to the transpose of a differential. Hence

$$\begin{aligned} \int_0^t \mathbf{i}_t\mathbf{L}\frac{\mathrm{d}\mathbf{i}}{\mathrm{d}t}\,\mathrm{d}t &= \frac{1}{2}\int_0^t \left\{\mathbf{i}_t\mathbf{L}\frac{\mathrm{d}\mathbf{i}}{\mathrm{d}t}+\frac{\mathrm{d}\mathbf{i}_t}{\mathrm{d}t}\mathbf{Li}\right\}\mathrm{d}t \\ &= \frac{1}{2}\int_0^t \frac{\mathrm{d}}{\mathrm{d}t}(\mathbf{i}_t\mathbf{Li})\,\mathrm{d}t = \frac{1}{2}\Big[\mathbf{i}_t\mathbf{Li}\Big]_0^{\mathbf{i}} \\ &= \tfrac{1}{2}\,\mathbf{i}_t\mathbf{Li} \end{aligned}$$

as before.

Torque Expressions

There are two alternative expressions for the torque, one being deduced from the voltage equation of p. 56, and the other from the commutator machine equation of p. 55, or from the transformed equation of p. 74, which are identical in form.

Derivation of the Torque Expression from the Equation $\mathbf{v} = \mathbf{Ri}+p(\mathbf{Li})$

In the system considered on p. 56, where the currents i^1, i^2, etc., were the actual currents, the inductances were functions of the angular position. The voltage equation of this system in matrix form is

$$\mathbf{v} = \mathbf{Ri}+\frac{\mathrm{d}}{\mathrm{d}t}(\mathbf{Li})$$

Consider such a system during the movement from an angular position θ to a position $\theta+\Delta\theta$, all currents being maintained constant by appropriate terminal voltages $\mathbf{v}$. Since the currents are constant, $\mathrm{d}\mathbf{i}/\mathrm{d}t$ is a null matrix, and the voltage equation reduces to

$$\mathbf{v} = \mathbf{Ri}+\frac{\mathrm{d}\mathbf{L}}{\mathrm{d}t}\mathbf{i}$$

For a system consisting of three windings this expands to

$$v_1 = R_{11}i^1+\frac{\mathrm{d}L_{11}}{\mathrm{d}t}i^1+\frac{\mathrm{d}M_{12}}{\mathrm{d}t}i^2+\frac{\mathrm{d}M_{13}}{\mathrm{d}t}i^3$$

$$v_2 = R_{22}i^2+\frac{\mathrm{d}M_{21}}{\mathrm{d}t}i^1+\frac{\mathrm{d}L_{22}}{\mathrm{d}t}i^2+\frac{\mathrm{d}M_{23}}{\mathrm{d}t}i^3$$

$$v_3 = R_{33}i^3+\frac{\mathrm{d}M_{31}}{\mathrm{d}t}i^1+\frac{\mathrm{d}M_{32}}{\mathrm{d}t}i^2+\frac{\mathrm{d}L_{33}}{\mathrm{d}t}i^3$$

Multiplying these equations by i^1, i^2, i^3 respectively and adding gives the total input power to the terminals as

$$\begin{aligned} i^1v_1+i^2v_2+i^3v_3 = {} & R_{11}(i^1)^2+(i^1)^2\frac{\mathrm{d}L_{11}}{\mathrm{d}t}+i^1i^2\frac{\mathrm{d}M_{12}}{\mathrm{d}t}+i^1i^3\frac{\mathrm{d}M_{13}}{\mathrm{d}t} \\ & +R_{22}(i^2)^2+i^2i^1\frac{\mathrm{d}M_{21}}{\mathrm{d}t}+(i^2)^2\frac{\mathrm{d}L_{22}}{\mathrm{d}t}+i^2i^3\frac{\mathrm{d}M_{23}}{\mathrm{d}t} \\ & +R_{33}(i^3)^2+i^3i^1\frac{\mathrm{d}M_{31}}{\mathrm{d}t}+i^3i^2\frac{\mathrm{d}M_{32}}{\mathrm{d}t}+(i^3)^2\frac{\mathrm{d}L_{33}}{\mathrm{d}t} \end{aligned}$$

Here again $\{R_{11}(i^1)^2+R_{22}(i^2)^2+R_{33}(i^3)^2\}$ is the power dissipated in the resistances. Since, however, the system is moving, the remainder includes both power being stored magnetically and the power being converted to mechanical form. The sum of the increase of stored energy and the energy converted is thus

$$\begin{aligned} \int_0^t \Bigg\{ & (i^1)^2\frac{\mathrm{d}L_{11}}{\mathrm{d}t}+(i^2)^2\frac{\mathrm{d}L^{22}}{\mathrm{d}t}+(i^3)^2\frac{\mathrm{d}L_{33}}{\mathrm{d}t} \\ & +2i^1i^2\frac{\mathrm{d}M_{12}}{\mathrm{d}t}+2i^2i^3\frac{\mathrm{d}M_{23}}{\mathrm{d}t}+2i^3i^1\frac{\mathrm{d}M_{31}}{\mathrm{d}t}\Bigg\}\,\mathrm{d}t \\ = {} & \int_{L_{11}}^{L_{11}+\Delta L_{11}}(i^1)^2\,\mathrm{d}L_{11}+\int_{L_{22}}^{L_{22}+\Delta L_{22}}(i^2)^2\,\mathrm{d}L_{22} \\ & +\int_{L_{33}}^{L_{33}+\Delta L_{33}}(i^3)^2\,\mathrm{d}L_{33}+\int_{M_{12}}^{M_{12}+\Delta M_{12}}2i^1i^2\,\mathrm{d}M_{12} \\ & +\int_{M_{23}}^{M_{23}+\Delta M_{23}}2i^2i^3\,\mathrm{d}M_{23}+\int_{M_{31}}^{M_{31}+\Delta M_{31}}2i^3i^1\,\mathrm{d}M_{31} \\ = {} & (i^1)^2\,\Delta L_{11}+(i^2)^2\,\Delta L_{22}+(i^3)^2\,\Delta L_{33} \\ & +2i^1i^2\,\Delta M_{12}+2i^2i^3\,\Delta M_{23}+2i^3i^1\,\Delta M_{31} \end{aligned}$$

However, the initial stored energy was

$$\tfrac{1}{2}(i^1)^2L_{11}+\tfrac{1}{2}(i^2)^2L_{22}+\tfrac{1}{2}(i^3)^2L_{33}+i^1i^2M_{12}+i^2i^3M_{23}+i^3i^1M_{31}$$

and the final stored energy is

$$\begin{aligned} & \tfrac{1}{2}(i^1)^2\,(L_{11}+\Delta L_{11})+\tfrac{1}{2}(i^2)^2\,(L_{22}+\Delta L_{22})+\tfrac{1}{2}(i^3)^2\,(L_{33}+\Delta L_{33}) \\ & \qquad +i^1i^2(M_{12}+\Delta M_{12})+i^2i^3(M_{23}+\Delta M_{23})+i^3i^1(M_{31}+\Delta M_{31}) \end{aligned}$$

so that the increase of stored energy is

$$\tfrac{1}{2}(i^1)^2\,\Delta L_{11}+\tfrac{1}{2}(i^2)^2\,\Delta L_{22}+\tfrac{1}{2}(i^3)^2\,\Delta L_{33}+i^1i^2\,\Delta M_{12}+i^2i^3\,\Delta M_{23}+i^3i^1\,\Delta M_{31}$$

This is exactly half the sum of the increase of stored energy and the energy converted to mechanical power. It follows that the energy converted to mechanical power is also

$$\tfrac{1}{2}(i^1)^2\,\Delta L_{11}+\tfrac{1}{2}(i^2)^2\,\Delta L_{22}+\tfrac{1}{2}(i^3)^2\,\Delta L_{33}+i^1i^2\,\Delta M_{12}+i^2i^3\,\Delta M_{32}+i^3i^1\,\Delta M_{31}$$

Since the angle turned through is $\Delta\theta$, the mean torque is

$$\tfrac{1}{2}(i^1)^2\frac{\Delta L_{11}}{\Delta\theta}+\tfrac{1}{2}(i^2)^2\frac{\Delta L_{22}}{\Delta\theta}+\tfrac{1}{2}(i^3)^2\frac{\Delta L_{33}}{\Delta\theta}+i^1i^2\frac{\Delta M_{12}}{\Delta\theta}+i^2i^3\frac{\Delta M_{23}}{\Delta\theta}+i^3i^1\frac{\Delta M_{31}}{\Delta\theta}$$

In the limit, as $\Delta\theta \to 0$, the torque is

$$\tfrac{1}{2}(i^1)^2\frac{\partial L_{11}}{\partial\theta}+\tfrac{1}{2}(i^2)^2\frac{\partial L_{22}}{\partial\theta}+\tfrac{1}{2}(i^3)^2\frac{\partial L_{33}}{\partial\theta}+i^1i^2\frac{\partial M_{12}}{\partial\theta}+i^2i^3\frac{\partial M_{23}}{\partial\theta}+i^3i^1\frac{\partial M_{31}}{\partial\theta}$$

$$=\tfrac{1}{2}\begin{array}{c}\begin{array}{ccc}1&2&3\end{array}\\\begin{bmatrix}i^1&i^2&i^3\end{bmatrix}\end{array}\begin{array}{c|ccc}&1&2&3\\\hline 1&\dfrac{\partial L_{11}}{\partial\theta}&\dfrac{\partial M_{12}}{\partial\theta}&\dfrac{\partial M_{13}}{\partial\theta}\\2&\dfrac{\partial M_{21}}{\partial\theta}&\dfrac{\partial L_{22}}{\partial\theta}&\dfrac{\partial M_{23}}{\partial\theta}\\3&\dfrac{\partial M_{31}}{\partial\theta}&\dfrac{\partial M_{32}}{\partial\theta}&\dfrac{\partial L_{33}}{\partial\theta}\end{array}\begin{array}{c|c}1&i^1\\2&i^2\\3&i^3\end{array}$$

$$=\tfrac{1}{2}\,\mathbf{i}_t\frac{\partial\mathbf{L}}{\partial\theta}\,\mathbf{i}$$

It is customary to use the symbol for partial differentiation on the ground that there may be other degrees of freedom. For example, the rotor may be free to move axially.

In the general case, the power equation is

$$\mathbf{i}_t\mathbf{v} = \mathbf{i}_t\mathbf{R}\mathbf{i}+\mathbf{i}_t\frac{d\mathbf{L}}{dt}\mathbf{i}$$

Since $\mathbf{i}_t\mathbf{R}\mathbf{i}$ is the power dissipated, $\mathbf{i}_t(d\mathbf{L}/dt)\mathbf{i}$ must be the sum of the power stored and the power converted to mechanical form.

If now the small finite displacement $\Delta\theta$ occurs in a time Δt, with a corresponding change of inductance $\Delta\mathbf{L}$, the initial stored energy is $\frac{1}{2}\mathbf{i}_t\mathbf{L}\mathbf{i}$ and the final stored energy is $\frac{1}{2}\mathbf{i}_t(\mathbf{L}+\Delta\mathbf{L})\mathbf{i} = \frac{1}{2}\mathbf{i}_t\mathbf{L}\mathbf{i}+\frac{1}{2}\mathbf{i}_t\Delta\mathbf{L}\mathbf{i}$. The increase in stored energy is thus $\frac{1}{2}\mathbf{i}_t\Delta\mathbf{L}\mathbf{i}$. Since Δt is a scalar, we may divide by it to obtain the rate of increase in stored energy as

$$\tfrac{1}{2}\mathbf{i}_t\frac{\Delta\mathbf{L}}{\Delta t}\mathbf{i} \to \tfrac{1}{2}\mathbf{i}_t\frac{d\mathbf{L}}{dt}\mathbf{i} \quad \text{as } \Delta t \text{ tends to zero.}$$

The power being stored is thus $\frac{1}{2}\mathbf{i}_t(d\mathbf{L}/dt)\mathbf{i}$, and consequently the power being converted is

$$\mathbf{i}_t\frac{d\mathbf{L}}{dt}\mathbf{i}-\tfrac{1}{2}\mathbf{i}_t\frac{d\mathbf{L}}{dt}\mathbf{i} = \tfrac{1}{2}\mathbf{i}_t\frac{d\mathbf{L}}{dt}\mathbf{i}$$

$$= \tfrac{1}{2}\mathbf{i}_t\frac{\partial\mathbf{L}}{\partial\theta}\frac{d\theta}{dt}\mathbf{i}$$

$$= \tfrac{1}{2}\mathbf{i}_t\frac{\partial\mathbf{L}}{\partial\theta}\dot{\theta}\mathbf{i}$$

Because the angular velocity $\dot{\theta}$ is a scalar, it is permissible to divide the power by it to obtain the two-pole torque T as $\frac{1}{2}\mathbf{i}_t(\partial\mathbf{L}/\partial\theta)\mathbf{i}$.

For a machine with p pole pairs, the power converted would be divided by the mechanical angular velocity $\dot{\theta}/\text{p}$ to obtain the torque as $(\text{p}/2)\,\mathbf{i}_t(\partial\mathbf{L}/\partial\theta)\mathbf{i}$, or $\text{p}T$ where T is the two-pole torque as above.

Torque, like stored energy, depends only upon the state of the system at the particular time, and is quite independent of the process whereby the system has been brought to this state. This expression is, therefore, valid under all conditions and is not in any way restricted by the sequence of steps used in its derivation.

The Transformation of $\partial\mathbf{L}/\partial\theta$

Once the elements of $\partial\mathbf{L}/\partial\theta$ have been determined for a particular case, the matrix $\partial\mathbf{L}/\partial\theta$ can be transformed to other systems of axes as required. Applying this to the impedance matrix of p. 69 we have

$$\mathbf{L} = \begin{array}{c|c|c|c|} & \alpha & \beta & D \\ \hline \alpha & L_d \cos^2\theta + L_q \sin^2\theta & (L_d - L_q)\sin\theta\cos\theta & L_{Dd}\cos\theta \\ \hline \beta & (L_d - L_q)\sin\theta\cos\theta & L_d \sin^2\theta + L_q \cos^2\theta & L_{Dd}\sin\theta \\ \hline D & L_{Dd}\cos\theta & L_{Dd}\sin\theta & L_{DD} \\ \hline \end{array}$$

Differentiating this gives us

$$\frac{\partial\mathbf{L}}{\partial\theta} = \begin{array}{c|c|c|c|} & \alpha & \beta & D \\ \hline \alpha & -(L_d - L_q)\sin 2\theta & (L_d - L_q)\cos 2\theta & -L_{Dd}\sin\theta \\ \hline \beta & (L_d - L_q)\cos 2\theta & (L_d - L_q)\sin 2\,\theta & L_{Dd}\cos\theta \\ \hline D & -L_{Dd}\sin\theta & L_{Dd}\cos\theta & 0 \\ \hline \end{array}$$

which may be transformed to d, q, D axes by means of the transformation matrix of p. 70 in the same way that impedance is transformed.

$$\frac{\partial\mathbf{L}}{\partial\theta}\mathbf{C} = \begin{array}{c|c|c|c|} & \alpha & \beta & D \\ \hline \alpha & -(L_d - L_q)\sin 2\theta & (L_d - L_q)\cos 2\theta & -L_{Dd}\sin\theta \\ \hline \beta & (L_d - L_q)\cos 2\theta & (L_d - L_q)\sin 2\theta & L_{Dd}\cos\theta \\ \hline D & -L_{Dd}\sin\theta & L_{Dd}\cos\theta & 0 \\ \hline \end{array} \; \begin{array}{c|c|c|c} & d & q & \text{I} \\ \hline \alpha & \cos\theta & \sin\theta & \\ \hline \beta & \sin\theta & -\cos\theta & \\ \hline D & 0 & 0 & \\ \hline \end{array}$$

$$= \begin{array}{c|c|c|c|} & d & q & D \\ \hline \alpha & -(L_d - L_q)\sin\theta & -(L_d - L_q)\cos\theta & -L_{Dd}\sin\theta \\ \hline \beta & (L_d - L_q)\cos\theta & -(L_d - L_q)\sin\theta & L_{Dd}\cos\theta \\ \hline D & 0 & -L_{Dd} & 0 \\ \hline \end{array}$$

Hence

$$\mathbf{C}_t \frac{\partial \mathbf{L}}{\partial \theta} \mathbf{C}$$

$$= \begin{array}{c|c|c|c|} & \alpha & \beta & \mathrm{D} \\ \hline \mathrm{d} & \cos\theta & \sin\theta & 0 \\ \hline \mathrm{q} & \sin\theta & -\cos\theta & 0 \\ \hline \mathrm{D} & 0 & 0 & 1 \\ \hline \end{array} \quad \begin{array}{c|c|c|c|} & \mathrm{d} & \mathrm{q} & \mathrm{D} \\ \hline \alpha & -(L_d-L_q)\sin\theta & -(L_d-L_q)\cos\theta & -L_{Dd}\sin\theta \\ \hline \beta & (L_d-L_q)\cos\theta & -(L_d-L_q)\sin\theta & L_{Dd}\cos\theta \\ \hline \mathrm{D} & 0 & -L_{Dd} & 0 \\ \hline \end{array}$$

$$= \begin{array}{c|c|c|c|} & \mathrm{d} & \mathrm{q} & \mathrm{D} \\ \hline \mathrm{d} & 0 & -(L_d-L_q) & 0 \\ \hline \mathrm{q} & -(L_d-L_q) & 0 & -L_{Dd} \\ \hline \mathrm{D} & 0 & -L_{Dd} & 0 \\ \hline \end{array}$$

In these terms the torque of a two-pole, two-phase, salient-pole machine is therefore

$$\frac{1}{2} \begin{array}{|c|c|c|} \mathrm{d} & \mathrm{q} & \mathrm{D} \\ \hline i^d & i^q & i^D \\ \hline \end{array} \quad \begin{array}{c|c|c|c|} & \mathrm{d} & \mathrm{q} & \mathrm{D} \\ \hline \mathrm{d} & & -(L_d-L_q) & \\ \hline \mathrm{q} & -(L_d-L_q) & & -L_{Dd} \\ \hline \mathrm{D} & & -L_{Dd} & \\ \hline \end{array} \quad \begin{array}{c|c|} \hline \mathrm{d} & i^d \\ \hline \mathrm{q} & i^q \\ \hline \mathrm{D} & i^D \\ \hline \end{array}$$

Derivation of the Torque Expression from the Equation $\mathbf{v} = \mathbf{Ri} + \mathbf{L}p\mathbf{i} + \mathbf{G}\dot{\theta}\mathbf{i}$

The voltage equation after transformation to "stationary" reference axes is given on p. 74 as

$$\mathbf{v} = \mathbf{Ri} + \mathbf{L}p\mathbf{i} + \mathbf{G}\dot{\theta}\mathbf{i}$$

and the matrix equation of the commutator machine given on p. 55 can be written in the same form, although in that case **G** is not necessarily related to **L**.

If this equation is multiplied by $\mathbf{i}_i$, the power equation is obtained as

$$\mathbf{i}_t\mathbf{v} = \mathbf{i}_t\mathbf{R}\mathbf{i}+\mathbf{i}_t\mathbf{L}p\mathbf{i}+\mathbf{i}_t\mathbf{G}\dot{\theta}\mathbf{i}$$

The left-hand side of this equation is the total power input to all terminals. The right-hand side consists of three terms: $\mathbf{i}_t\mathbf{R}\mathbf{i}$, which is the sum of all the I^2R losses; $\mathbf{i}_t\mathbf{L}p\mathbf{i}$, which is the power being stored, as deduced on p. 83; and $\mathbf{i}_t\mathbf{G}\dot{\theta}\mathbf{i}$. It follows that $\mathbf{i}_t\mathbf{G}\dot{\theta}\mathbf{i}$, being neither dissipated nor stored, must be the power converted to mechanical form. Dividing it by the scalar angular velocity $\dot{\theta}$ gives the two-pole torque T as $\mathbf{i}_t\mathbf{G}\mathbf{i}$.

For a machine with p pole-pairs, $\mathbf{i}_t\mathbf{G}\dot{\theta}\mathbf{i}$ would be divided by the mechanical angular velocity $\dot{\theta}/\mathrm{p}$ to obtain the torque as $\mathrm{p}\mathbf{i}_t\mathbf{G}\mathbf{i}$ or $\mathrm{p}T$, where T is the two-pole torque as above.

These expressions can be used for the torque wherever the axes are "stationary", i.e. at rest relative to one another, or have been derived from such axes by a transformation involving only constant elements. Such a transformation results in a new impedance matrix of which the elements are, like those of the matrix from which it was derived, constant. Since most problems are solved to find the currents in terms of stationary axes in the first instance, the expression $\mathbf{i}_t\mathbf{G}\mathbf{i}$ is much more often used than the expression $\frac{1}{2}\mathbf{i}_t(\partial\mathbf{L}/\partial\theta)\mathbf{i}$. The latter is directly applicable only to the actual winding axes themselves or to axes at rest relative to the windings which they represent. It may, however, also be used for stationary axes if $\partial\mathbf{L}/\partial\theta$ is first determined in the actual winding axes and then transformed to the stationary axes in the usual form $\mathbf{C}_t(\partial\mathbf{L}/\partial\theta)\mathbf{C}$, as in the example on p. 88.

For the machine considered on pp. 65–73 the torque is

$$\mathbf{i}_t\mathbf{G}\mathbf{i} = \begin{array}{c} \begin{array}{ccc} \mathrm{d} & \mathrm{q} & \mathrm{D} \end{array} \\ \left[\begin{array}{ccc} i^{\mathrm{d}} & i^{\mathrm{q}} & i^{\mathrm{D}} \end{array}\right] \\ \\ \\ \end{array} \begin{array}{c} \\ \mathrm{d} \\ \mathrm{q} \\ \mathrm{D} \end{array} \begin{array}{c} \begin{array}{ccc} \mathrm{d} & \mathrm{q} & \mathrm{D} \end{array} \\ \left[\begin{array}{ccc} & L_{\mathrm{q}} & \\ -L_{\mathrm{d}} & & -L_{\mathrm{Dd}} \\ & & \end{array}\right] \end{array} \begin{array}{c} \\ \mathrm{d} \\ \mathrm{q} \\ \mathrm{D} \end{array} \begin{array}{c} \\ \left[\begin{array}{c} i^{\mathrm{d}} \\ i^{\mathrm{q}} \\ i^{\mathrm{D}} \end{array}\right] \end{array}$$

It is interesting to compare this expression with that on p. 89, derived from $\partial\mathbf{L}/\partial\theta$ after transformation to d, q, D axes.

Although quite obviously $\mathbf{G} \neq \frac{1}{2}(\partial\mathbf{L}/\partial\theta)$, both these torque expressions reduce to $-(L_d - L_q)i^d i^q - L_{Dd} i^q i^D$. This result serves to remind us that there is no division in matrix algebra and that $\mathbf{i}_t$ and $\mathbf{i}$, being singular, have no inverses. It is therefore fundamentally incorrect to attempt to divide both sides of the true equation $\mathbf{i}_t\mathbf{G}\mathbf{i} = \mathbf{i}_t \frac{1}{2}(\partial\mathbf{L}/\partial\theta)\mathbf{i}$ by $\mathbf{i}_t$ and $\mathbf{i}$ in an endeavour to prove that $\mathbf{G}$ and $\frac{1}{2}(\partial\mathbf{L}/\partial\theta)$ are equal, which is obviously not true.

This minor difficulty can be explained as follows: any matrix, and therefore $\mathbf{G}$, can be expressed as the sum of a symmetric and a skew-symmetric matrix:

$$\mathbf{G} = \begin{array}{c|c|c|c|} & \mathrm{d} & \mathrm{q} & \mathrm{D} \\ \hline \mathrm{d} & & L_q & \\ \hline \mathrm{q} & -L_d & & -L_{Dd} \\ \hline \mathrm{D} & & & \\ \hline \end{array}$$

$$= \frac{1}{2}\ \begin{array}{c|c|c|c|} & \mathrm{d} & \mathrm{q} & \mathrm{D} \\ \hline \mathrm{d} & & -(L_d - L_q) & \\ \hline \mathrm{q} & -(L_d - L_q) & & -L_{Dd} \\ \hline \mathrm{D} & & -L_{Dd} & \\ \hline \end{array}$$

$$+ \frac{1}{2}\ \begin{array}{c|c|c|c|} & \mathrm{d} & \mathrm{q} & \mathrm{D} \\ \hline \mathrm{d} & & (L_d + L_q) & \\ \hline \mathrm{q} & -(L_d + L_q) & & -L_{Dd} \\ \hline \mathrm{D} & & L_{Dd} & \\ \hline \end{array}$$

$$= \mathbf{G}_{sym} + \mathbf{G}_{skew} \text{ say.}$$

Hence $$T = \mathbf{i}_t\mathbf{G}\mathbf{i} = \mathbf{i}_t\mathbf{G}_{sym}\mathbf{i} + \mathbf{i}_t\mathbf{G}_{skew}\mathbf{i}$$

Now a product such as $\mathbf{i}_t\mathbf{G}_{skew}\mathbf{i}$, which is known as a "skew-symmetric quadratic form", is always zero.† This being so, $\mathbf{i}_t\mathbf{G}\mathbf{i} = \mathbf{i}_t\mathbf{G}_{sym}\mathbf{i}$ and it will be noted that $\mathbf{G}_{sym}$ is in fact equal to $\frac{1}{2}(\partial\mathbf{L}/\partial\theta)$.

The torque may therefore be derived either from $\frac{1}{2}\mathbf{i}_t(\partial\mathbf{L}/\partial\theta)\mathbf{i}$ or from $\mathbf{i}_t\mathbf{G}\mathbf{i}$, whichever is the more convenient for the particular case.

The Mean Steady-state Torque in A.C. Machines

The preceding analysis has been conducted in terms of instantaneous or d.c. currents, i.e. real currents in the mathematical sense. If steady-state a.c. conditions are being considered in terms of complex numbers, it will be necessary to replace $\mathbf{i}_t$ by its conjugate $\mathbf{i}_t^*$ throughout, and to take the real part of $\mathbf{i}_t^*\mathbf{G}\mathbf{i}$ as can be shown by repeating the analysis noting that p will be replaced by $j\omega$ and that the power equation is now obtained from the voltage equation by pre-multiplying by $\mathbf{i}_t^*$ and taking real parts.

When the torque is to be determined in terms of the terminal voltages, the currents will all have as a common denominator the determinant Δ of the impedance matrix. Since Δ is a scalar, the Δ^* in the denominator of $\mathbf{i}_t^*$ and the Δ in the denominator of $\mathbf{i}$ can be brought together as $\Delta^*\Delta$.

It is helpful to remember that this product is always wholly real and equal to $|\Delta|^2$, the square of the magnitude of Δ. This makes it much easier to pick out the real part of $\mathbf{i}_t^*\mathbf{G}\mathbf{i}$ which may be written Re $\mathbf{I}_t^*\mathbf{G}\mathbf{I}$.

Direction of Torque

It has been shown that $\mathbf{i}_t\mathbf{G}\dot{\theta}\mathbf{i}$ represents the electrical power conr verted to mechanical form. If it is positive, the mechanical *output* power is thus positive and the electrodynamic torque and angula, velocity are in the same direction. From the mathematical viewpoint- if $\mathbf{i}_t\mathbf{G}\dot{\theta}\mathbf{i}$ is positive, $(\mathbf{i}_t\mathbf{G}\mathbf{i})$ and $\dot{\theta}$ are either both positive or both negative. If $\dot{\theta}$ is positive, i.e. counterclockwise, the torque is also counterclockwise and $(\mathbf{i}_t\mathbf{G}\mathbf{i})$ is positive. Similarly, if $\dot{\theta}$ is negative, i.e. clock-

† This may be demonstrated by multiplying out the product in this particular case, or, better still, for a general skew-symmetric matrix.

wise, the torque is also clockwise and $(\mathbf{i}_t\mathbf{G}\mathbf{i})$ is negative. Thus positive $(\mathbf{i}_t\mathbf{G}\mathbf{i})$ denotes counterclockwise torque and negative $(\mathbf{i}_t\mathbf{G}\mathbf{i})$ denotes clockwise torque.

The same conclusion is reached by considering the generating condition in which $\mathbf{i}_t\mathbf{G}\dot{\theta}\mathbf{i}$ is negative, $(\mathbf{i}_t\mathbf{G}\mathbf{i})$ and $\dot{\theta}$ are of opposite sign and the torque and angular velocity are in opposite directions.

The above reasoning applies specifically when the currents are "real", i.e. instantaneous or steady-state d.c., but consideration of Re $\mathbf{I}_t^*\mathbf{G}\dot{\theta}\mathbf{I}$ leads to a similar interpretation of the sign of $\mathrm{Re}(\mathbf{I}_t^*\mathbf{G}\mathbf{I})$ for steady-state a.c. conditions treated in complex form.

It can also be shown that positive $\frac{1}{2}\mathbf{i}_t(\partial\mathbf{L}/\partial\theta)\mathbf{i}$ indicates torque in the same direction as positive $\dot{\theta}$.

The convention assumed for the angle θ that counterclockwise is positive thus applies also to all the torque expressions.

CHAPTER 7

Linear Transformations in Circuits and Machines

REFERENCE has already been made on pp. 24–36, 38, 44, 62, 70 to the use of matrix methods for effecting linear transformations in electric circuit equations. In electrical machine analysis it is frequently necessary to change the variables to another related set for the purpose of either

(a) obtaining the equations of a different machine, or of the same machine differently connected, or of a machine connected to another machine or circuit element, or
(b) simplifying the process of obtaining the solution of a problem. In such cases it may be necessary to transform the solution back into terms of the original variables and, even so, obtain the required solution more easily and quickly than could have been done without the intermediate transformation.

It is convenient to group together in this chapter a number of linear transformations commonly used in machine analysis.

Any arbitrary set of linear equations may be used to define a linear transformation provided that they are independent, but in general such an arbitrary transformation would be pointless. A transformation may be selected on purely algebraic grounds, e.g. to give a matrix of constants in place of a matrix of functions, or to give a symmetric or a diagonal matrix in place of an asymmetric one. To select a transformation of this type requires some experience of matrix algebra. Frequently, however, a transformation corresponds to a physical change to the system which it would be possible, at least in principle, to perform. An example of this is the transformation on pp. 60 and 70 which is equivalent to replacing the slip-rings of a synchronous machine by a commutator, thereby making the machine into a metadyne. The reason for performing this transformation is, however, to obtain a set of equations with constant coefficients. It must be emphasized that it is not necessary for a transformation to correspond to any physical equivalent, and if a transformation used for purely mathematical reasons has a physical significance this must be regarded as fortuitous, but advantageous, since the associated physical image is helpful to those with a distaste for or distrust of abstract mathematics. In fact not only do most of the transformations used here have physical parallels, but they were derived by physical considerations long before matrix algebra was applied to electrical machine analysis.

In some cases there is a wider application than that first visualized. For example, the symmetrical component transformation, originating in ideas of contra-rotating m.m.f.s in electrical machines, and applied to steady-state currents in complex form, can also be applied to instantaneous, i.e. "real" currents, thereby leading to complex instantaneous currents in the transformed system. Any visualization of a complex instantaneous current is difficult, if not impossible. This, however, does not invalidate in any way analysis in which this transformation is applied to instantaneous values.

Whilst the rules of matrix algebra do not necessitate any relationship between the transformation of current and that of voltage, it has already been mentioned on p. 33 that for electrical circuits it is desirable that the two transformations should be so related that the power is the same in terms of both sets of variables. In electrical machine analysis this is even more necessary if powers and torques are to be calculated. If this "invariance of power" under transformation is not maintained, it will be necessary to determine powers and torques only from currents expressed in the original system of reference axes. If, then, the currents have been determined in a transformed system, it will be necessary to transform them back to the original system before calculating power or torque or to determine a special expression for the power or torque applicable only to that particular case. If, however, the transformations used conform to the requirements for invariant power, power and torque can be calculated directly from the transformed currents.

In polyphase machines there is the option of keeping the total power and torque of the machine constant under transformation or of keeping the power per phase constant. These two are obviously different when the transformation is from a three-phase to a two-phase system. In traditional analysis of polyphase machines such as the three-phase induction motor, it is customary to consider one phase only. This may also be done in matrix analysis and it is then convenient to keep the power per phase invariant. It is not appropriate, however, for more advanced analysis in which acceleration and oscillations are considered, since the moment of inertia is then included in the matrix, and the torques in both electrical and mechanical equations must relate to the whole machine. This entails keeping the *total* power invariant under transformation. Invariance of total power is also preferable according to the philosophy of tensor analysis, which is the underlying basis of this approach to electrical machine theory.

As has been shown in pp. 34–36, the condition for invariance of power is that $\mathbf{v}' = \mathbf{C}_t^*\mathbf{v}$ when $\mathbf{i} = \mathbf{C}\mathbf{i}'$. In addition to keeping power invariant under transformation, it is sometimes possible to have identical transformations for current and voltage. This is desirable in the interest of simplicity. For example, in the transformation from three-phase to two-phase, it results in the phase impedance under balanced conditions having the same magnitude in both cases.†

For this condition to exist we have

$$\mathbf{i} = \mathbf{C}\mathbf{i}' \quad \text{and} \quad \mathbf{v} = \mathbf{C}\mathbf{v}'$$

Pre-multiplying the voltage equation by $\mathbf{C}^{-1}$ gives,

$$\mathbf{C}^{-1}\mathbf{v} = \mathbf{C}^{-1}\mathbf{C}\mathbf{v}' = \mathbf{v}'$$

But for the power to be invariant $\mathbf{v}' = \mathbf{C}_t^*\mathbf{v}$, hence

$$\mathbf{C}^{-1}\mathbf{v} = \mathbf{C}_t^*\mathbf{v}$$

† See p. 136.

For this to be true for all values of the independent elements of **v**, $\mathbf{C}^{-1} = \mathbf{C}_t^*$.

If **C** is real, this condition reduces to $\mathbf{C}^{-1} = \mathbf{C}_t$. If $\mathbf{C}^{-1} = \mathbf{C}_t$, **C** is orthogonal; if $\mathbf{C}^{-1} = \mathbf{C}_t^*$, **C** is hermitian orthogonal, or unitary.

If **C** is of the form $\begin{vmatrix} P & Q \\ R & S \end{vmatrix}$ where P, Q, R, and S are complex,

$$\mathbf{C}_t = \begin{vmatrix} P & R \\ Q & S \end{vmatrix} \quad \text{and} \quad \mathbf{C}^{-1} = \frac{1}{PS-QR} \begin{vmatrix} S & -Q \\ -R & P \end{vmatrix}$$

Therefore, if $\mathbf{C}^{-1} = \mathbf{C}_t^*$,

$$P^*(PS-QR) = S \qquad Q^*(PS-QR) = -R$$
$$S^*(PS-QR) = P \qquad R^*(PS-QR) = -Q$$

It follows that $P^*P = S^*S$ and $Q^*Q = R^*R$ and that $|P| = |S|$ and $|Q| = |R|$ and hence that $|PS-QR| = 1$.

For orthogonal matrices, all the elements of **C** are real and $P = \pm S$, $Q = \pm R$, and $PS-QR = \pm 1$.

Resolution of Rotor M.M.F.

In machine analysis it is frequently only rotor currents which are transformed. In physical terms the transformed system of rotor currents must at all times produce an m.m.f. identical with that of the original system of rotor currents in respect of both magnitude and direction. If this condition were not fulfilled, the effect of the rotor currents on the stator windings would be different in the two cases, and this is obviously inadmissible.

Resolution of the instantaneous rotor m.m.f. along any convenient axes, usually d and q, is thus a simple way of deriving transformations and also of recalling them. Considerable use of this approach will be made in the following pages.

Where appropriate, of course, it would be possible to resolve the stator m.m.f. in a similar manner.

The use of this technique is strictly justified only where it is reasonable to assume that the m.m.f.s and fluxes are sinusoidally distributed.

Transformation between Two Sets of Stationary Axes (Brush-shifting Transformation)

If a commutator machine has brushes which are not at 90° electrically to one another, or which are not on the axes of symmetry of the stator (magnetic axes if salient, winding axes if non-salient), a transformation is necessary to relate such a machine to one which has its brushes on

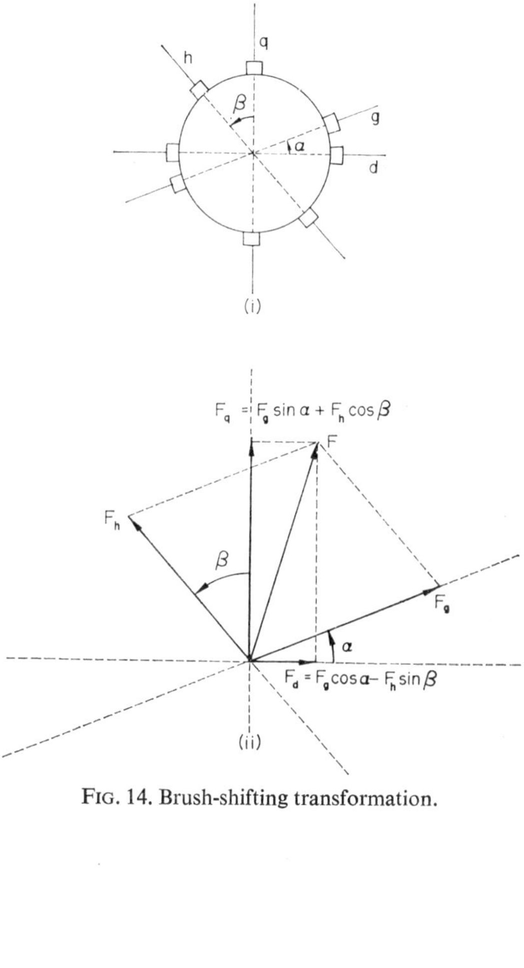

FIG. 14. Brush-shifting transformation.

these axes of symmetry, since the impedance matrix of the latter is known.

Suppose a set of brushes g makes an angle α with the d axis and another set h makes an angle β with the q axis, as shown in Fig. 14 (i).

Resolving the m.m.f.s and cancelling the number of turns, which is the same for all circuits, gives

$$i^{d} = i^{g}\cos\alpha - i^{h}\sin\beta$$

$$i^{q} = i^{g}\sin\alpha + i^{h}\cos\beta$$

In matrix form this is

$$\begin{array}{c|c|} d & i^{d} \\ \hline q & i^{q} \end{array} = \begin{array}{c|c|c|} & g & h \\ \hline d & \cos\alpha & -\sin\beta \\ \hline q & \sin\alpha & \cos\beta \end{array} \begin{array}{c|c|} g & i^{g} \\ \hline h & i^{h} \end{array}$$

or

$$\mathbf{C} = \begin{array}{c|c|c|} & g & h \\ \hline d & \cos\alpha & -\sin\beta \\ \hline q & \sin\alpha & \cos\beta \end{array}$$

If the axes g and h are at right angles to one another, $\alpha = \beta$ and the transformation matrix becomes

$$\mathbf{C} = \begin{array}{c|c|c|} & g & h \\ \hline d & \cos\alpha & -\sin\alpha \\ \hline q & \sin\alpha & \cos\alpha \end{array}$$

The inverse of this is

$$\mathbf{C}^{-1} = \begin{array}{c|c|c|} & d & q \\ \hline g & \cos\alpha & \sin\alpha \\ \hline h & -\sin\alpha & \cos\alpha \end{array}$$

Thus $\mathbf{C}^{-1} = \mathbf{C}_t$ and $\mathbf{C}$ is then orthogonal.

This gives an indication of the reason for calling such matrices orthogonal: when this transformation is applied to a set of orthogonal axes (d, q) it leads to another set of orthogonal axes (g, h).

Pre-multiplication of a vector by this last transformation matrix may be regarded as the operation of rotating the vector through an angle α in the counterclockwise direction or, alternatively, of rotating the axes through an angle α in the clockwise direction g, h to d, q. The second interpretation is appropriate for the brush-shifting transformaion.

The Equivalence of Three-phase and Two-phase Systems

The phasor (time-vector) diagrams of the three-phase and two-phase systems are shown in Fig. 15 and the space vector diagrams and windings in Fig. 16.

A balanced three-phase system of currents in a balanced three-phase winding sets up an m.m.f. the space fundamental of which is constant

FIG. 15. (i) Three-phase and (ii) two-phase phasor diagrams.

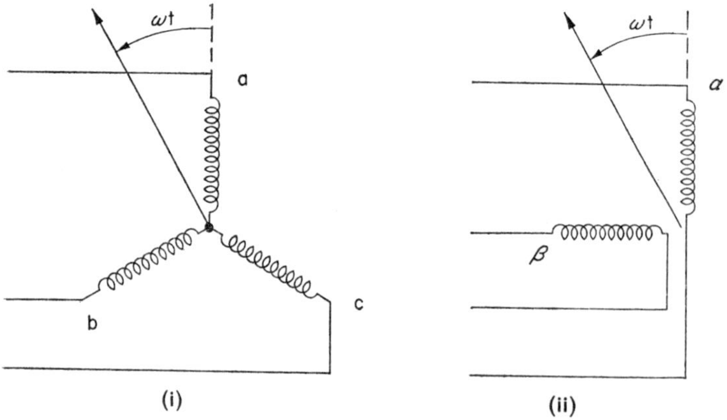

FIG. 16. (i) Three-phase and (ii) two-phase space vector diagrams.

in magnitude and rotates at a constant angular velocity. Let the currents be

$$i^{a} = \hat{I}\cos\omega t$$
$$i^{b} = \hat{I}\cos(\omega t - 2\pi/3)$$
$$i^{c} = \hat{I}\cos(\omega t - 4\pi/3)$$

In the three-phase windings these will produce an m.m.f. of magnitude $3\hat{I}N/2$ making an angle ωt in a counterclockwise direction with the axis of the a-phase winding, where N is the effective number of turns, i.e. allowing for distribution factor, and chording, etc.

A balanced two-phase system of currents in a balanced two-phase winding also sets up an m.m.f., the space fundamental of which is of constant magnitude and rotates at a constant angular velocity.

Let the currents be

$$i^{\alpha} = \hat{I}\cos\omega t$$
$$i^{\beta} = \hat{I}\cos(\omega t - \pi/2) = \hat{I}\sin\omega t$$

In the two-phase winding these will produce an m.m.f. of magnitude $\hat{I}N$ making an angle ωt in a counterclockwise direction with the axis of the α phase winding.

With the chosen currents the m.m.f.s of the three-phase and two-phase systems are thus in the same direction provided that the axes of the a- and α-phase windings are coincident. They could also be made equal in magnitude by making the appropriate change to either the magnitude of the two-phase current or to the number of turns of the two-phase winding, or to both current and turns.

It is clear that if the opposite phase sequence, namely c, b, a and β, α, were used, say in the forms

$$i^{a} = \hat{I}\cos\omega t \qquad i^{\alpha} = \hat{I}\cos\omega t$$
$$i^{b} = \hat{I}\cos(\omega t - 4\pi/3) \qquad i^{\beta} = -\hat{I}\sin\omega t$$
$$i^{c} = \hat{I}\cos(\omega t - 2\pi/3)$$

both m.m.f. waves would rotate in a clockwise direction and the two systems could again be equivalent. By super-position it follows that the two systems can also be equivalent when the currents contain components, not necessarily equal, of both positive phase-sequence (a, b, c and α, β) and negative phase-sequence (c, b, a and β, α). This equivalence is thus sufficiently general to justify more detailed consideration as a mathematical transformation whereby a three-phase system can be replaced by an equivalent two-phase system for analytical purposes.

The Transformation from Three-phase to Two-phase Axes

(a, b, c to α, β, o)

Let the a-phase winding axis and the α-phase winding axis be coincident as shown in Fig. 16.

Let N_3 be the effective number of turns per phase of the three-phase winding and N_2 be the effective number of turns per phase of the two-phase winding. Resolving the three-phase m.m.f.s along the α and β axis directions, and equating the three-phase and two-phase values,

$$N_2 i^{\alpha} = N_3 i^{a} + N_3 i^{b} \cos 2\pi/3 + N_3 i^{c} \cos 4\pi/3$$
$$N_2 i^{\beta} = N_3 i^{b} \sin 2\pi/3 + N_3 i^{c} \sin 4\pi/3$$

whence

$$N_2 i^{\alpha} = N_3 i^{a} - \tfrac{1}{2} N_3 i^{b} - \tfrac{1}{2} N_3 i^{c}$$
$$N_2 i^{\beta} = (\sqrt{3}/2) N_3 i^{b} - (\sqrt{3}/2) N_3 i^{c}$$

For completeness we need a third variable in addition to i^{α} and i^{β} and independent of them. Let it be i^{o}, defined by

$$N_2 i^{o} = k N_3 i^{a} + k N_3 i^{b} + k N_3 i^{c}, \quad \text{where } k \text{ is to be determined later.}$$

The current i^{o} cannot be associated *physically* with a two-phase system, which obviously has only two independent currents. We have now reached the point where the transformation is a purely mathematical operation, without physical significance unless i^{o} happens to be zero. Since this condition is fulfilled in most practical cases, the transformation is usually physically realizable. For purposes of analysis, however, it is immaterial whether this is so or not. The reason for the choice of this particular third variable is the very fact that it is in most cases zero. In a star-connected winding without neutral connection, this is necessarily so since $i^{a} + i^{b} + i^{c} = 0$; in delta-connected windings, it is usually so, because there is rarely any net voltage to cause a current to circulate around the delta.

In matrix form the equation defining the transformation is

$$\begin{array}{c|c|} & \\ \alpha & i^{\alpha} \\ \beta & i^{\beta} \\ \mathrm{o} & i^{\mathrm{o}} \end{array} = \frac{N_3}{N_2} \begin{array}{c|c|c|c|} & \mathrm{a} & \mathrm{b} & \mathrm{c} \\ \alpha & 1 & -1/2 & -1/2 \\ \beta & & \sqrt{3}/2 & -\sqrt{3}/2 \\ \mathrm{o} & k & k & k \end{array} \begin{array}{c|c|} & \\ \mathrm{a} & i^{\mathrm{a}} \\ \mathrm{b} & i^{\mathrm{b}} \\ \mathrm{c} & i^{\mathrm{c}} \end{array}$$

which is of the form $\mathbf{i}' = \mathbf{C}^{-1}\mathbf{i}$ if the three-phase system is regarded as the "old" system and the two-phase system as the "new".

If the above $\mathbf{C}^{-1}$ matrix is inverted, we obtain

$$\mathbf{C} = \frac{2}{3}\frac{N_2}{N_3} \begin{array}{c|c|c|c|} & \alpha & \beta & \mathrm{o} \\ \mathrm{a} & 1 & & 1/(2k) \\ \mathrm{b} & -1/2 & \sqrt{3}/2 & 1/(2k) \\ \mathrm{c} & -1/2 & -\sqrt{3}/2 & 1/(2k) \end{array}$$

so that

$$\mathbf{C}_t = \frac{2}{3}\frac{N_2}{N_3} \begin{array}{c|c|c|c|} & \mathrm{a} & \mathrm{b} & \mathrm{c} \\ \alpha & 1 & -1/2 & -1/2 \\ \beta & & \sqrt{3}/2 & -\sqrt{3}/2 \\ \mathrm{o} & 1/(2k) & 1/(2k) & 1/(2k) \end{array}$$

Now it is desirable,† although not essential, that the transformations for current and voltage should be the same, as well as consistent with the invariance of power. It was shown on p. 97 that the condition for this to be so is that $\mathbf{C}_t = \mathbf{C}^{-1}$. This condition is fulfilled here if

$$\frac{2}{3}\frac{N_2}{N_3} = \frac{N_3}{N_2} \quad \text{and} \quad k = 1/(2k)$$

i.e. if $$N_2/N_3 = \sqrt{(3/2)} \quad \text{and} \quad k = 1/\sqrt{2}.$$

† See p. 137.

The *orthogonal* transformation matrix is thus

$$\mathbf{C} = \sqrt{\frac{2}{3}}\;\begin{array}{c|c|c|c|} & \alpha & \beta & \mathrm{o} \\ \hline a & 1 & & 1/\sqrt{2} \\ \hline b & -1/2 & \sqrt{3}/2 & 1/\sqrt{2} \\ \hline c & -1/2 & -\sqrt{3}/2 & 1/\sqrt{2} \\ \hline \end{array}$$

$$= \begin{array}{c|c|c|c|} & \alpha & \beta & \mathrm{o} \\ \hline a & \sqrt{2}/\sqrt{3} & & 1/\sqrt{3} \\ \hline b & -1/\sqrt{6} & 1/\sqrt{2} & 1/\sqrt{3} \\ \hline c & -1/\sqrt{6} & -1/\sqrt{2} & 1/\sqrt{3} \\ \hline \end{array}$$

and

$$\mathbf{C}_t = \mathbf{C}^{-1} = \sqrt{\frac{2}{3}}\;\begin{array}{c|c|c|c|} & a & b & c \\ \hline \alpha & 1 & -1/2 & -1/2 \\ \hline \beta & & \sqrt{3}/2 & -\sqrt{3}/2 \\ \hline \mathrm{o} & 1/\sqrt{2} & 1/\sqrt{2} & 1/\sqrt{2} \\ \hline \end{array}$$

$$= \begin{array}{c|c|c|c|} & a & b & c \\ \hline \alpha & \sqrt{2}/\sqrt{3} & -1/\sqrt{6} & -1/\sqrt{6} \\ \hline \beta & & 1/\sqrt{2} & -1/\sqrt{2} \\ \hline \mathrm{o} & 1/\sqrt{3} & 1/\sqrt{3} & 1/\sqrt{3} \\ \hline \end{array}$$

The transformation thus corresponds to rewinding the three-phase machine with $\sqrt{(2/3)}$ times as many turns before re-connecting it as a two-phase machine. By this means both the rated phase-current and the rated phase-voltage are increased in the ratio $\sqrt{(3/2)}$. The power per phase is thus increased in the ratio 3/2, but, because the number of phases has been changed in the ratio 2/3, the total power is unchanged.

It is interesting to note that the resistances per phase of the two-phase and three-phase windings are equal.† The total number of turns having been changed in the ratio $\sqrt{(2/3)}$, the cross-sectional area of the conductor will have been increased in the ratio $\sqrt{(3/2)}$, but the number of turns per phase, and hence the total length of conductor per phase, will also have been increased in the same ratio $\sqrt{(3/2)}$.

The same transformation would have been obtained if the opposite rotation to that of Fig. 16 had been chosen, i.e. if b and c of Fig. 16(i) were interchanged and β shown to the right of α instead of to the left in Fig. 16(ii). These two changes would have to be made simultaneously so that the directions of rotation of the two m.m.f. waves remained alike.

To take the a-phase and α-phase windings as having the same axis is an arbitrary, but usual, choice, which is preferred to all others because of the special relationship between the a-phase and α-phase currents and voltages which then results under balanced conditions (See p. 137.)

If there is no current on the o-axis, **C** may be used omitting the last column, thus enabling i^a, i^b, and i^c to be found from i^α and i^β, although the matrix giving i^α and i^β in terms of i^a, i^b, and i^c could not then be inverted.

If it is preferred, these transformations can be expressed in terms of the trigonometric functions which were used in their derivation:

$\mathbf{C} = \sqrt{(2/3)}$

	α	β	o
a	$\cos 0$	$\sin 0$	$1/\sqrt{2}$
b	$\cos 2\pi/3$	$\sin 2\pi/3$	$1/\sqrt{2}$
c	$\cos 4\pi/3$	$\sin 4\pi/3$	$1/\sqrt{2}$

† The relationship between the two-phase and three-phase inductances cannot be expressed so simply. See p. 138.

$$\mathbf{C}^{-1} = \mathbf{C}_t = \sqrt{(2/3)}\; \begin{array}{c|c|c|c|} & a & b & c \\ \hline \alpha & \cos 0 & \cos 2\pi/3 & \cos 4\pi/3 \\ \beta & \sin 0 & \sin 2\pi/3 & \sin 4\pi/3 \\ o & 1/\sqrt{2} & 1/\sqrt{2} & 1/\sqrt{2} \\ \hline \end{array}$$

As a result of this transformation, it is unnecessary to consider three-phase machines as such. The simpler case of the two-phase machine can be analysed, together with the o-axis quantities if required, and the results transformed to the three-phase equivalents if necessary. This is discussed in p. 132.

The Transformation from Three-phase Axes to Symmetrical Component Axes (a, b, c to p, n, o)

The elements of the previous transformation matrices were all real, those which follow have complex elements. The idea underlying symmetrical components originated in the study of single-phase induction motors in which an alternating m.m.f. can be visualized as being composed of two components, each of half its maximum magnitude, rotating at equal angular velocities in opposite directions. Such contra-rotating m.m.f.s could be produced in a machine with a balanced three-phase winding by two equal balanced systems of three-phase currents differing only in that the phase sequence of one is a, b, c and that of the other a, c, b. The m.m.f.s of such a system will coincide on, say, the a-phase axis and thus give a resultant m.m.f. always on that axis but alternating in magnitude.

It is more convenient here to regard this as a straightforward case of linear transformation from three variables i^a, i^b, i^c to three other variables i^p, i^n, i^o according to the equations

$$\begin{aligned} i^a &= i^{pa}+i^{na}+i^{oa} = i^p+i^n+i^o \\ i^b &= i^{pb}+i^{nb}+i^{ob} = h^2 i^p+h i^n+i^o \\ i^c &= i^{pc}+i^{nc}+i^{oc} = h i^p+h^2 i^n+i^o \end{aligned}$$

where h is the 120° operator discussed on p. 28, and i^{pa} is the positive-sequence component of the a-phase current, commonly called the

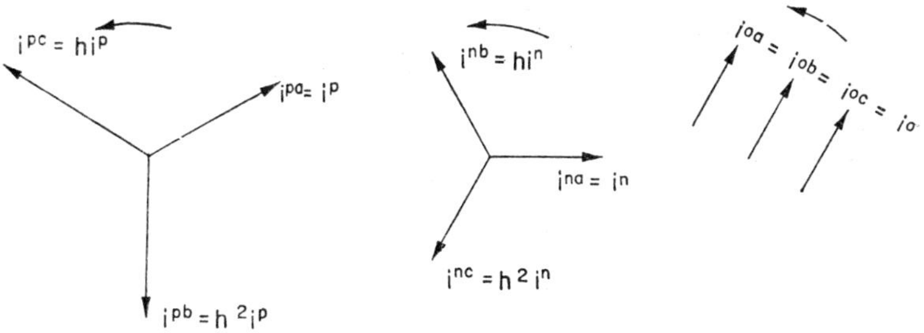

FIG. 17. Three-phase symmetrical components.

positive-sequence current and which is represented here by i^p. i^{pb} is the positive-sequence component of the b-phase current which lags $2\pi/3$ behind that of the a-phase. A lag of $2\pi/3$ is identical with a lead of $4\pi/3$ and is thus represented by h^2 and $i^{pb} = h^2 i^{pa} = h^2 i^p$. Correspondingly the negative-sequence component of the b-phase current leads that of the a-phase by $2\pi/3$ so that $i^{nb} = hi^{na} = hi^n$. The zero-sequence components i^{oa}, i^{ob} and i^{oc} are all in phase and hence are all represented by i^o.

The phasor relationships of these "symmetrical" components are shown in Fig. 17.

The transformation from phase values to conventional symmetrical component quantities is thus defined by the matrix equation

$$\begin{array}{c|c|} & \\ a & i^a \\ b & i^b \\ c & i^c \end{array} = \begin{array}{c|c|c|c|} & p & n & o \\ a & 1 & 1 & 1 \\ b & h^2 & h & 1 \\ c & h & h^2 & 1 \end{array} \begin{array}{c|c|} & \\ p & i^p \\ n & i^n \\ o & i^o \end{array}$$

The determinant of this transformation matrix is $j3\sqrt{3}$ and its inverse is

$$\frac{1}{3}\begin{array}{c|c|c|c|} & a & b & c \\ p & 1 & h & h^2 \\ n & 1 & h^2 & h \\ o & 1 & 1 & 1 \end{array}$$

h and h^2 are conjugates, being $(-\frac{1}{2}+j\sqrt{3}/2)$ and $(-\frac{1}{2}-j\sqrt{3}/2)$ respectively, hence the transposed conjugate is

	a	b	c
p	1	h	h^2
n	1	h^2	h
o	1	1	1

The inverse thus differs from the transposed conjugate only by the presence of the factor $\frac{1}{3}$. Bearing in mind that if the power is invariant, the transformation of the voltage is determined by the conjugate of the transpose of the current transformation, this means that the transformation of current and voltage are the same except for the factor $\frac{1}{3}$. To comply with the requirement of invariant power and also to give identical transformations for current and voltage, it is necessary to divide the transformation matrix by the cube root† of the magnitude of its determinant. Since the determinant is $j3\sqrt{3}$, the transformation matrix above is divided by $\sqrt{3}$ giving the *unitary* transformation matrix

$\mathbf{C} = \frac{1}{\sqrt{3}}$

	p	n	o
a	1	1	1
b	h^2	h	1
c	h	h^2	1

$\mathbf{C}^{-1} = \mathbf{C}_t^* = \frac{1}{\sqrt{3}}$

	a	b	c
p	1	h	h^2
n	1	h^2	h
o	1	1	1

In this form, the zero-sequence axis is clearly identical with the o-axis defined on p. 104, and, as already indicated, zero-sequence components do not normally arise in machine analysis, but do occur when there is a circuit to the star-point, either made deliberately, or, more likely, as the result of a fault.

† Cube root since the matrix, and hence the determinant, has three rows and columns.

It is sometimes more convenient to express this transformation in terms of exponentials:

$$\mathbf{C} = \frac{1}{\sqrt{3}} \begin{array}{c|c|c|c|} & \text{p} & \text{n} & \text{o} \\ \hline \text{a} & 1 & 1 & 1 \\ \hline \text{b} & \varepsilon^{j4\pi/3} & \varepsilon^{j2\pi/3} & 1 \\ \hline \text{c} & \varepsilon^{j2\pi/3} & \varepsilon^{j4\pi/3} & 1 \\ \hline \end{array}$$

The particular importance of this transformation to symmetrical components is that it transforms an impedance matrix of the form

	a	b	c
a	Z	X_f	X_r
b	X_r	Z	X_f
c	X_f	X_r	Z

into a diagonal matrix. This has already been shown on pp. 27–32 for the transformation as defined on p. 104. As an exercise the transformation **C** as re-defined above may be applied to this form of impedance matrix.

The symmetrical component transformation can be determined directly from this impedance matrix without invoking physical concepts, by techniques more advanced than the matrix algebra covered in this book, namely diagonalization by means of the eigen vectors.

The Transformation from Two-phase Axes to Symmetrical Component Axes (α, β, o to p, n, o)

Since the zero-sequence components are not transformed in this instance, we are concerned only with the transformation from α, β to p, n axes.

This operation can be treated in two ways; either from first principles, or as two successive transformations from two-phase to three-phase followed by three-phase to symmetrical components.

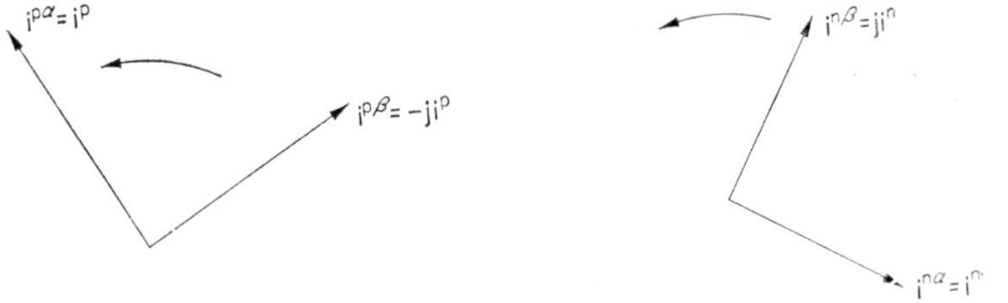

FIG. 18. Two-phase symmetrical components.

In exactly the same way that the three-phase quantities are defined in terms of three balanced systems, two-phase quantities can be defined in terms of two balanced systems, positive-sequence two-phase and negative-sequence two-phase. The zero-sequence has no equivalent in a two-phase system. The phasor diagrams are as shown in Fig. 18.

The equations are

$$i^{\alpha} = i^{p\alpha} + i^{n\alpha} = i^{p} + i^{n}$$
$$i^{\beta} = i^{p\beta} + i^{n\beta} = -ji^{p} + ji^{n}$$

In matrix form this is

$$\begin{array}{c|c|} \alpha & i^{\alpha} \\ \hline \beta & i^{\beta} \end{array} = \begin{array}{c|c|c|} & \text{p} & \text{n} \\ \alpha & 1 & 1 \\ \hline \beta & -j & j \end{array} \begin{array}{c|c|} \text{p} & i^{p} \\ \hline \text{n} & i^{n} \end{array}$$

which leads to the inverse relation

$$\begin{array}{c|c|} \text{p} & i^{p} \\ \hline \text{n} & i^{n} \end{array} = \frac{1}{2} \begin{array}{c|c|c|} & \alpha & \beta \\ \text{p} & 1 & j \\ \hline \text{n} & 1 & -j \end{array} \begin{array}{c|c|} \alpha & i^{\alpha} \\ \hline \beta & i^{\beta} \end{array}$$

The transposed conjugate of the transformation matrix is, however,

$$\begin{array}{c|c|c|} & \alpha & \beta \\ \text{p} & 1 & j \\ \hline \text{n} & 1 & -j \end{array}$$

In this case, therefore, the inverse differs from the transposed conjugate by the presence of the factor $\frac{1}{2}$. To obtain invariant power with identical transformations for current and voltage will thus necessitate our dividing the initial transformation matrix by $\sqrt{2}$ obtaining

$$\mathbf{C} = \frac{1}{\sqrt{2}} \begin{array}{c|cc|} & \mathrm{p} & \mathrm{n} \\ \hline \alpha & 1 & 1 \\ \beta & -j & j \\ \hline \end{array} \qquad \mathbf{C}^{-1} = \mathbf{C}_t^* = \frac{1}{\sqrt{2}} \begin{array}{c|cc|} & \alpha & \beta \\ \hline \mathrm{p} & 1 & j \\ \mathrm{n} & 1 & -j \\ \hline \end{array}$$

If preferred this unitary transformation can also be expressed in terms of exponentials:

$$\mathbf{C} = \frac{1}{\sqrt{2}} \begin{array}{c|cc|} & \mathrm{p} & \mathrm{n} \\ \hline \alpha & 1 & 1 \\ \beta & \varepsilon^{j3\pi/2} & \varepsilon^{j\pi/2} \\ \hline \end{array}$$

This transformation may now be compared with that obtained by multiplying the two-phase to three-phase and the three-phase to symmetrical component transformations:

$$\begin{array}{c|c|} \alpha & i^{\alpha} \\ \beta & i^{\beta} \\ \end{array} = \begin{array}{c|ccc|} & \mathrm{a} & \mathrm{b} & \mathrm{c} \\ \hline \alpha & \sqrt{\frac{2}{3}} & -\sqrt{\frac{1}{6}} & -\sqrt{\frac{1}{6}} \\ \beta & & \sqrt{\frac{1}{2}} & -\sqrt{\frac{1}{2}} \\ \hline \end{array} \begin{array}{c|c|} \mathrm{a} & i^{\mathrm{a}} \\ \mathrm{b} & i^{\mathrm{b}} \\ \mathrm{c} & i^{\mathrm{c}} \\ \end{array}$$

$$= \begin{array}{c|ccc|} & \mathrm{a} & \mathrm{b} & \mathrm{c} \\ \hline \alpha & \sqrt{\frac{2}{3}} & -\sqrt{\frac{1}{6}} & -\sqrt{\frac{1}{6}} \\ \beta & & \sqrt{\frac{1}{2}} & -\sqrt{\frac{1}{2}} \\ \hline \end{array} \frac{1}{\sqrt{3}} \begin{array}{c|cc|} & \mathrm{p} & \mathrm{n} \\ \hline \mathrm{a} & 1 & 1 \\ \mathrm{b} & h^2 & h \\ \mathrm{c} & h & h^2 \\ \hline \end{array} \begin{array}{c|c|} \mathrm{p} & i^{\mathrm{p}} \\ \mathrm{n} & i^{\mathrm{n}} \\ \end{array}$$

$$= \frac{1}{\sqrt{2}} \begin{array}{c|cc|} & \mathrm{p} & \mathrm{n} \\ \hline \alpha & 1 & 1 \\ \beta & -j & j \\ \hline \end{array} \begin{array}{c|c|} \mathrm{p} & i^{\mathrm{p}} \\ \mathrm{n} & i^{\mathrm{n}} \\ \end{array}$$

If the two-phase symmetrical component transformation is applied to a two-by-two circulant matrix (which is also symmetric), the resulting matrix is *not* diagonal but is hermitian. The transformation required to diagonalize such a matrix is quite different, namely

$$\begin{array}{|c|c|}\hline 1 & 1 \\ \hline 1 & -1 \\ \hline\end{array}$$

This situation arises because the "two-phase" system is not a true two-phase, for which the angle between phases is π, and which is known as bi-phase. The so-called two-phase system has an angle $\pi/2$ between phases and may be regarded as half a four-phase system, or, preferably, as a four-phase system in which opposite phases have been connected in series opposition.

For the same reason, the typical two-phase impedance matrix is not a circulant, but is the sum of a scalar matrix and a skew-symmetric matrix, as may be seen by transforming the three-phase circulant matrix on p. 109 to two-phase. This type of matrix *is* reduced to diagonal form by the two-phase to symmetrical component transformation.

Steady-state and Instantaneous Symmetrical Components

When balanced voltages or currents are transformed to symmetrical components, the negative-sequence and zero-sequence components are found to be zero *provided that the phase values are expressed in complex form*:

$$\frac{1}{\sqrt{3}}\;\begin{array}{c|c|c|c|}\multicolumn{1}{c}{} & \multicolumn{1}{c}{\mathrm{a}} & \multicolumn{1}{c}{\mathrm{b}} & \multicolumn{1}{c}{\mathrm{c}} \\ \cline{2-4} \mathrm{p} & 1 & h & h^2 \\ \cline{2-4} \mathrm{n} & 1 & h^2 & h \\ \cline{2-4} \mathrm{o} & 1 & 1 & 1 \\ \cline{2-4}\end{array}\;\begin{array}{c|c|}\cline{2-2} \mathrm{a} & V \\ \cline{2-2} \mathrm{b} & h^2V \\ \cline{2-2} \mathrm{c} & hV \\ \cline{2-2}\end{array} = \frac{1}{\sqrt{3}}\;\begin{array}{c|c|}\cline{2-2} \mathrm{p} & V(1+h^3+h^3) \\ \cline{2-2} \mathrm{n} & V(1+h+h^2) \\ \cline{2-2} \mathrm{o} & V(1+h^2+h) \\ \cline{2-2}\end{array}$$

$$= \frac{1}{\sqrt{3}}\;\begin{array}{c|c|}\cline{2-2} \mathrm{p} & 3V \\ \cline{2-2} \mathrm{n} & 0 \\ \cline{2-2} \mathrm{o} & 0 \\ \cline{2-2}\end{array} = \begin{array}{c|c|}\cline{2-2} \mathrm{p} & \sqrt{3}V \\ \cline{2-2} \mathrm{n} & 0 \\ \cline{2-2} \mathrm{o} & 0 \\ \cline{2-2}\end{array}$$

$$\text{and}\quad \frac{1}{\sqrt{2}}\begin{array}{c|cc|}
 & \alpha & \beta \\ \hline
\mathrm{p} & 1 & j \\
\mathrm{n} & 1 & -j \\ \hline
\end{array}\;
\begin{array}{c|c|}
\alpha & V \\
\beta & -jV \\
\end{array}
= \frac{1}{\sqrt{2}}\begin{array}{c|c|}
\mathrm{p} & 2V \\
\mathrm{n} & 0 \\
\end{array}
= \begin{array}{c|c|}
\mathrm{p} & \sqrt{2}V \\
\mathrm{n} & 0 \\
\end{array}$$

This was to be expected from the phasor diagrams Figs. 17 and 18 and from the corresponding development of the equations.

If, however, balanced voltages or currents expressed as *instantaneous values* are transformed to symmetrical components, both positive- and negative-sequence terms are present. For example

$$\frac{1}{\sqrt{3}}\begin{array}{c|ccc|}
 & \mathrm{a} & \mathrm{b} & \mathrm{c} \\ \hline
\mathrm{p} & 1 & h & h^2 \\
\mathrm{n} & 1 & h^2 & h \\
\mathrm{o} & 1 & 1 & 1 \\ \hline
\end{array}\;
\begin{array}{c|c|}
\mathrm{a} & \hat{V}\cos\omega t \\
\mathrm{b} & \hat{V}\cos(\omega t-2\pi/3) \\
\mathrm{c} & \hat{V}\cos(\omega t-4\pi/3) \\
\end{array}$$

$$= \frac{\sqrt{3}}{2}\begin{array}{c|c|}
\mathrm{p} & \hat{V}(\cos\omega t+j\sin\omega t) \\
\mathrm{n} & \hat{V}(\cos\omega t-j\sin\omega t) \\
\mathrm{o} & 0 \\
\end{array}$$

$$\text{and}\quad \frac{1}{\sqrt{2}}\begin{array}{c|cc|}
 & \alpha & \beta \\ \hline
\mathrm{p} & 1 & j \\
\mathrm{n} & 1 & -j \\ \hline
\end{array}\;
\begin{array}{c|c|}
\alpha & \hat{V}\cos\omega t \\
\beta & \hat{V}\sin\omega t \\
\end{array}
= \frac{1}{\sqrt{2}}\begin{array}{c|c|}
\mathrm{p} & \hat{V}(\cos\omega t+j\sin\omega t) \\
\mathrm{n} & \hat{V}(\cos\omega t-j\sin\omega t) \\
\end{array}$$

This result might also have been expected, because, in the absence of zero-sequence components, the sum of the instantaneous positive- and negative-sequence components must be real when $i^{a} = (1/\sqrt{3})(i^{p}+i^{n}+i^{o})$ and $i^{\alpha} = (1/\sqrt{2})(i^{p}+i^{n})$ are real. The instantaneous positive- and negative-sequence components must accordingly be complex conjugates. They can be represented diagrammatically by vectors which rotate in opposite directions and which coincide when on the real axis.†

Because of the extra complication, instantaneous symmetrical components should be used only when the particular problem demands it. (See, for example, p. 222.)

The effect of transformation to symmetrical components may be summarized as follows:

(a) The impedance matrix becomes symmetric or even diagonal.
(b) If the voltages and currents are balanced:
  (i) in complex form only one component is present,
  (ii) in instantaneous form the positive-sequence and negative-sequence components are complex conjugates and the analysis need be done for only one of them.

In many problems the symmetrical component transformation thus leads to a considerable reduction in the amount of algebra needed to obtain a solution.

Transformation from Two-phase Rotating Axes to Stationary Axes (α, β, o to d, q, o)

Since the zero-sequence quantities are not transformed, we are concerned only with the transformation from α, β to d, q. This transformation has already been derived in p. 60 by considering how differential equations with constant coefficients might be obtained. It will be reconsidered here from the viewpoint of resolution of m.m.f.s. Whilst the transformations of this chapter have so far had constant elements, those of this transformation are functions of the relative position of the rotor and stator and have the effect of replacing a moving system by a stationary one, with the addition of series-connected voltage sources. Moving and stationary here refer to the axes of the non-salient member (here assumed to be the rotor) and describe them relatively to the other member. We could equally well have a

† In fact, the positive-sequence vector, rotating counterclockwise, is identical, apart from a constant factor, with the rotating vector of the vector diagram from which the normal phasor diagram was derived.

stationary reference system in which all the axes were at rest relative to the rotor, i.e. were moving at the same speed as the rotor relative to the stator. This latter has its applications, but will not be considered further here.

The term "stationary" is also applied to axes which move relative to both stator and rotor, provided that they are *all stationary relative to one another*. Such axes are treated on pp. 260–263, where they are used in the consideration of small oscillations of induction machines.

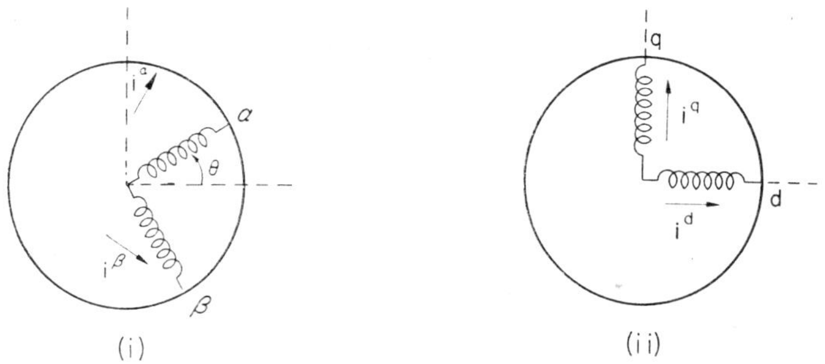

FIG. 19. Rotating and stationary axes.

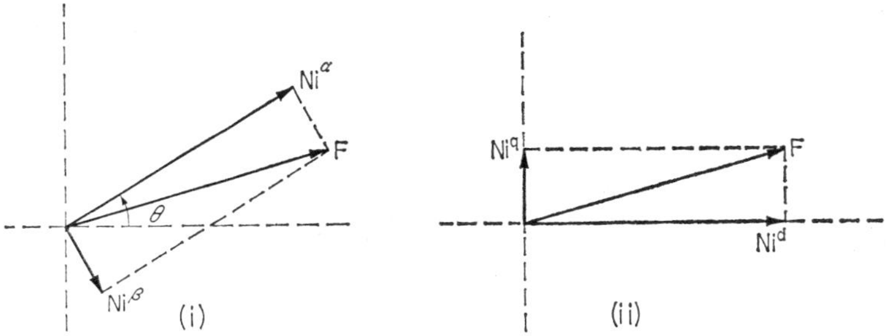

FIG. 20. Space vectors of rotating and stationary axes.

At the instant when the α phase of the rotating winding makes an angle θ with the stationary d winding, as shown in Fig. 19, the m.m.f. space vectors are as shown in Fig. 20. Since the α and β windings are balanced, and assuming that the d and q windings are identical with the α and β windings except for the absence of rotation, we can resolve the m.m.f.s along the d and q axes and divide by the effective number of turns per phase to get

$$i^{d} = \cos\theta i^{\alpha} + \sin\theta i^{\beta}$$
$$i^{q} = \sin\theta i^{\alpha} - \cos\theta i^{\beta}$$

The transformation is thus defined by

$$\mathbf{C}^{-1} = \begin{array}{c|c|c|} & \alpha & \beta \\ \hline d & \cos\theta & \sin\theta \\ \hline q & \sin\theta & -\cos\theta \\ \hline \end{array}$$

The determinant of this matrix is -1.
Its inverse is

$$\mathbf{C} = \begin{array}{c|c|c|} & d & q \\ \hline \alpha & \cos\theta & \sin\theta \\ \hline \beta & \sin\theta & -\cos\theta \\ \hline \end{array}$$

Since the transpose of this is equal to $\mathbf{C}^{-1}$, the matrix is orthogonal. More than this, however, **C** is symmetric and is therefore equal to its own transpose. In this particular case,† therefore,

$$\mathbf{C} = \mathbf{C}_t = \mathbf{C}^{-1}$$

Whereas the orthogonal matrix on p. 99 represents a rotation, this matrix represents rotation and reflection.‡ It is this particular combination of operations which makes it possible for it to be equal to its own inverse.

The transformation for voltage is therefore the same as that for current.

† This transformation is defined differently in this book from the corresponding transformation used by Kron, primarily to get the correct phase sequences with a positive $d\theta/dt$. It is fortunate that the change also avoids the necessity for distinguishing between the transformation and its inverse.

‡ See ref. 8, pp. 35–37.

If zero-sequence current exists, it is not transformed. The transformation matrix is therefore extended by an additional row and an additional column containing a solitary unity as their common element:

$$\mathbf{C} = \begin{array}{c|c|c|c|} & \text{d} & \text{q} & \text{o} \\ \hline \alpha & \cos\theta & \sin\theta & \\ \hline \beta & \sin\theta & -\cos\theta & \\ \hline \text{o} & & & 1 \\ \hline \end{array}$$

This matrix is also orthogonal and being symmetric, is its own transpose and inverse.

Transformation from Three-phase Rotating Axes to Stationary Axes (a, b, c to d, q, o)

The transformation from three-phase axes a, b, c to stationary axes d, q, o can now be deduced.

$$\begin{array}{c|c|} \text{a} & i^{\text{a}} \\ \hline \text{b} & i^{\text{b}} \\ \hline \text{c} & i^{\text{c}} \\ \hline \end{array} = \sqrt{\tfrac{2}{3}} \begin{array}{c|c|c|c|} & \alpha & \beta & \text{o} \\ \hline \text{a} & \cos 0 & \sin 0 & 1/\sqrt{2} \\ \hline \text{b} & \cos 2\pi/3 & \sin 2\pi/3 & 1/\sqrt{2} \\ \hline \text{c} & \cos 4\pi/3 & \sin 4\pi/3 & 1/\sqrt{2} \\ \hline \end{array} \begin{array}{c|c|} \alpha & i^{\alpha} \\ \hline \beta & i^{\beta} \\ \hline \text{o} & i^{\text{o}} \\ \hline \end{array}$$

$$= \sqrt{\tfrac{2}{3}} \begin{array}{c|c|c|c|} & \alpha & \beta & \text{o} \\ \hline \text{a} & \cos 0 & \sin 0 & 1/\sqrt{2} \\ \hline \text{b} & \cos 2\pi/3 & \sin 2\pi/3 & 1/\sqrt{2} \\ \hline \text{c} & \cos 4\pi/3 & \sin 4\pi/3 & 1/\sqrt{2} \\ \hline \end{array} \begin{array}{c|c|c|c|} & \text{d} & \text{q} & \text{o} \\ \hline \alpha & \cos\theta & \sin\theta & \\ \hline \beta & \sin\theta & -\cos\theta & \\ \hline \text{o} & & & 1 \\ \hline \end{array} \begin{array}{c|c|} \text{d} & i^{\text{d}} \\ \hline \text{q} & i^{\text{q}} \\ \hline \text{o} & i^{\text{o}} \\ \hline \end{array}$$

$$= \sqrt{\tfrac{2}{3}} \begin{array}{c|c|c|c|} & \text{d} & \text{q} & \text{o} \\ \hline \text{a} & \cos\theta & \sin\theta & 1/\sqrt{2} \\ \hline \text{b} & \cos(\theta-2\pi/3) & \sin(\theta-2\pi/3) & 1/\sqrt{2} \\ \hline \text{c} & \cos(\theta-4\pi/3) & \sin(\theta-4\pi/3) & 1/\sqrt{2} \\ \hline \end{array} \begin{array}{c|c|} \text{d} & i^{\text{d}} \\ \hline \text{q} & i^{\text{q}} \\ \hline \text{o} & i^{\text{o}} \\ \hline \end{array}$$

It will be found that this transformation matrix is also orthogonal, i.e. has an inverse equal to its transpose and a determinant of magnitude unity. However, because it is not symmetric, it is not equal to its own inverse which is

$$\mathbf{C}^{-1} = \mathbf{C}_t = \sqrt{\tfrac{2}{3}}\; \begin{array}{c|c|c|c|} & a & b & c \\ \hline d & \cos\theta & \cos(\theta-2\pi/3) & \cos(\theta-4\pi/3) \\ \hline q & \sin\theta & \sin(\theta-2\pi/3) & \sin(\theta-4\pi/3) \\ \hline o & 1/\sqrt{2} & 1/\sqrt{2} & 1/\sqrt{2} \\ \hline \end{array}$$

This transformation could equally well have been deduced directly by resolution of the m.m.f.s of the three-phase windings.

The Transformation from Stationary Axes to Forward and Backward Axes (d, q, o to f, b, o)

The transformation

$$\mathbf{C} = \frac{1}{\sqrt{2}}\; \begin{array}{c|c|c|} & p & n \\ \hline \alpha & 1 & 1 \\ \hline \beta & -j & j \\ \hline \end{array}$$

gives symmetrical positive- and negative-sequence components from phase values. A transformation having the same elements can be applied to stationary axes values to give forward and backward components f, b:

$$\mathbf{C} = \frac{1}{\sqrt{2}}\; \begin{array}{c|c|c|} & f & b \\ \hline d & 1 & 1 \\ \hline q & -j & j \\ \hline \end{array}$$

That these are not the same follows from the fact that the first is applied to α, β quantities directly, whereas the second is applied after the transformation from α, β, to d, q axes.

If the original system is three-phase and it is necessary to include the zero-sequence axis, this axis is retained without transformation in all four systems of axes, α, β, o; p, n, o; d, q, o, and f, b, o, leading to the three-by-three transformation matrix

$$\mathbf{C} = \frac{1}{\sqrt{2}} \begin{array}{c|ccc} & \text{f} & \text{b} & \text{o} \\ \hline \text{d} & 1 & 1 & \\ \text{q} & -j & j & \\ \text{o} & & & \sqrt{2} \end{array}$$

The Transformation of Stator Winding Axes

So far transformations of rotor winding axes only have been considered. Some of these transformations are clearly equally applicable to the stator winding axes, e.g. three-phase to two-phase, three-phase to symmetrical components and two-phase to symmetrical components. Others, such as a, b, c to d, q and α, β to d, q, have no equivalent in the stator axes when all axes D, Q, d, q are at rest relative to the stator. It is possible to take all axes at rest relative to the rotor, e.g. in the analysis of the performance of a revolving-field synchronous machine. This possibility was indicated on pp. 64–65 and 78. These particular transformations then apply to the stator winding axes and not to the rotor winding axes. This alternative is not considered in this book since, as already stated, it leads to identical results except for the sign (or conventional positive direction) of θ and $\dot{\theta}$.

Because the first two capital letters of the Greek alphabet are indistinguishable from those of the Roman alphabet, the capital forms of α and β will be denoted by A′ and B′ respectively, when we have occasion to refer specifically to two-phase stator windings.

Physical Interpretation of Various Sets of Axes

Rotor

a, b, c and α, β axes are the actual rotor winding axes or are obtained from the actual rotor winding axes by a transformation involving only constant real elements. These axes are, therefore, at rest relative to the rotor and rotate with it. They therefore rotate relative to the stator at the same angular velocity $\dot{\theta}$ as the rotor itself.

d, q axes are at rest relative to the stator and hence rotate relative to the rotor at the rotor angular velocity $\dot{\theta}$, but in the opposite direction.

p, n and f, b axes are complex and consequently neither pair represents any realizable set of physical axes. However, when they are used for steady-state a.c. conditions, the m.m.f.s due to the various (complex) currents can be identified as follows:

The m.m.f.s due to positive-sequence and negative-sequence rotor currents rotate relative to the rotor with an angular velocity equal to the angular frequency of the rotor current, the positive-sequence m.m.f. in the direction a, b, c (or α, β) and the negative sequence in the direction c, b, a (or β, α). The motion of these m.m.f.s relative to the stator obviously depends on the angular velocity of the rotor. If the rotor rotates at synchronous speed in a counterclockwise direction the positive-sequence m.m.f. is at rest relative to the stator and the negative-sequence m.m.f. rotates at twice synchronous speed relative to the stator, in a counterclockwise direction.

The m.m.f.s due to forward and backward currents rotate relative to the stator with angular velocity equal to the angular frequency of the rotor current referred to the stator. This angular frequency will, of course, normally be the stator angular frequency. These two m.m.f.s rotate in opposite directions as their names indicate, in counterclockwise and clockwise directions respectively.

Stator

Stator axes A, B, C; A′, B′ and D, Q are all at rest relative to the stator. In fact the sets A′, B′ and D, Q may be taken as identical, D corresponding to A′ and Q to B′, if counterclockwise rotation from A′ to B′ is acceptable. As indicated on p. 101 it is advisable to take the sequence of the three-phase axes A, B, C in the same direction as that of the two-phase axes A′, B′, so that one form of the three-phase to two-phase transformation may be taken as standard.

If A′, B′ and D, Q axes are identical, it follows that P, N and F, B axes are also identical. The positive-sequence (or forward) and negative-sequence (or backward) m.m.f.s rotate relative to the stator with angular velocity equal to the stator angular frequency in the direction A, B, C or A′, B′ and C, B, A or B′, A′ respectively. In general then, P(F) and f axes are at rest relative to each other and so also are N (B, backward) and b.

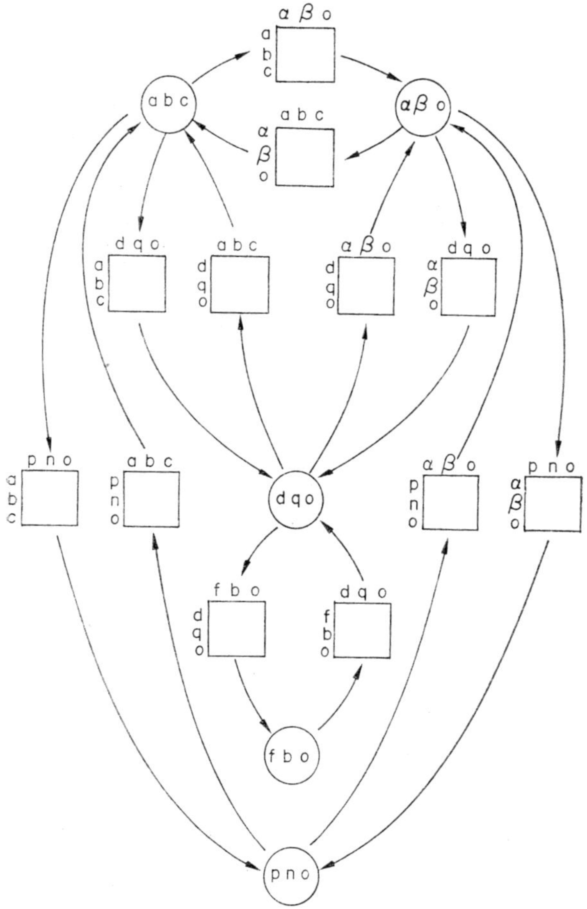

FIG. 21. Rotor (armature) transformations.

It must be remembered that the relative rotations referred to above are those of space vectors and have no relation to the rotation of vectors representing alternating quantities. The latter must, by definition, rotate in the same direction to make possible the addition of the various components as on pp. 107 and 110.

It is *not*, of course, necessary to comprehend the physical significance of the various axes and m.m.f.s in order to use them satisfactorily. It is, however, desirable to appreciate the relationships between them. These are set out in Figs 21 and 22 for rotor and stator respectively. In these diagrams the circles represent the various sets of axes and the squares the transformation matrices denoted by their indices.

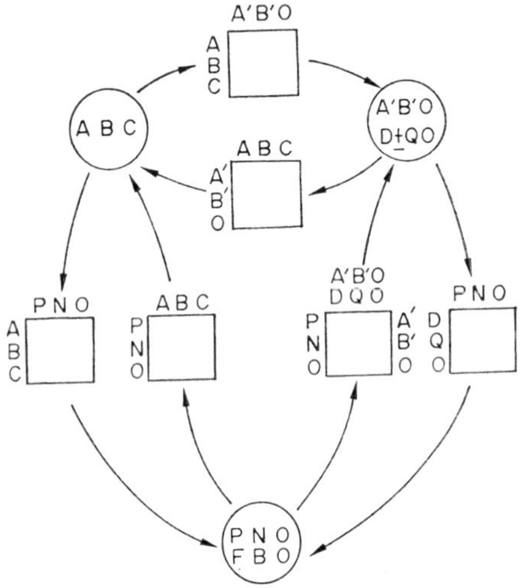

FIG. 22. Stator transformations.

CHAPTER 8

The Application of Matrix Techniques to Routine Performance Calculations

The Establishment of the Transient Impedance Matrix

The first step in machine performance calculation is to establish the "transient impedance" matrix for the specified machine. This matrix may be regarded as representing the operation which, if performed on the current, gives the voltage, or, alternatively, as the expression which gives the Laplace transform of the voltage when post-multiplied by the Laplace transform of the current. This process is begun by writing down by inspection, or by simple rules, the transient impedance of a "primitive" machine, that is one in which:

(a) there are no interconnections between windings;

(b) there are two stationary rotor axes at right angles to each other and arranged so that:

  (i) if the stator has salient poles, the axes of the rotor (armature) circuits lie on the axes of symmetry of the stator; or

  (ii) if the air-gap is uniform, the axis of at least one of the rotor (armature) circuits lies on the axis of one of the stator windings,

(c) if the stator has salient poles, the stator windings have axes coincident with the axes of symmetry.

In other words one in which the reference axes are D, Q, d, q, all stationary relative to the stator.

It is assumed here that saliency is associated only with the stator. If, however, the rotor is salient and the stator non-salient, the axes must be taken stationary relative to the rotor and coincident with its axes of symmetry. The resulting equations will be identical in form with those of the corresponding machine with a salient-pole stator.

If the specified machine has a balanced winding, the resistances of the two axes of this winding are equal. If the air-gap is uniform, i.e. both members are non-salient, the self-inductances on the two axes of a balanced winding are equal, and if, moreover, both stator and rotor windings are balanced, the two mutual inductances are also equal.

The primitive machine must have an adequate number of windings on both stator and rotor. In general this means a D and a Q winding for each actual stator winding not itself on either the D or Q axes. For each actual winding on the D or Q axes a single winding on the appropriate axis will suffice. Both d and q windings are required for each balanced set of rotor windings.

Transformations are then performed on the primitive system as follows. If the specified machine has

(a) pairs of commutator brushes which are not at right angles to one another and/or which result in the axes of the armature circuits not coinciding with the axes of symmetry of the stator core if salient, or of the stator winding if the gap be uniform, or
(b) salient-poles and stator windings with axes not coincident with the axes of symmetry of the poles,

a transformation of the "brush-shifting" type must be performed to give the actual winding arrangement of the specified machine.

A further transformation must be performed if there are any interconnections between windings. It is essential that the windings to be connected are expressed in terms of either the actual winding axes, or of axes which have been obtained from the actual winding axes by identical transformations.

Series and parallel connection must be considered separately.

(a) If windings are connected in series, the same current flows in them all, and this indicates the required transformation. The separate identity of the series-connected windings, and also of their terminal voltages, is then lost, the latter being replaced in the voltage matrix by their sum. This is usually of little consequence, since if the individual terminal voltages are required, they can easily be found when all the currents have been determined.

(b) If windings are connected in parallel, their terminal voltages are equal. The concept of duality indicates that transformation would replace the paralleled windings by a single equivalent winding carrying a current equal to the sum of the currents in the individual windings. The identity of these individual currents is then lost. If the paralleled windings form part of a larger network, so that their terminal voltage is not known, this inconvenience may have to be suffered in order to express the constraints between the currents. If, however, the terminal voltage of the paralleled windings is known, as for example with a d.c. shunt motor, it may be better not to perform any transformation but to determine all currents directly, merely making all the relevant elements of the voltage matrix equal to the known terminal voltage.

The matrix resulting from these operations is the transient impedance matrix **Z** of the specified machine in terms of stationary reference axes. In the case of stationary windings and of armature circuits formed by stationary commutator brushes, these reference axes are those of the actual circuits and the currents and voltages are those actually existing. In the case of armature windings connected through slip-rings, or short-circuited otherwise than by commutator brushes (as in revolving-armature synchronous machines and induction machines) the reference axes are not those of the actual windings and the currents and voltages are not those actually existing. In almost all cases, however, it is easiest to express the equations in terms of the stationary axes d, q in the first instance. In some cases it is convenient to solve the problem in these axes, in which case the terminal data (voltage or current) is first transformed to d, q axes if it is not already in terms of these axes. Finally, the results are transformed back to the given axes if necessary, as shown for the synchronous machine in pp. 201–2. In other cases the solution of the voltage equation is easier if a further transformation is first performed, e.g. as with the induction machine to P, N, f, b axes in pp. 165 and 180 or to synchronous axes A, B, a, b in pp. 260–3. In these cases the terminal data is transformed to the new axes and the solution transformed back from them to the given axes.

For the calculation of torque, the **G** matrix can be written down by inspection as the coefficients of $\dot{\theta}$ in the transient impedance matrix **Z**.

These transient impedance matrices permit solutions to be obtained for transient electrical conditions and for instantaneous and r.m.s. values in steady-state a.c. conditions and for steady-state d.c. conditions. If only steady-state solutions are required, the p may be replaced in the impedance matrix by:

(i) zero, if all the exciting functions (i.e. applied voltages and currents) are constant *when referred to the stationary axes*;
(ii) $j\omega$, if all the exciting functions are sinusoidal and of the same angular frequency ω *when referred to the stationary axes.*

This latter substitution leads to solutions in terms of r.m.s. phasor quantities.

In synchronous machines appropriate allowance must be made for the difference in time between the alignment of the α and D winding axes (i.e. when $\theta = 0$) and the time zero of t determined by the phase of the terminal voltage. Mathematically this arises when $\dot{\theta} = \omega$ is integrated giving $\theta = \omega t +$constant. This constant is not necessarily, or even usually, zero. It is, of course, the load angle or an angle simply related to the load angle. It will be denoted in this work by ψ, since it is not usually apparent, when it is introduced, whether it is or is not the load angle, for which the usual symbol is δ.

Phasor Diagrams and Equivalent Circuits

Phasor diagrams and equivalent circuits are widely used in the older methods of machine analysis and it is interesting to reflect upon their uses.

Phasor Diagrams

Phasor diagrams are useful:

(i) as *aides-mémoire* in writing down the equations of a machine;
(ii) in that they lead to locus diagrams, which epitomize the performance;
(iii) in that they permit graphical solutions of some problems which are tedious arithmetically.

These advantages are not so attractive when the analysis is performed directly in terms of equations with the aid of matrix algebra. However, phasor diagrams can be deduced from the matrix equations and this is done for synchronous machines in Chapter 10 partly to show the procedure and partly to show that the present analysis leads to the same phasor diagrams as do the earlier methods.

Phasor diagrams cannot be deduced directly from stationary reference axes equations when the actual axes are moving. It must be noted that in such cases the equations are often purely algebraic and not phasor equations, and the currents and voltages in these terms are instantaneous or "d.c." quantities. The equations must be in terms of the actual winding axes, when the phase angles will appear in the equations.

Equivalent Circuits

Equivalent circuits are useful:

(i) as *aides-mémoire* in writing down equations;
(ii) for use on network analysers for the solution of problems which are too tedious to do arithmetically.

Again the first of these uses is less attractive when the analysis is performed in terms of matrices than when traditional methods are used, whilst the second has been largely superseded by digital techniques.

Some equivalent circuits for machines involve active elements.† This is particularly undesirable for circuits intended for use with a

† For example, Morrill's equivalent circuit of the capacitor motor. See ref. 9.

network analogue. It is characteristic of networks formed entirely of passive elements that the impedance matrix is necessarily symmetric since L_{xy} and L_{yx} are equal. To derive an equivalent circuit consisting only of passive elements it is therefore necessary to transform the variables so that the transformed impedance matrix is symmetric and contains only constant elements, i.e. which are not functions of time.

Interconnections between Machines or between Machines and other Circuit Elements

Interconnection of machines or machines and other circuit elements is represented as follows. After obtaining the transient impedance matrices of the separate machines and other circuit elements, one writes these as sub-matrices on the principal diagonal of a compound matrix. The "mutual" sub-matrices are composed wholly of zeros. The compound matrix, when written out in full, thus has as many rows and columns as all the component matrices together. A transformation matrix is then derived according to the current relationships arising from the interconnections and this transformation is applied to the combined transient impedance matrix written in full.†

It is important to realize that interconnections can, in general, be made only between the actual winding axes of the machines and not between hypothetical axes obtained by transformation. For example, if the rotors of two induction motors are interconnected,‡ the matrices must be in terms of a_1, b_1, c_1 and a_2, b_2, c_2 or α_1, β_1, and α_2, β_2, *not* between d_1, q_1 and d_2, q_2. It may be possible to perform a transformation to other axes after the interconnection, and, having done so, it may then be possible to see how it could have been done before—but it is most unwise to attempt to do it before the interconnection has been performed.

† An example of this is given on pp. 202–4.

‡ See pp. 292–300.

Closed Circuits

When a circuit is closed its terminal voltage is zero. This is not necessarily the same as saying that the terminal voltage of a winding is zero (i.e. that a winding is short-circuited), since the winding may be only part of that particular circuit. If a closed circuit exists, the voltage across a winding forming only part of that circuit can be found only by first determining the current in it and the other circuits and multiplying by the impedance of either the winding itself or of the remainder of its circuit—not forgetting the mutual inductance terms.

Specific Types of Problem

Whilst it is obviously not possible to foresee all types of problem which can arise in the consideration of electrical machines, most problems fall within the five categories which follow. In all cases it is first necessary to determine the transient impedance matrix $\mathbf{Z}$ of the machine or system involved in terms of stationary reference axes. The subsequent procedure depends upon the particular problem.

(*a*) *Given the terminal voltages of all windings carrying current, to find the currents*

If the stationary reference axes of $\mathbf{Z}$ are the actual winding axes it is merely necessary to invert $\mathbf{Z}$, when the currents are given by $\mathbf{i} = \mathbf{Z}^{-1}\mathbf{v}$.

If, however, the stationary reference axes are not the actual winding axes it is first necessary to transform the given voltages, which will be in terms of the actual winding axes, into the terminal voltages of the stationary reference axis system by $\mathbf{v}' = \mathbf{C}_t^*\mathbf{v}$. After inverting $\mathbf{Z}'$—the impedance in terms of stationary reference axes—the currents in terms of stationary reference axes are obtained by $\mathbf{i}' = \mathbf{Z}'^{-1}\mathbf{v}'$. Lastly, the actual currents are obtained by $\mathbf{i} = \mathbf{C}\mathbf{i}'$.

Whilst this latter procedure may seem unduly long, it is not possible to attach any meaning to an attempt to invert the impedance in terms of the actual moving reference axes since inversion is used to solve the Laplace transform equations, and this method of solution cannot be used if the coefficients are not constant.

Some labour may sometimes be saved where circuits are closed and consequently have zero terminal voltage. Since the elements of the corresponding column of $\mathbf{Z}^{-1}$ will multiply zeros in the voltage matrix, there is no need to determine such columns of $\mathbf{Z}^{-1}$. If, however, the impedance matrix is not in terms of the actual winding axes, zeros in the transformed voltage $\mathbf{v}'$ will not necessarily occur when there are zeros in the actual terminal voltage matrix $\mathbf{v}$, and vice versa.

Again, if some windings of the machine are open-circuited, and are consequently not carrying current, their terminal voltages are usually unknown. This in itself does not occasion any difficulty, because the rows and columns of the matrices corresponding to an open-circuited winding may be omitted, since its presence has no effect on the currents or voltages of other windings. There is no difficulty therefore when the axes are the actual winding axes. When, however, the actual axes are moving, open-circuit of an actual winding will not necessarily correspond to zero current in any transformed winding. Moreover, if the open circuiting of a winding leaves the machine so that it does not comply with the limitations detailed on p. 76, the problem cannot be solved by routine techniques in terms of stationary reference axes.†

(*b*) *Given the terminal voltages of all windings carrying current, to find the terminal voltages of the open-circuited windings*

It is first necessary here to find the currents as described in the preceding section, omitting the open-circuited winding axes. Again the limitation that after this omission the machine must still comply with the requirements of pp. 76–77 must not be overlooked.

Once the currents are known, the voltages are found by $\mathbf{v} = \mathbf{Zi}$ using the complete impedance matrix. In performing the multiplication $\mathbf{Zi}$ it is obviously not necessary, except for checking purposes, to calculate the voltages which were given in the data.

† See, for example, the analysis of the single-phase alternator pp. 305–8, and refs. 13 and 14.

(c) Given the terminal voltages of some windings and the currents in the others, to find the remaining currents and voltages

This type of problem can be solved by partitioning. The order of the reference axes in the matrix is arranged so that all the known voltages come before the unknown voltages and all the known currents after the unknown ones. If $\mathbf{v}_1$ is the matrix of known voltages and $\mathbf{v}_2$ that of the unknown ones, and $\mathbf{i}^1$ the unknown currents and $\mathbf{i}^2$ the known currents, the impedance matrix is partitioned accordingly into $\mathbf{Z}_{11}$, $\mathbf{Z}_{12}$, $\mathbf{Z}_{21}$ and $\mathbf{Z}_{22}$ so that the voltage equation is

$$\begin{bmatrix} \mathbf{v}_1 \\ \hline \mathbf{v}_2 \end{bmatrix} = \begin{bmatrix} \mathbf{Z}_{11} & \mathbf{Z}_{12} \\ \hline \mathbf{Z}_{21} & \mathbf{Z}_{22} \end{bmatrix} \begin{bmatrix} \mathbf{i}^1 \\ \hline \mathbf{i}^2 \end{bmatrix}$$

from which

$$\begin{aligned} \mathbf{v}_1 &= \mathbf{Z}_{11}\mathbf{i}^1 + \mathbf{Z}_{12}\mathbf{i}^2 \\ \mathbf{Z}_{11}\mathbf{i}^1 &= \mathbf{v}_1 - \mathbf{Z}_{12}\mathbf{i}^2 \\ \mathbf{i}^1 &= \mathbf{Z}_{11}^{-1}[\mathbf{v}_1 - \mathbf{Z}_{12}\mathbf{i}^2] \end{aligned}$$

so that the unknown currents $\mathbf{i}^1$ can be found in terms of the known voltages $\mathbf{v}_1$ and the known currents $\mathbf{i}^2$.

The unknown voltages could be found by a similar process after inverting the matrix but it is easier to use the equation

$$\mathbf{v}_2 = \mathbf{Z}_{21}\mathbf{i}^1 + \mathbf{Z}_{22}\mathbf{i}^2$$

after $\mathbf{i}^1$ has been determined. Expressed directly in terms of $\mathbf{v}_1$ and $\mathbf{i}_2$:

$$\begin{aligned} \mathbf{v}_2 &= \mathbf{Z}_{21}\mathbf{Z}_{11}^{-1}[\mathbf{v}_1 - \mathbf{Z}_{12}\mathbf{i}^2] + \mathbf{Z}_{22}\mathbf{i}^2 \\ &= \mathbf{Z}_{21}\mathbf{Z}_{11}^{-1}\mathbf{v}_1 + [\mathbf{Z}_{22} - \mathbf{Z}_{21}\mathbf{Z}_{11}^{-1}\mathbf{Z}_{12}]\mathbf{i}^2 \end{aligned}$$

(d) Given the terminal voltages, to find the torque in terms of the speed or phase angle

It is first necessary to find the currents as described above.

The $\mathbf{G}$ matrix may be written down from the stationary axis impedance matrix and the torque calculated as Re $\mathbf{i}_t^*\mathbf{G}\mathbf{i}$ in which $\mathbf{i}_t$, $\mathbf{G}$, and $\mathbf{i}$ are in terms of D, Q, d, q axes, whether or not these are the actual winding axes. There is no need to transform back to the actual winding axes when the currents have been determined in other axes.

There is not usually any advantage in calculating the torque from $\frac{1}{2}\mathbf{i}_t^*(\partial\mathbf{L}/\partial\theta)\mathbf{i}$ in the type of problems covered in this book.

The speed or phase angle will arise naturally in the current matrix according to whether the machine is operating asynchronously or synchronously, so that the torque is expressed as a function of the speed or angle respectively.

(*e*) *Given the terminal voltages to find the speed and/or currents at a given torque*

Since the torque is given as a function of speed (or phase angle) it is necessary to equate the torque expression to the given torque and solve for the speed—this last operation may not be possible algebraically, and resort to numerical or graphical methods may be necessary.

Once the speed has been found the currents can be determined.

The Analysis of Three-phase Machines

The direct approach to the analysis of three-phase machines would be to determine the impedance matrix in terms of the three-phase windings, shown diagrammatically in Fig. 23, in a manner similar to

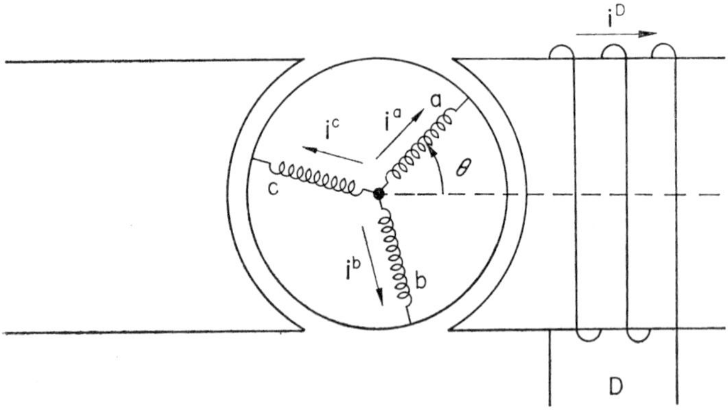

FIG. 23. Balanced threep-hase salient-pole machine.

that used for the two-phase machine on pp. 65–69. This can be done,[†] but requires considerable understanding of the various components of flux linking the windings. Difficulty arises in particular with flux, other than the space-fundamental flux, due to current in one phase of the armature winding which links another phase of that winding.

An alternative approach is to use the impedance matrix already derived for the two-phase machine with axes α, β, to add another row and column for the zero-sequence axis and then to use the transformation derived on p. 104, between three-phase axes a, b, c and the α, β, o system. Although this may appear simpler, it merely replaces the original difficulty by another one, namely that of determining the elements of the zero-sequence row and column of the impedance matrix in α, β, o axes. The components of flux involved in this are, however, more easily identified. The result of this approach is a three-phase impedance matrix expressed in terms of the normal machine parameters X_d, X_q, and X_o, whereas direct derivation of the three-phase impedance matrix involves other quantities related linearly to these reactances.

The Effects of Zero-sequence Currents

Zero-sequence currents, by definition, are of the same instantaneous magnitude in all the three winding phases a, b, c. Consideration of the air-gap m.m.f. of a winding having the usual 60° phase groups[‡] shows that such currents produce no space-fundamental m.m.f. but only third space-harmonic and its higher multiples. The resulting third space-harmonic flux will have no or negligible linkage with the field or damper windings and will be very little affected by the relative position of armature and field. It may, therefore, be taken as independent of θ.

There will, of course, be end-winding leakage fiuxes and slot-leakage fluxes. If the winding is chorded, however, the slots containing coil-sides of different winding phases require particular consid-

† See ref. 10, pp. 10–16.
‡ See ref. 3, p. 234.

eration. With positive- or negative-sequence currents, the two coil-sides of such slots carry, in opposite directions, currents differing in time-phase by 120°, which corresponds to a net difference of 60°. Thus chording has some effect on slot leakage flux under such conditions. With zero-sequence current the coil-sides of such slots carry, at all times, identical currents in opposite directions. Consequently their contribution to slot-leakage flux is then very small indeed. The slot-leakage contribution to zero-sequence reactance thus varies widely according to the proportion of slots having coil-sides of different winding phases, i.e. according to the amount of chording.†

It is clear that, provided that the three windings a, b, c are balanced, none of the components of flux due to zero-sequence current can lead to positive- or negative-sequence voltage, nor can positive- or negative-sequence currents produce zero-sequence voltages.

We can sum up the above deductions mathematically by saying that the elements in the zero-sequence row and column of the impedance matrix in α, β, o or d, q, o or p, n, o axes will all be zero except for that on the principal diagonal, which will have a resistance equal to the actual winding resistance and an inductance which is effectively independent of θ and, therefore, neglecting saturation changes, a constant. This element of the impedance matrix may thus be written $R_a+L_o p$, in which L_o will be small, and, with chorded windings, perhaps much smaller than the ordinary armature leakage inductance l_a.

The two-phase impedance matrix, given on p. 69, with the addition of the zero-sequence impedance in the form $R_a+L_o p$, is

	α	β	o	D
α	$R_a+p(L_d\cos^2\theta+L_q\sin^2\theta)$	$p(L_d-L_q)\sin\theta\cos\theta$		$pL_{Dd}\cos\theta$
β	$p(L_d-L_q)\sin\theta\cos\theta$	$R_a+p(L_d\sin^2\theta+L_q\cos\theta^2)$		$pL_{Dd}\sin\theta$
o			$R_a+L_o p$	
D	$pL_{Dd}\cos\theta$	$pL_{Dd}\sin\theta$		$R_{DD}+L_{DD}p$

† See ref. 11, p. 1205.

This can be transformed to three-phase axes by the transformation matrix

$C = \sqrt{\frac{2}{3}}$

	a	b	c	D
α	1	$-\frac{1}{2}$	$-\frac{1}{2}$	
β		$\sqrt{3}/2$	$-\sqrt{3}/2$	
o	$1/\sqrt{2}$	$1/\sqrt{2}$	$1/\sqrt{2}$	
D				$\sqrt{3}/\sqrt{2}$

Since this matrix array contains only constant elements, the transformation is of the simple form $\mathbf{C_t Z C}$, without any intermediate differentiations, although considerable algebra and trigonometry is involved.

The resulting impedance matrix† is shown on p. 136.

When transformed to d, q, o, D axes, both two-phase and three-phase impedance matrices lead to

	d	q	o	D
d	$R_a+L_d p$	$L_q\dot{\theta}$		$L_{Dd}p$
q	$-L_d\dot{\theta}$	$R_a+L_q p$		$-L_{Dd}\dot{\theta}$
o			$R_a+L_o p$	
D	$L_{Dd}p$			$R_{DD}+L_{DD}p$

Comparison of the three impedance matrices in α, β, o, D; a, b, c, D; and d, q, o, D axes makes it obvious why it is preferable, in most cases,‡ when determining the performance of a three-phase machine, to:

(a) start with a two-phase system rather than to attempt to deduce the three-phase impedance matrix directly from physical considerations;

† This matrix may be compared with those of ref. 2, vol. 38, p. 590, and ref. 12, and the equations of ref. 10, pp. 10–16.

‡ Only in cases of such problems as that considered in ref. 14 and in a simplified form in pp. 305–8 is there no advantage in d, q axes.

	a	b	c	D
a	R_a $+p(\frac{1}{3})(L_d+L_q+L_o)$ $+p(\frac{1}{3})(L_d-L_q)\cos 2\theta$	$-p(\frac{1}{6})(L_d+L_q-2L_o)$ $+p(\frac{1}{3})(L_d-L_q)\cos(2\theta-2\pi/3)$	$-p(\frac{1}{6})(L_d+L_q-2L_o)$ $+p(\frac{1}{3})(L_d-L_q)\cos(2\theta-4\pi/3)$	$p\sqrt{(\frac{2}{3})}L_{Dd}\cos\theta$
b	$-p(\frac{1}{6})(L_d+L_q-2L_o)$ $+p(\frac{1}{3})(L_d-L_q)\cos(2\theta-2\pi/3)$	R_a $+p(\frac{1}{3})(L_d+L_q+L_o)$ $+p(\frac{1}{3})(L_d-L_q)\cos(2\theta-4\pi/3)$	$-p(\frac{1}{6})(L_d+L_q-2L_o)$ $+p(\frac{1}{3})(L_d-L_q)\cos 2\theta$	$p\sqrt{(\frac{2}{3})}L_{Dd}\cos(\theta-2\pi/3)$
c	$-p(\frac{1}{6})(L_d+L_q-2L_o)$ $+p(\frac{1}{3})(L_d-L_q)\cos(2\theta-4\pi/3)$	$-p(\frac{1}{6})(L_d+L_q-2L_o)$ $+p(\frac{1}{3})(L_d-L_q)\cos 2\theta$	R_a $+p(\frac{1}{3})(L_d+L_q+L_o)$ $+p(\frac{1}{3})(L_d-L_q)\cos(2\theta-2\pi/3)$	$p\sqrt{(\frac{2}{3})}L_{Dd}\cos(\theta-4\pi/3)$
D	$p\sqrt{(\frac{2}{3})}L_{Dd}\cos\theta$	$p\sqrt{(\frac{2}{3})}L_{Dd}\cos(\theta-2\pi/3)$	$p\sqrt{(\frac{2}{3})}L_{Dd}\cos(\theta-4\pi/3)$	$R_{DD}+pL_{DD}$

(b) to transform the rotor axes, α, β, of the two-phase system to stationary axes d, q as in pp. 70–72. This need be done only on the first occasion that the particular machine is considered;

(c) to transform the data, from whatever axes it is given in, to d, q, axes; to solve the problem in d, q, axes or others derived from them, and then to transform the solution back into the desired reference axes if necessary.

In the absence of any zero-sequence voltages and currents, the problem is thus solved in only two rotor axes instead of the three of the a, b, c system, in addition to avoiding the complication of time-varying inductances.

From the transformation matrices on p. 117 it can be seen that

$$v_\alpha = \cos\theta v_d + \sin\theta v_q \quad \text{and}$$
$$i^\alpha = \cos\theta i^d + \sin\theta i^q$$

whereas

$$v_a = \sqrt{(\tfrac{2}{3})}\{\cos\theta v_d + \sin\theta v_q + (1/\sqrt{2})v_o\}$$

and

$$i^a = \sqrt{(\tfrac{2}{3})}\{\cos\theta i^d + \sin\theta i^q + (1/\sqrt{2})i^o\}$$

In the absence of zero-sequence components, therefore,

$$v_a = \sqrt{(\tfrac{2}{3})}v_\alpha \quad \text{and} \quad i^a = \sqrt{(\tfrac{2}{3})}i^\alpha$$

so that

$$v_a/i^a = v_\alpha/i^\alpha$$

The relationship between v_a and i^a being the same as that between v_α and i^α implies that the numerical values of the parameters, such as L_d, L_q, X_S, X_M, X_r, of the two systems are the same. It also means that, in the absence of zero-sequence components, a solution for the α phase of a two-phase system applies equally to the a phase of a three-phase system without the necessity of performing the transformation. This is particularly convenient under balanced conditions when the performance can be completely expressed in terms of the voltage and current of one phase alone.

These results arise from the particular choice of three- to two-phase transformation and more specifically from the choice of a transformation which is the same for both current and voltage i.e. orthogonal.

It is important to note that although the values of L_d, L_q, etc., in the two-phase axes are the same as the values in three-phase axes, there is a significant difference. Reference to the two-phase impedance matrix on p. 134 shows that L_d and L_q are the values of the inductance of the α phase when $\theta = 0$ and $\theta = \pi/2$ respectively *with all other windings open-circuit*. It is clear, however, from the impedance matrix on p. 136 that the inductances of the a phase of a three-phase machine for these two conditions are $(\frac{2}{3})L_d + (\frac{1}{3})L_o$ and $(\frac{2}{3})L_q + (\frac{1}{3})L_o$ respectively. A more detailed consideration of this impedance matrix shows that when $\theta = 0$, the voltage v_a is $(R_a + L_d p)i^a$ *provided that there is also current in the* b *and* c *phases such that* $p(i^a + i^b + i^c) = 0$. L_d is thus an effective self-inductance of the a phase for $\theta = 0$, when there is no zero-sequence current. Similarly, L_q is an the effective self-inductance when $\theta = \pi/2$ under the same condition.

When the α phase of the two-phase machine is considered, it is, of course, irrelevant whether there is current in the β-phase winding, since the two windings are at right angles to one another and on the axes of symmetry of the salient-pole field. The relationships can be summarized by saying that L_d and L_q are the effective inductances of the a phase in the absence of zero-sequence current and the actual indu ctances of the α phase, when $\theta = 0$ and $\theta = \pi/2$ respectively.

In single-phase line-to-line operation of a three-phase machine, the form of the two-phase equation will differ according to whether i^a, i^b, or i^c is taken as zero. As a further result of the particular choice of three-phase to two-phase transformation, $i^\alpha = 0$ if $i^a = 0$, so that this is to be preferred to either $i^b = 0$ or $i^c = 0$.

CHAPTER IX

D.C. and Single-phase Commutator Machines

The Series Commutator Machine

The arrangement of the series commutator machine is shown in Fig. 24, and the corresponding primitive machine in Fig. 25(i). However, it is preferable to consider first the more complicated primitive machine of Fig. 25(ii) of which the impedance matrix is

	d	q	F
d	$R_a + L_a p$	$L_a\dot{\theta}$	Mp
q	$-L_a\dot{\theta}$	$R_a + L_a p$	$-M\dot{\theta}$
F	Mp		$R_F + L_F p$

Since there is no d circuit the corresponding row and column may be omitted leaving

	a	F
a	$R_a + L_a p$	$-M\dot{\theta}$
F		$R_F + L_F p$

in which the sole armature circuit is designated a instead of q. The reason for reverting to the primitive machine with the d winding included is apparent from the presence of the $-M\dot{\theta}$ element in the a row of the last matrix. Without the d row and column of the machine of Fig. 25(ii) the sign, the magnitude, and possibly the existence of this element might be in doubt.

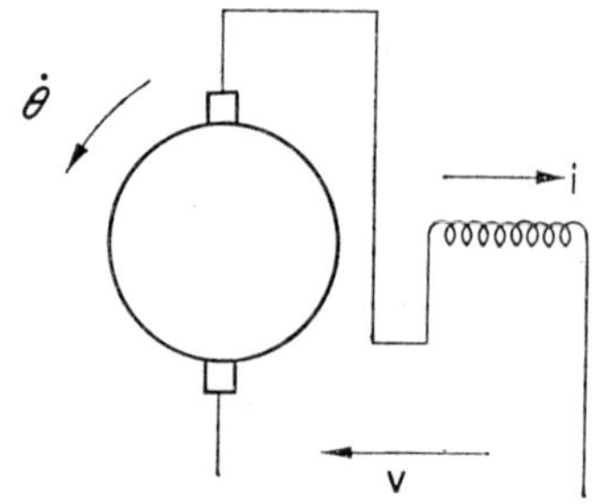

FIG. 24. Series commutator machine.

M is the mutual inductance measured between the field winding and direct-axis brushes when fitted, if the flux is sinusoidally distributed. If this last condition is not fulfilled, M is just a constant with the dimensions of inductance relating generated voltage to speed and field current. In this case, if the d axis were still present in the equa-

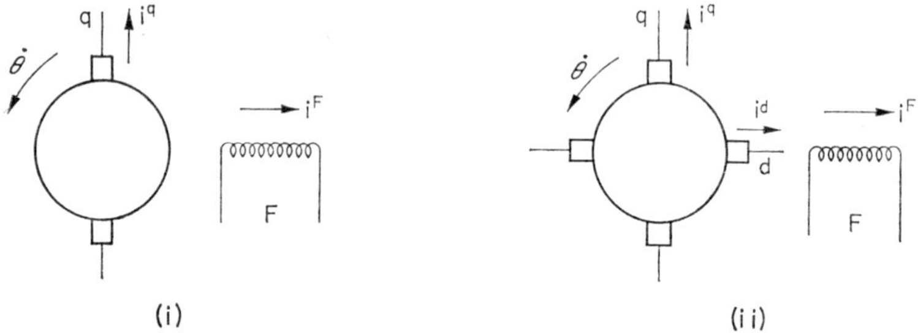

FIG. 25. Primitive commutator machines.

tions, it would be necessary to make a distinction between the true mutual inductance and this constant, as in the matrix of p. 55.

The interconnection to form the series machine connected as in Fig. 24 is simply defined by

$$\begin{aligned} i^a &= i \\ i^F &= i \end{aligned} \qquad \text{or} \qquad \begin{matrix} \text{a} \\ \text{F} \end{matrix} \begin{bmatrix} i^a \\ i^F \end{bmatrix} = \begin{matrix} \text{a} \\ \text{F} \end{matrix} \begin{bmatrix} 1 \\ 1 \end{bmatrix} \begin{bmatrix} i \end{bmatrix}$$

The impedance matrix of the connected machine is therefore

$$\mathbf{Z}' = \begin{array}{cc} \text{a} & \text{F} \\ \hline \multicolumn{1}{|c|}{1} & \multicolumn{1}{c|}{1} \\ \hline \end{array} \quad \begin{array}{c|c|c|} & \text{a} & \text{F} \\ \hline \text{a} & R_a+L_a p & -M\dot{\theta} \\ \hline \text{F} & & R_F+L_F p \\ \hline \end{array} \quad \begin{array}{c|c|} \hline \text{a} & 1 \\ \hline \text{F} & 1 \\ \hline \end{array}$$

$$= \boxed{(R_a+R_F)+(L_a+L_F)p-M\dot{\theta}}$$

whence $\mathbf{G}' = \boxed{-M}$

If $R = R_a+R_F$ and $L = L_a+L_F$ the voltage equation is

$$v = (R+Lp-M\dot{\theta})i$$

This equation is equally valid for d.c. or a.c. single-phase series machines.

For the d.c. machine p may be replaced by zero for steady-state conditions, leading to

$$V = RI-M\dot{\theta}I \qquad \text{and} \qquad I = \frac{V}{R-M\dot{\theta}}$$

The torque is $\mathbf{i}_t'\mathbf{G}'\mathbf{i}' = -IMI = -MI^2 = -\dfrac{V^2M}{(R-M\dot{\theta})^2}$ which, being negative, shows that the armature will rotate in the clockwise direction with the connection as in Fig. 24. This being so, $\dot{\theta}$ is normally negative and consequently $-M\dot{\theta}I$ is positive, and is, of course, the "back e.m.f.", which may be represented by E, giving the familiar d.c. motor equation $V = RI+E$. The value of M must correspond to the degree of saturation in the machine at the current I.

For steady-state conditions in the a.c. machine the p of the transient impedance matrix must be replaced by $j\omega$, leading to

$$\begin{aligned} V &= (R+j\omega L-M\dot{\theta})I \\ &= \{(R-M\dot{\theta})+j\omega L\}I \end{aligned}$$

as a phasor equation, from which

$$I = \frac{V}{(R-M\dot{\theta})+j\omega L}$$
$$= \frac{V(R-M\dot{\theta})-jV\omega L}{(R-M\dot{\theta})^2+\omega^2L^2}$$

The mean torque is given by Re $\mathbf{I}_t'^*\mathbf{G}'\mathbf{I}'$

$$= \text{Re}\left[\left\{\frac{V(R-M\dot{\theta})+jV\omega L}{(R-M\dot{\theta})^2+\omega^2L^2}\right\}\{-M\}\left\{\frac{V(R-M\dot{\theta})-jV\omega L}{(R-M\dot{\theta})^2+\omega^2L^2}\right\}\right]$$
$$= \frac{-V^2M}{(R-M\dot{\theta})^2+\omega^2L^2}$$

In the above ωL may be written X, and M as X_M/ω.

Alternatively, we may consider the instantaneous conditions in the a.c. machine by putting $v = \hat{V}\sin\omega t$. The Laplace transform equation is then

$$\frac{\hat{V}\omega}{(s^2+\omega^2)} = (R-M\dot{\theta}+Ls)\bar{i}$$

or

$$\bar{i} = \frac{\hat{V}\omega}{(s^2+\omega^2)(R-M\dot{\theta}+Ls)}$$
$$= \frac{As+B\omega}{(s^2+\omega^2)}+\frac{C}{(R-M\dot{\theta}+Ls)}$$

where

$$A = -\frac{\hat{V}\omega L}{(R-M\dot{\theta})^2+\omega^2L^2} \qquad B = \frac{\hat{V}(R-M\dot{\theta})}{(R-M\dot{\theta})^2+\omega^2L^2}$$

and

$$C = \frac{\hat{V}\omega L^2}{(R-M\dot{\theta})^2+\omega^2L^2}$$

The term containing C represents the unidirectional transient current which would result if the voltage $\hat{V}\sin\omega t$ were suddenly applied at time zero with the motor running with an angular velocity $\dot{\theta}$.

The inverse transform of the remainder is the instantaneous current under steady-state conditions and is

$$\frac{\hat{V}(R-M\dot{\theta})\sin\omega t-\hat{V}\omega L\cos\omega t}{(R-M\dot{\theta})^2+\omega^2L^2}=\frac{\hat{V}\sin(\omega t-\phi)}{\sqrt{\{(R-M\dot{\theta})^2+\omega^2L^2\}}}$$

where $\cos\phi = (R-M\dot{\theta})/\sqrt{\{(R-M\dot{\theta})^2+\omega^2L^2\}}$.

The instantaneous torque is $\mathbf{i}'_t\mathbf{G}'\mathbf{i}'$

$$=\frac{\hat{V}\sin(\omega t-\phi)}{\sqrt{\{(R-M\dot{\theta})^2+\omega^2L^2\}}}(-M)\frac{\hat{V}\sin(\omega t-\phi)}{\sqrt{\{(R-M\dot{\theta})^2+\omega^2L^2\}}}$$

$$=\frac{-\hat{V}^2M\sin^2(\omega t-\phi)}{(R-M\dot{\theta})^2+\omega^2L^2}$$

$$=\frac{-\hat{V}^2M\{1-\cos 2(\omega t-\phi)\}}{2\{(R-M\dot{\theta})^2+\omega^2L^2\}}$$

Since $\hat{V} = \sqrt{2}V$, the mean of this expression is the same as that derived on p. 142. The remainder is the variation of the torque with time showing that the torque is pulsating at double the supply frequency, between a value equal to twice that of the mean torque and zero.

The Shunt Commutator Machine

The primitive of the shunt commutator machine is shown in Fig. 26 where a denotes the armature (quadrature) circuit and F the field winding on the direct axis. The impedance matrix of this machine is

$\mathbf{Z} =$

	a	F
a	R_a+L_ap	$-M\dot{\theta}$
F		R_F+L_Fp

to the M of which the remarks on p. 140 apply.

The connected machine is shown in Fig. 27. However, since the relation between i^a and i^F depends on the terminal conditions, it is better not to transform the impedance matrix of the primitive machine at this stage.

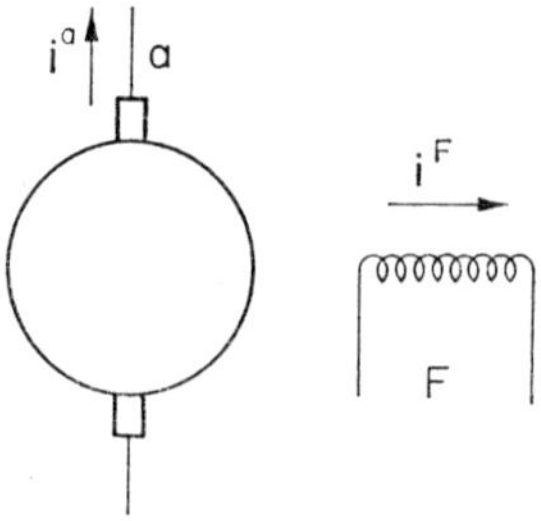

FIG. 26. Primitive commutator machine.

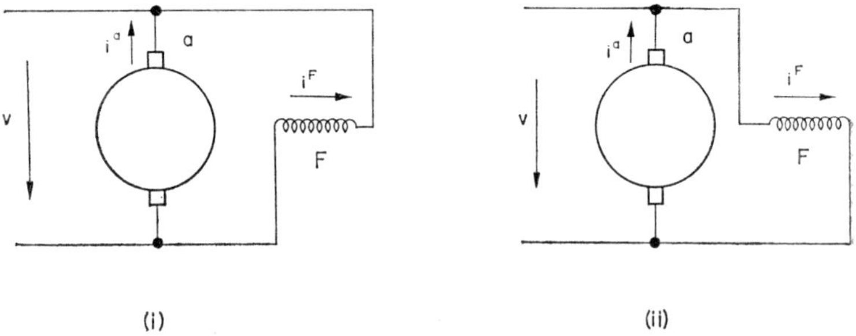

FIG. 27. Shunt commutator machine.

Two cases of this machine are of interest: the d.c. shunt motor in which the terminal voltages of the two windings are both equal and known; and the d.c. shunt generator in which they are equal but unknown. It is convenient to use different connections for motor and generator as shown in Fig. 27(i) and (ii) respectively.

The D.C. Shunt Motor

The steady-state impedance matrix is

$$\mathbf{Z} = \begin{array}{c|c|c|} & \mathrm{a} & \mathrm{F} \\ \hline \mathrm{a} & R_{\mathrm{a}} & -M\dot{\theta} \\ \hline \mathrm{F} & & R_{\mathrm{F}} \\ \hline \end{array}$$

and the voltage equation of the armature is

$$V_{\mathrm{a}} = R_{\mathrm{a}}I^{\mathrm{a}} - M\dot{\theta}I^{\mathrm{F}}$$

and of the field winding,

$$V_F = R_F I^F$$

Again the first of these equations is commonly written

$$V = RI + E$$

where E is the back e.m.f. and is equal to $-M\dot{\theta}I^F$.

The torque is $\mathbf{i}_t\mathbf{G}\mathbf{i}$

$$= \begin{array}{cc} \scriptstyle a & \scriptstyle F \\ \hline \multicolumn{1}{|c|}{I^a} & \multicolumn{1}{c|}{I^F} \\ \hline \end{array} \quad \begin{array}{c|c|c|} & \scriptstyle a & \scriptstyle F \\ \cline{2-3} \scriptstyle a & & -M \\ \cline{2-3} \scriptstyle F & & \\ \cline{2-3} \end{array} \quad \begin{array}{c|c|} \cline{2-2} \scriptstyle a & I^a \\ \cline{2-2} \scriptstyle F & I^F \\ \cline{2-2} \end{array}$$

$$= -I^a M I^F = I^a E/\dot{\theta}.$$

Inverting the impedance matrix leads to the current equation

$$\begin{array}{c|c|} \cline{2-2} \scriptstyle a & I^a \\ \cline{2-2} \scriptstyle F & I^F \\ \cline{2-2} \end{array} = \frac{1}{R_a R_F} \begin{array}{c|c|c|} & \scriptstyle a & \scriptstyle F \\ \cline{2-3} \scriptstyle a & R_F & M\dot{\theta} \\ \cline{2-3} \scriptstyle F & & R_a \\ \cline{2-3} \end{array} \quad \begin{array}{c|c|} \cline{2-2} \scriptstyle a & V_a \\ \cline{2-2} \scriptstyle F & V_F \\ \cline{2-2} \end{array}$$

Since $V_a = V_F = V$ say, in the shunt motor connected as in Fig. 27(i),

$$\begin{array}{c|c|} \cline{2-2} \scriptstyle a & I^a \\ \cline{2-2} \scriptstyle F & I^F \\ \cline{2-2} \end{array} = \frac{V}{R_a R_F} \begin{array}{c|c|} \cline{2-2} \scriptstyle a & R_F + M\dot{\theta} \\ \cline{2-2} \scriptstyle F & R_a \\ \cline{2-2} \end{array}$$

In these terms, therefore, the two-pole torque T is

$$-\frac{V^2(R_F + M\dot{\theta})}{R_a R_F^2} M$$

The implication that the torque varies as the square of the terminal voltage is true only so long as M is constant, i.e. in the unsaturated condition. At normal voltages M is a function of I^F and hence of V.

The D.C. Shunt Generator on Open Circuit

Because there is no other load on the armature, $i^a = i^F$ if the directions are as shown in Fig. 27(ii), and

$$\mathbf{C} = \begin{array}{c|c|} & \\ \mathrm{a} & 1 \\ \mathrm{F} & 1 \\ \end{array}$$

$$\mathbf{Z}' = \mathbf{C}_t\mathbf{Z}\mathbf{C} = \begin{array}{cc} \mathrm{a} & \mathrm{F} \\ \hline 1 & 1 \\ \hline \end{array} \quad \begin{array}{c|c|c|} & \mathrm{a} & \mathrm{F} \\ \hline \mathrm{a} & R_a+L_a p & -M\dot{\theta} \\ \hline \mathrm{F} & & R_F+L_F p \\ \hline \end{array} \quad \begin{array}{c|c|} \hline \mathrm{a} & 1 \\ \hline \mathrm{F} & 1 \\ \hline \end{array}$$

$$= \boxed{(R_a+R_F)+(L_a+L_F)p-M\dot{\theta}}$$

$$\mathbf{v}' = \mathbf{C}_t\mathbf{v} = \begin{array}{cc} \mathrm{a} & \mathrm{F} \\ \hline 1 & 1 \\ \hline \end{array} \quad \begin{array}{c|c|} \hline \mathrm{a} & v_a \\ \hline \mathrm{F} & v_F \\ \hline \end{array} = \boxed{v_a+v_F} = \boxed{0}$$

since the circuit is closed.

Let the current at time zero be i_o, then the Laplace transform voltage equation is

$$0 = \{(R_a+R_F)+(L_a+L_F)s-M\dot{\theta}\}\bar{i}-(L_a+L_F)i_o$$

and

$$\bar{i} = \frac{(L_a+L_F)i_o}{(L_a+L_F)s+(R_a+R_F-M\dot{\theta})}$$

$$= \frac{i_o}{s+\dfrac{R_a+R_F-M\dot{\theta}}{L_a+L_F}}$$

whence

$$i = i_o \varepsilon^{-\{(R_a+R_F-M\dot{\theta})/(L_a+L_F)\}t}$$

Three possible cases arise:

$(R_a + R_F) > M\dot{\theta}$—the current decays to zero

$(R_a + R_F) = M\dot{\theta}$—the current remains constant

$(R_a + R_F) < M\dot{\theta}$—the current increases without limit

In practice the initial current i_o is provided by the residual magnetism and if $(R_a + R_F) < M\dot{\theta}$ in the unsaturated region, the machine "builds up", i.e. the current and armature voltage increase. When, however, the value of current reaches the saturated region, M begins to decrease and when $M\dot{\theta}$ has decreased to $(R_a + R_F)$, there is no further increase of current and voltage.

If $(R_a + R_F) > M\dot{\theta}$ in the unsaturated region the resistance is said to be greater than the "critical resistance" ($= M\dot{\theta}$) and the increase in voltage above that corresponding to residual flux is negligible. Correspondingly $(R_a + R_F)/M$ is called the "critical" speed.†

The D.C. Shunt Generator on Resistance Load

As explained on p. 128, for the interconnection of machine and load, the machine impedance matrix has an additional row and column for the external load resistance R_L:

	a	F	L
a	$R_a + L_a p$	$-M\dot{\theta}$	
F		$R_F + L_F p$	
L			R_L

The current relations deduced from Fig. 28(i) are

$$\begin{aligned} i^a &= i^a \\ i^F &= \quad i^F \\ i^L &= i^a - i^F \end{aligned}$$

† See ref. 3, p. 304.

The connection matrix is therefore

$$\mathbf{C} = \begin{array}{c|c|c|} & \text{a} & \text{F} \\ \hline \text{a} & 1 & \\ \hline \text{F} & & 1 \\ \hline \text{L} & 1 & -1 \\ \hline \end{array}$$

The impedance matrix of the system is therefore $\mathbf{Z}' = \mathbf{C}_t\mathbf{Z}\mathbf{C}$

$$= \begin{array}{c|c|c|c|} & \text{a} & \text{F} & \text{L} \\ \hline \text{a} & 1 & & 1 \\ \hline \text{F} & & 1 & -1 \\ \hline \end{array} \begin{array}{c|c|c|c|} & \text{a} & \text{F} & \text{L} \\ \hline \text{a} & R_a + L_a p & -M\dot{\theta} & \\ \hline \text{F} & & R_F + L_F p & \\ \hline \text{L} & & & R_L \\ \hline \end{array} \begin{array}{c|c|c|} & \text{a} & \text{F} \\ \hline \text{a} & 1 & \\ \hline \text{F} & & 1 \\ \hline \text{L} & 1 & -1 \\ \hline \end{array}$$

$$= \begin{array}{c|c|c|} & \text{a} & \text{F} \\ \hline \text{a} & R_a + L_a p + R_L & -M\dot{\theta} - R_L \\ \hline \text{F} & -R_L & R_F + L_F p + R_L \\ \hline \end{array}$$

The voltage is

$$\mathbf{v}' = \mathbf{C}_t\mathbf{v} = \begin{array}{c|c|c|c|} & \text{a} & \text{F} & \text{L} \\ \hline \text{a} & 1 & & 1 \\ \hline \text{F} & & 1 & -1 \\ \hline \end{array} \begin{array}{c|c|} \hline \text{a} & v_a \\ \hline \text{F} & v_F \\ \hline \text{L} & v_L \\ \hline \end{array} = \begin{array}{c|c|} \hline \text{a} & v_a + v_L \\ \hline \text{F} & v_F - v_L \\ \hline \end{array}$$

Bearing in mind the positive directions defined by the current arrows in Fig. 28(i) it can be seen that these voltages are both zero, as they must be, since both circuits are closed. The analysis is in fact now in terms of two meshes, as shown in Fig. 28(ii) and the voltages around the meshes are necessarily zero.

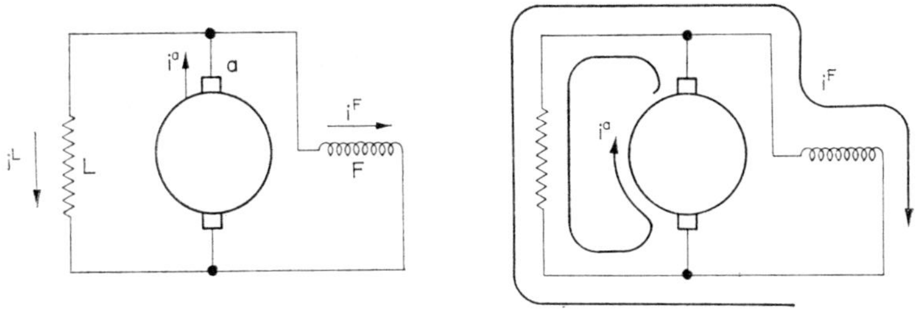

(i) (ii)

FIG. 28. Shunt commutator machine on load.

The voltage equation $\mathbf{v}' = \mathbf{Z}'\mathbf{i}'$ is thus

$$\begin{array}{c|c|} \mathrm{a} & 0 \\ \hline \mathrm{F} & 0 \end{array} = \begin{array}{c|c|c|} & \mathrm{a} & \mathrm{F} \\ \mathrm{a} & R_a + L_a p + R_L & -M\dot{\theta} - R_L \\ \hline \mathrm{F} & -R_L & R_F + L_F p + R_L \end{array} \begin{array}{c|c|} \mathrm{a} & i^a \\ \hline \mathrm{F} & i^F \end{array}$$

and the Laplace transform voltage epuation is

$$\begin{array}{c|c|} \mathrm{a} & 0 \\ \hline \mathrm{F} & 0 \end{array} = \begin{array}{c|c|c|} & \mathrm{a} & \mathrm{F} \\ \mathrm{a} & R_a + L_a s + R_L & -M\dot{\theta} - R_L \\ \hline \mathrm{F} & -R_L & R_F + L_F s + R_L \end{array} \begin{array}{c|c|} \mathrm{a} & \bar{i}^a \\ \hline \mathrm{F} & \bar{i}^F \end{array}$$

$$- \begin{array}{c|c|c|} & \mathrm{a} & \mathrm{F} \\ \mathrm{a} & L_a & \\ \hline \mathrm{F} & & L_F \end{array} \begin{array}{c|c|} \mathrm{a} & i^a_o \\ \hline \mathrm{F} & i^F_o \end{array}$$

or

$$\begin{array}{c|c|} \mathrm{a} & L_a i^a_o \\ \hline \mathrm{F} & L_F i^F_o \end{array} = \begin{array}{c|c|c|} & \mathrm{a} & \mathrm{F} \\ \mathrm{a} & R_a + L_a s + R_L & -M\dot{\theta} - R_L \\ \hline \mathrm{F} & -R_L & R_F + L_F s + R_L \end{array} \begin{array}{c|c|} \mathrm{a} & \bar{i}^a \\ \hline \mathrm{F} & \bar{i}^F \end{array}$$

where i^a_o and i^F_o are the initial armature and field currents respectively.

The determinant of the impedance matrix is

$$\begin{aligned}\Delta &= \{(R_a+R_L)+L_a s\}\{(R_F+R_L)+L_F s\}-R_L(M\dot{\theta}+R_L)\\ &= (L_a L_F)s^2+\{L_F(R_L+R_a)+L_a(R_F+R_L)\}s\\ &\quad +(R_a R_F+R_a R_L+R_L R_F-R_L M\dot{\theta})\end{aligned}$$

The inverse of the impedance matrix is

$$\frac{1}{\Delta}\;\begin{array}{c|c|c|} & \text{a} & \text{F} \\ \hline \text{a} & (R_F+R_L)+L_F s & M\dot{\theta}+R_L \\ \hline \text{F} & R_L & (R_a+R_L)+L_a s \\ \hline \end{array}$$

The currents are therefore given by

$$\begin{array}{c|c|} \text{a} & \bar{i}^a \\ \hline \text{F} & \bar{i}^F \\ \hline \end{array} = \frac{1}{\Delta}\;\begin{array}{c|c|c|} & \text{a} & \text{F} \\ \hline \text{a} & (R_F+R_L)+L_F s & M\dot{\theta}+R_L \\ \hline \text{F} & R_L & (R_a+R_L)+L_a s \\ \hline \end{array}\;\begin{array}{c|c|} \text{a} & L_a i_o^a \\ \hline \text{F} & L_F i_o^F \\ \hline \end{array}$$

The stability of the system is dependent upon the roots of the equation $\Delta = 0$. For stability it is necessary that no root has a positive real part and a necessary, but not a sufficient, condition for this is that all the coefficients of Δ are of the same sign.† It is apparent that the coefficients of s^2 and s are positive. The constant term may be positive, negative or zero. For it to be zero,

$$\begin{aligned}M\dot{\theta} &= \frac{R_a R_F+R_a R_L+R_L R_F}{R_L}\\ &= (R_a+R_F)+R_a R_F/R_L\end{aligned}$$

If $M\dot{\theta}$ exceeds this value the constant term will be negative, and the currents and voltages will increase until M decreases with saturation sufficiently to satisfy this equation. If $M\dot{\theta}$ is less, there will be no build up. If R_L is put equal to infinity, this condition reduces to that previously determined for the generator on no load.

† For a full consideration of the coefficients of a polynomial in relation to its zeros, i.e. in relation to stability, Routh's (Hurwitz's) criterion should be studied. For this see ref. 13, or books on circuit theory.

Parameters of D.C. Machines

Shunt and Separately Excited D.C. Machines

From the impedance matrix given on p. 144, the simple shunt or separately excited d.c. machine has a steady-state voltage equation

$$\begin{array}{c|c|} & \\ \hline a & V_a \\ \hline F & V_F \\ \hline \end{array} = \begin{array}{c|c|c|} & a & F \\ \hline a & R_a & -M\dot{\theta} \\ \hline F & & R_F \\ \hline \end{array} \begin{array}{c|c|} & \\ \hline a & I^a \\ \hline F & I_F \\ \hline \end{array}$$

If the armature is stationary, $\dot{\theta} = 0$, and the values of R_a and R_F can be found as V_a/I^a and V_F/I^F respectively, although for accuracy, V_a should be measured on the commutator and not at the terminals. If this is not done, the brush contact voltage drop will be included in the measurement of V_a.

If the machine is separately driven with $I^a = 0$, i.e. with the armature circuit open, and a constant current I^F in the field winding, the armature terminal voltage is $V_a = -M\dot{\theta}I^F$.† This is, of course, the e.m.f. E of classical theory measured by the open-circuit test.‡ The magnitude of M is obtained as $V_a/(\dot{\theta}I^F)$.

The value of M varies widely with saturation, i.e. with the value of I^F, as shown in Fig. 29. The constant part of the graph of M corresponds to the straight part of the graph of e.m.f. against exciting current. The value of M on load might also vary significantly with I^a. That is, "armature reaction" might reduce the e.m.f. on load compared to the no-load e.m.f. for the same excitation. This effect could be taken into account by a test with the machine on load in a manner similar to that described on p. 153 for the series machine.§

The inductances L_a and L_F of the transient impedance matrix of p. 143 are simply the inductances measured at the armature and field

† The negative sign arises from the conventions chosen and is not relevant to the present interest.

‡ See ref. 3, p. 293. § See ref. 3, pp. 295–6.

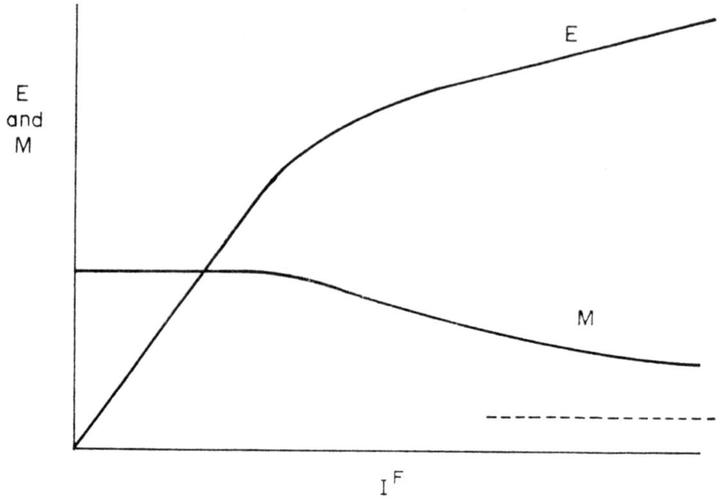

FIG. 29. Open-circuit characteristic.

terminals with the armature stationary. They can be measured approximately by steady-state a.c. or by uni-directional transient tests.†
In practice the value of L_a will be affected by the value of the field current I^F and the value of L_F will be affected by I^a. It is possible in each case to conduct the test with current in the other circuit in order to take into account this interaction due to saturation of pole face and pole tip.

D.C. Series Machine

It is possible to disconnect the field and armature windings of a series d.c. machine and to test it as described above for the shunt machine. In practice, however, it is more convenient to run it as a motor varying the applied voltage and load torque to keep the speed constant over the range of current.

† A.C. tests will give values of inductance including the effects of any coupled current paths through coils or field structure, and hence lower values, in general, than unidirectional tests. The test which corresponds with the condition to be analysed should be chosen. See refs. 7, 14, and 15.

From the matrix on p. 141 the steady-state impedance is seen to be $(R_a+R_F)-M\dot{\theta}$.

Hence $V = \{(R_a+R_F)-M\dot{\theta}\}I$, $E = -M\dot{\theta}I = V-(R_a+R_F)I$, and $M = \{V/I-(R_a+R_F)\}/\dot{\theta}$.

This method has the advantage over the open-circuit test in that, although error is increased by finding M in terms of a difference (which in practice involves the brush contact voltage drop ignored here), the effect of armature reaction is included. However, in many d.c. series motor applications the armature and field currents are, at times, different.† For such cases it is necessary to supply the armature and field windings with these different currents during the test to allow correctly for armature reaction. The effect of armature reaction is also dependent upon whether the machine is motoring or generating, so that a complete test requires the machine to be operated under both conditions.

The Repulsion Motor

Kron appreciated from the beginning that the repulsion motor was a particularly useful machine to demonstrate the method of analysis, although it is now of very little practical importance. It consists of a commutator machine with a pair of short-circuited brushes displaced from the neutral axis and a single-phase stator winding, as shown in Fig. 30. The primitive machine is shown in Fig. 31.

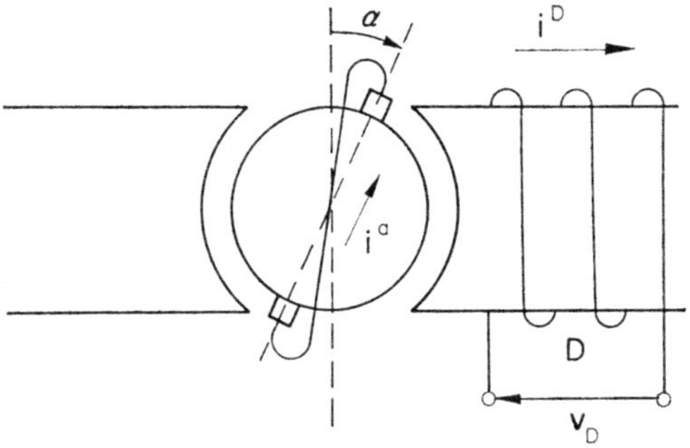

FIG. 30. Repulsion motor.

† See ref. 16, pp. 142–52.

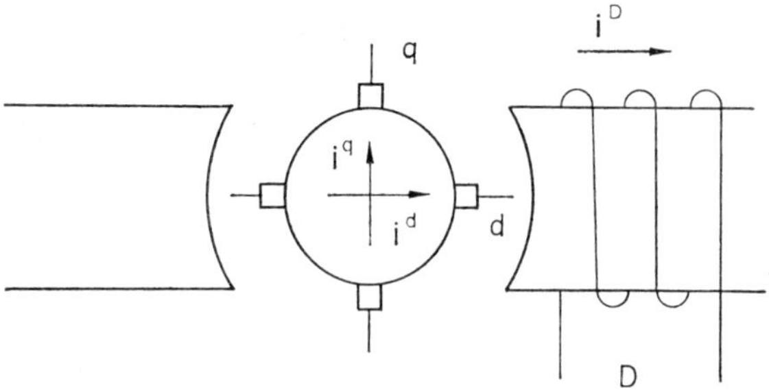

FIG. 31. Primitive machine.

The brush-shift transformation required is

$$i^{d} = \sin \alpha i^{a}$$
$$i^{q} = \cos \alpha i^{a}$$

and since the stator winding axis is not to be transformed, the transformation matrix is

$\mathbf{C} =$

	D	a
D	1	
d		$\sin \alpha$
q		$\cos \alpha$

and its transpose is

$\mathbf{C}_t =$

	D	d	q
D	1		
a		$\sin \alpha$	$\cos \alpha$

The impedance matrix **Z** of the primitive machine is that of the metadyne, namely

	D	d	q
D	$R_D + L_D p$	Mp	
d	Mp	$R_a + L_d p$	$L_q \dot{\theta}$
q	$-M\dot{\theta}$	$-L_d \dot{\theta}$	$R_a + L_q p$

That of the repulsion motor is therefore $\mathbf{Z}' = \mathbf{C}_t\mathbf{Z}\mathbf{C}$

=

	D	d	q
D	1		
a		$\sin\alpha$	$\cos\alpha$

	D	d	q
D	R_D+L_Dp	Mp	
d	Mp	R_a+L_dp	$L_q\dot{\theta}$
q	$-M\dot{\theta}$	$-L_d\dot{\theta}$	R_a+L_qp

	D	a
D	1	
d		$\sin\alpha$
q		$\cos\alpha$

=

	D	a
D	R_D+L_Dp	$Mp\sin\alpha$
a	$Mp\sin\alpha-M\dot{\theta}\cos\alpha$	$(R_a+L_dp)\sin^2\alpha+(R_a+L_qp)\cos^2\alpha$ $+(L_q-L_d)\dot{\theta}\sin\alpha\cos\alpha$

=

	D	a
D	R_D+L_Dp	$Mp\sin\alpha$
a	$M(p\sin\alpha-\dot{\theta}\cos\alpha)$	$(R_a+L_qp)+(L_d-L_q)\{p\sin^2\alpha-\dot{\theta}\sin\alpha\cos\alpha\}$

=

	D	a
D	R_D+L_Dp	$Mp\sin\alpha$
a	$M(p\sin\alpha-\dot{\theta}\cos\alpha)$	$(R_a+L_qp)+\frac{1}{2}(L_d-L_q)\{p(1-\cos 2\alpha)-\dot{\theta}\sin 2\alpha\}$

Although it is not essential to do so, it is convenient here to restrict further consideration to a machine with $L_d = L_q$ in order to reduce the length of the expressions.

If $L_d = L_q = L_a$ the impedance matrix becomes

	D	a
D	R_D+L_Dp	$Mp\sin\alpha$
a	$M(p\sin\alpha-\dot{\theta}\cos\alpha)$	R_a+L_ap

If we were interested only in the stator current, it would be preferable to eliminate the armature current; however, we will proceed to calculate both currents and the torque. Inverting the impedance matrix gives the currents as

$$\begin{array}{c|c|} \mathrm{D} & i^{\mathrm{D}} \\ \hline \mathrm{a} & i^{\mathrm{a}} \end{array} = \frac{1}{\Delta}\; \begin{array}{c|c|c|} & \mathrm{D} & \mathrm{a} \\ \hline \mathrm{D} & R_{\mathrm{a}}+L_{\mathrm{a}}p & -Mp\sin\alpha \\ \hline \mathrm{a} & -M(p\sin\alpha-\dot{\theta}\cos\alpha) & R_{\mathrm{D}}+L_{\mathrm{D}}p \end{array} \; \begin{array}{c|c|} \mathrm{D} & v_{\mathrm{D}} \\ \hline \mathrm{a} & v_{\mathrm{a}} \end{array}$$

Since the armature brushes are short-circuited $v_{\mathrm{a}} = 0$ and

$$\begin{array}{c|c|} \mathrm{D} & i^{\mathrm{D}} \\ \hline \mathrm{a} & i^{\mathrm{a}} \end{array} = \frac{v_{\mathrm{D}}}{\Delta}\; \begin{array}{c|c|} \mathrm{D} & R_{\mathrm{a}}+L_{\mathrm{a}}p \\ \hline \mathrm{a} & -M(p\sin\alpha-\dot{\theta}\cos\alpha) \end{array}$$

where $\Delta = (L_{\mathrm{D}}L_{\mathrm{a}}-M^2\sin^2\alpha)p^2+(R_{\mathrm{D}}L_{\mathrm{a}}+R_{\mathrm{a}}L_{\mathrm{D}}$
$$+M^2\dot{\theta}\sin\alpha\cos\alpha)p+R_{\mathrm{D}}R_{\mathrm{a}}$$

Steady-state Performance in Complex Terms

Considering first steady-state conditions with an r.m.s. applied voltage V_{D}, we may replace p by $j\omega$ to get

$$\begin{array}{c|c|} \mathrm{D} & I^{\mathrm{D}} \\ \hline \mathrm{a} & I^{\mathrm{a}} \end{array} = \frac{V_{\mathrm{D}}}{\Delta}\; \begin{array}{c|c|} \mathrm{D} & R_{\mathrm{a}}+jX_{\mathrm{a}} \\ \hline \mathrm{a} & X_{\mathrm{M}}(\mathrm{v}\cos\alpha-j\sin\alpha) \end{array}$$

where $X_{\mathrm{M}} = \omega M$, $X_{\mathrm{a}} = \omega L_{\mathrm{a}}$, $X_{\mathrm{D}} = \omega L_{\mathrm{D}}$, $\mathrm{v} = \dot{\theta}/\omega$ and now

$$\Delta = \{R_{\mathrm{D}}R_{\mathrm{a}}-(X_{\mathrm{D}}X_{\mathrm{a}}-X_{\mathrm{M}}^2\sin^2\alpha)\}+j\{R_{\mathrm{D}}X_{\mathrm{a}}+R_{\mathrm{a}}X_{\mathrm{D}}+X_{\mathrm{M}}^2\mathrm{v}\sin\alpha\cos\alpha\}$$

From the transient impedance matrix

$$\mathbf{G} = \begin{array}{c|c|c|} & \mathrm{D} & \mathrm{a} \\ \hline \mathrm{D} & & \\ \hline \mathrm{a} & -M\cos\alpha & \end{array} = \frac{1}{\omega}\; \begin{array}{c|c|c|} & \mathrm{D} & \mathrm{a} \\ \hline \mathrm{D} & & \\ \hline \mathrm{a} & -X_{\mathrm{M}}\cos\alpha & \end{array}$$

The steady-state two-pole torque T is therefore

$$\operatorname{Re}\frac{1}{\omega}\begin{array}{c}\begin{array}{cc}\text{D} & \text{a}\end{array}\\ \begin{array}{|c|c|}\hline I^{*\text{D}} & I^{\text{a}*}\\ \hline\end{array}\end{array}\begin{array}{c|c|c|}\multicolumn{1}{c}{} & \multicolumn{1}{c}{\text{D}} & \multicolumn{1}{c}{\text{a}}\\ \cline{2-3}\text{D} & & \\ \cline{2-3}\text{a} & -X_{\text{M}}\cos\alpha & \\ \cline{2-3}\end{array}\begin{array}{c|c|}\cline{2-2}\text{D} & I^{\text{D}}\\ \cline{2-2}\text{a} & I^{\text{a}}\\ \cline{2-2}\end{array}$$

$$= \operatorname{Re}\left\{-\frac{1}{\omega}I^{\text{a}*}I^{\text{D}}X_{\text{M}}\cos\alpha\right\}$$

$$= \operatorname{Re}\left\{-\frac{X_{\text{M}}^2\cos\alpha}{\omega}\,\frac{V_{\text{D}}^*V_{\text{D}}}{\Delta^*\Delta}(\text{v}\cos\alpha+j\sin\alpha)(R_{\text{a}}+jX_{\text{a}})\right\}$$

$$= -\frac{X_{\text{M}}^2\cos\alpha}{\omega}\,\frac{V_{\text{D}}^*V_{\text{D}}}{\Delta^*\Delta}(\text{v}R_{\text{a}}\cos\alpha-X_{\text{a}}\sin\alpha)$$

$$= \frac{1}{\omega}\,\frac{V_{\text{D}}^2X_{\text{M}}^2\cos\alpha(X_{\text{a}}\sin\alpha-\text{v}R_{\text{a}}\cos\alpha)}{\{R_{\text{D}}R_{\text{a}}-(X_{\text{D}}X_{\text{a}}-X_{\text{M}}^2\sin^2\alpha)\}^2+\{R_{\text{D}}X_{\text{a}}+R_{\text{a}}X_{\text{D}}+X_{\text{M}}^2\text{v}\sin\alpha\cos\alpha\}^2}$$

When $\text{v} = 0$, this torque is positive, i.e. in the counterclockwise direction for positive α. The armature thus rotates in a counterclockwise direction with a displacement of the brushes in a clockwise direction and vice versa.

Steady-state Instantaneous Currents and Torque

The Laplace transform equation for the currents is

$$\begin{array}{|c|}\hline \bar{\imath}^{\text{D}}\\ \hline \bar{\imath}^{\text{a}}\\ \hline\end{array} = \frac{\hat{V}s}{(s^2+\omega^2)\{(L_{\text{D}}L_{\text{a}}-M^2\sin^2\alpha)s^2+(R_{\text{D}}L_{\text{a}}+R_{\text{a}}L_{\text{D}}+M^2\dot{\theta}\sin\alpha\cos\alpha)s+R_{\text{D}}R_{\text{a}}\}}\begin{array}{c|c|}\cline{2-2}\text{D} & R_{\text{a}}+L_{\text{a}}s\\ \cline{2-2}\text{a} & -M(s\sin\alpha-\dot{\theta}\cos\alpha)\\ \cline{2-2}\end{array}$$

assuming that the applied voltage V_{D} is $\hat{V}\cos\omega t$.

It is apparent that

$$(L_{\text{D}}L_{\text{a}}-M^2\sin^2\alpha)s^2+(R_{\text{D}}L_{\text{a}}+R_{\text{a}}L_{\text{D}}+M^2\dot{\theta}\sin\alpha\cos\alpha)s+R_{\text{D}}R_{\text{a}}$$

cannot be factorized simply and that consequently the transient solution can be obtained easily only for numerical cases. For steady-state conditions we may assume that it factorizes into

$$(L_D L_a - M^2 \sin^2 \alpha)(s+\lambda)(s+\mu)$$

It is clear that λ and μ cannot be conjugate imaginaries from the presence of the term $(R_D L_a + R_a L_D + M^2\dot{\theta} \sin \alpha \cos \alpha)s$, which is only zero for particular values of $\dot{\theta}$ and α. In general, therefore, the currents associated with these terms are transient currents only. Knowledge of the relative orders of magnitude of the coefficients indicates that the zeros cannot be complex, but it is not necessary to know this for our present purpose.

The transform of the stator current is

$$\begin{aligned}\bar{\imath}^D &= \frac{\hat{V}s(R_a + L_a s)}{(s^2+\omega^2)(L_D L_a - M^2 \sin^2 \alpha)(s+\lambda)(s+\mu)} \\ &= \frac{\hat{V}}{(L_D L_a - M^2 \sin^2 \alpha)} \frac{L_a s^2 + R_a s}{(s^2+\omega^2)(s+\lambda)(s+\mu)} \\ &= \frac{\hat{V}}{(L_D L_a - M^2 \sin^2 \alpha)} \left[\frac{As+B\omega}{s^2+\omega^2} + \frac{C}{s+\lambda} + \frac{D}{s+\mu}\right]\end{aligned}$$

in which we do not know λ or μ, but are not interested in them or in C and D.

If we put $s = j\omega$ in the two expressions we get

$$j\omega(R_a + j\omega L_a) = (j\omega+\lambda)(j\omega+\mu)(j\omega A + B\omega)$$

Although we do not know the value of λ and μ, we do know that of $(s+\lambda)(s+\mu)$ and hence that of $(j\omega+\lambda)(j\omega+\mu)$, so that

$$j\omega(R_a + j\omega L_a) = \frac{\left[\begin{aligned}&\{R_D R_a - \omega^2 L_D L_a + \omega^2 M^2 \sin^2 \alpha\} \\ &+ j\omega\{R_D L_a + R_a L_D + M^2\,\dot{\theta} \sin \alpha \cos \alpha\}\end{aligned}\right]}{L_D L_a - M^2 \sin^2 \alpha} (j\omega A + B\omega)$$

whence

$$A = \frac{\left[\begin{aligned}&R_a\{R_D R_a - \omega^2 L_D L_a + \omega^2 M^2 \sin^2 \alpha\} \\ &+ \omega^2 L_a\{R_D L_a + R_a L_D + M^2\dot{\theta} \sin \alpha \cos \alpha\}\end{aligned}\right]}{\left[\begin{aligned}&\{R_D R_a - \omega^2 L_D L_a + \omega^2 M^2 \sin^2 \alpha\}^2 \\ &+ \omega^2\{R_D L_a + R_a L_D + M^2\dot{\theta} \sin \alpha \cos \alpha\}^2\end{aligned}\right]} (L_D L_a - M^2 \sin^2 \alpha)$$

and

$$B = \frac{\begin{bmatrix} -\omega L_a\{R_D R_a - \omega^2 L_D L_a + \omega^2 M^2 \sin^2 \alpha\} \\ +\omega R_a\{R_D L_a + R_a L_D + M^2 \dot{\theta} \sin \alpha \cos \alpha\} \end{bmatrix}}{\begin{bmatrix} \{R_D R_a - \omega^2 L_D L_a + \omega^2 M^2 \sin^2 \alpha\}^2 \\ +\omega^2\{R_D L_a + R_a L_D + M^2 \dot{\theta} \sin \alpha \cos \alpha\}^2 \end{bmatrix}} (L_D L_a - M^2 \sin^2 \alpha\}$$

Similarly,

$$\bar{i}^a = \frac{\hat{V}}{(L_D L_a - M^2 \sin^2 \alpha)} \frac{M(-s^2 \sin \alpha + s\dot{\theta} \cos \alpha)}{(s^2+\omega^2)(s+\lambda)(s+\mu)}$$

$$= \frac{\hat{V}}{(L_D L_a - M^2 \sin^2 \alpha)} \left[\frac{Ps+Q\omega}{s^2+\omega^2} + \frac{S}{s+\lambda} + \frac{T}{s+\mu} \right]$$

in which we are not interested in S or T. In a similar manner to before we find that

$$P = \frac{\begin{bmatrix} M\dot{\theta} \cos \alpha\{R_D R_a - \omega^2 L_D L_a + \omega^2 M^2 \sin^2 \alpha\} \\ -\omega^2 M \sin \alpha\{R_D L_a + R_a L_D \\ + M^2 \dot{\theta} \sin \alpha \cos \alpha\} \end{bmatrix}}{\begin{bmatrix} \{R_D R_a - \omega^2 L_D L_a + \omega^2 M^2 \sin^2 \alpha\}^2 \\ +\omega^2\{R_D L_a + R_a L_D + M^2 \dot{\theta} \sin \alpha \cos \alpha\}^2 \end{bmatrix}} (L_D L_a - M^2 \sin^2 \alpha)$$

$$Q = \frac{\begin{bmatrix} \omega M \sin \alpha\{R_D R_a - \omega^2 L_D L_a + \omega^2 M^2 \sin^2 \alpha\} \\ +\omega M\dot{\theta} \cos \alpha\{R_D L_a + R_a L_D \\ + M^2 \dot{\theta} \sin \alpha \cos \alpha\} \end{bmatrix}}{\begin{bmatrix} \{R_D R_a - \omega^2 L_D L_a + \omega^2 M^2 \sin^2 \alpha\}^2 \\ +\omega^2\{R_D L_a + R_a L_D + M^2 \dot{\theta} \sin \alpha \cos \alpha\}^2 \end{bmatrix}} (L_D L_a - M^2 \sin^2 \alpha)$$

The instantaneous steady-state currents are therefore

$$i^D = \frac{\hat{V}}{L_D L_a - M^2 \sin^2 \alpha} (A \cos \omega t + B \sin \omega t)$$

$$i^a = \frac{\hat{V}}{L_D L_a - M^2 \sin^2 \alpha} (P \cos \omega t + Q \sin \omega t)$$

The instantaneous two-pole torque T is

$$\begin{array}{cc} \text{D} & \text{a} \\ \hline i^{\text{D}} & i^{\text{a}} \end{array} \quad \begin{array}{c|c|c|} & \text{D} & \text{a} \\ \text{D} & & \\ \hline \text{a} & -M\cos\alpha & \end{array} \quad \begin{array}{c|c|} \text{D} & i^{\text{D}} \\ \hline \text{a} & i^{\text{a}} \end{array}$$

$$= -i^{\text{a}} i^{\text{D}} M\cos\alpha$$

$$= -\frac{\hat{V}^2 M\cos\alpha}{(L_{\text{D}}L_{\text{a}} - M^2\sin^2\alpha)^2}\{(A\cos\omega t + B\sin\omega t)(P\cos\omega t + Q\sin\omega t)\}$$

$$= -\frac{\hat{V}^2 M\cos\alpha}{(L_{\text{D}}L_{\text{a}} - M^2\sin^2\alpha)^2}\{AP\cos^2\omega t + BQ\sin^2\omega t + (AQ+BP)\sin\omega t\cos\omega t\}$$

$$= -\frac{\hat{V}^2 M\cos\alpha}{2(L_{\text{D}}L_{\text{a}} - M^2\sin^2\alpha)^2}\{(AP+BQ) + (AP-BQ)\cos 2\omega t + (AQ+BP)\sin 2\omega t\}$$

$$= -\frac{\hat{V}^2 M\cos\alpha}{2(L_{\text{D}}L_{\text{a}} - M^2\sin^2\alpha)^2}\{(AP+BQ) + \sqrt{[(A^2+B^2)(P^2+Q^2)]} \times \sin(2\omega t + \psi)\}$$

where ψ is arc sin $(AP-BQ)/\sqrt{[(A^2+B^2)(P^2+Q^2)]}$.

The mean torque thus has a magnitude

$$-\frac{\hat{V}^2 M\cos\alpha}{2(L_{\text{D}}L_{\text{a}} - M^2\sin^2\alpha)^2}(AP+BQ)$$

which on substitution of the values of A, B, P, and Q can be reduced to

$$\frac{1}{2}\;\frac{\hat{V}^2 M^2\cos\alpha\{\omega^2 L_{\text{a}}\sin\alpha - \text{v}\omega R_{\text{a}}\cos\alpha\}}{\{R_{\text{D}}R_{\text{a}} - \omega^2 L_{\text{D}}L_{\text{a}} + \omega^2 M^2\sin^2\alpha\}^2 + \omega^2\{R_{\text{D}}L_{\text{a}} + R_{\text{a}}L_{\text{D}} + \text{v}M^2\omega\sin\alpha\cos\alpha\}^2}$$

$$= \frac{1}{2\omega}\;\frac{\hat{V}^2 X_{\text{M}}^2\cos\alpha\{X_{\text{a}}\sin\alpha - \text{v}R_{\text{a}}\cos\alpha\}}{\{R_{\text{D}}R_{\text{a}} - X_{\text{D}}X_{\text{a}} + X_{\text{M}}^2\sin^2\alpha\}^2 + \{R_{\text{D}}X_{\text{a}} + R_{\text{a}}X_{\text{D}} + \text{v}X_{\text{M}}^2\sin\alpha\cos\alpha\}^2}$$

$$= \frac{1}{\omega}\;\frac{V_{\text{D}}^2 X_{\text{M}}^2\cos\alpha\{X_{\text{a}}\sin\alpha - \text{v}R_{\text{a}}\cos\alpha\}}{\{R_{\text{D}}R_{\text{a}} - (X_{\text{D}}X_{\text{a}} - X_{\text{M}}^2\sin^2\alpha)\}^2 + \{R_{\text{D}}X_{\text{a}} + R_{\text{a}}X_{\text{D}} + \text{v}X_{\text{M}}^2\sin\alpha\cos\alpha\}^2}$$

which is identical to the expression for the mean torque obtained from the complex currents on p. 157. But from the expression for the instantaneous torque we can obtain, in addition, the magnitude of the alternating component of the torque as

$$\frac{\hat{V}^2 M \cos\alpha}{2(L_D L_a - M^2 \sin^2\alpha)^2}\sqrt{[(A^2+B^2)(P^2+Q^2)]}$$

Substituting for A, B, P, and Q reduces this to

$$\frac{1}{2}\,\frac{\omega \hat{V}^2 M^2 \cos\alpha\,\sqrt{[(R_a^2+\omega^2 L_a^2)(\sin^2\alpha+\mathrm{v}^2\cos^2\alpha)]}}{\{R_D R_a-\omega^2 L_D L_a+\omega^2 M^2\sin^2\alpha\}^2+\omega^2\{R_D L_a+R_a L_D+\mathrm{v}M^2\omega\sin\alpha\cos\alpha\}^2}$$

$$=\frac{1}{\omega}\,\frac{\hat{V}_D^2 X_M^2\cos\alpha\,\sqrt{[(R_a^2+X_a^2)(\sin^2\alpha+\mathrm{v}^2\cos^2\alpha)]}}{\{R_D R_a-X_D X_a+X_M^2\sin^2\alpha\}^2+\{R_D X_a+R_a X_D+\mathrm{v}X_M^2\sin\alpha\cos\alpha\}^2}$$

The ratio of alternating torque to mean torque is thus

$$\frac{\sqrt{[(R_a^2+X_a^2)(\sin^2\alpha+\mathrm{v}^2\cos^2\alpha)]}}{X_a\sin\alpha-\mathrm{v}R_a\cos\alpha}$$

$$=\sqrt{\left(\frac{R_a^2\sin^2\alpha+X_a^2\sin^2\alpha+R_a^2\mathrm{v}^2\cos^2\alpha+X_a^2\mathrm{v}^2\cos^2\alpha}{X_a^2\sin^2\alpha+R_a^2\mathrm{v}^2\cos^2\alpha-2X_aR_a\mathrm{v}\sin\alpha\cos\alpha}\right)}$$

Since in general v changes sign with α, it can be seen that the numerator of this expression is always bigger than the denominator, i.e. the magnitude of the alternating component is bigger than the magnitude of the mean torque and consequently during part of the cycle the net torque is negative. It is also interesting to note that although the mean torque changes sign as α changes sign when the machine is stationary, when it is running the torque changes sign at $\alpha = \arctan \mathrm{v}R_a/X_a$.

CHAPTER 10

The Steady-state Performance of Polyphase Machines

The Balanced Polyphase Induction Machine

The equations of this type of machine have already been deduced in terms of stationary reference axes on p. 63, and expressed in matrix form with a change in their order are

$$\begin{array}{c|c|}
 & \\
D & v_D \\
Q & v_Q \\
d & v_d \\
q & v_q
\end{array}
=
\begin{array}{c|c|c|c|c|}
 & D & Q & d & q \\
D & R_S+L_Sp & & Mp & \\
Q & & R_S+L_Sp & & Mp \\
d & Mp & M\dot{\theta} & R_r+L_rp & L_r\dot{\theta} \\
q & -M\dot{\theta} & Mp & -L_r\dot{\theta} & R_r+L_rp
\end{array}
\begin{array}{c|c|}
 & \\
D & i^D \\
Q & i^Q \\
d & i^d \\
q & i^q
\end{array}$$

in which the indices relate to the windings arranged as in Fig. 32.

Because the rotor windings of an induction machine are short-circuited, their terminal voltages are zero, in both three- and two-phase axes. In terms of d and q reference axes the rotor voltages are therefore

$$\begin{array}{c|c|}
d & v_d \\
q & v_q
\end{array}
=
\begin{array}{c|c|c|}
 & \alpha & \beta \\
d & \cos\theta & \sin\theta \\
q & \sin\theta & -\cos\theta
\end{array}
\begin{array}{c|c|}
\alpha & 0 \\
\beta & 0
\end{array}
=
\begin{array}{c|c|}
d & 0 \\
q & 0
\end{array}$$

This result is obvious and is an example of the general rule that, with transformations of the type used in this work, a matrix which

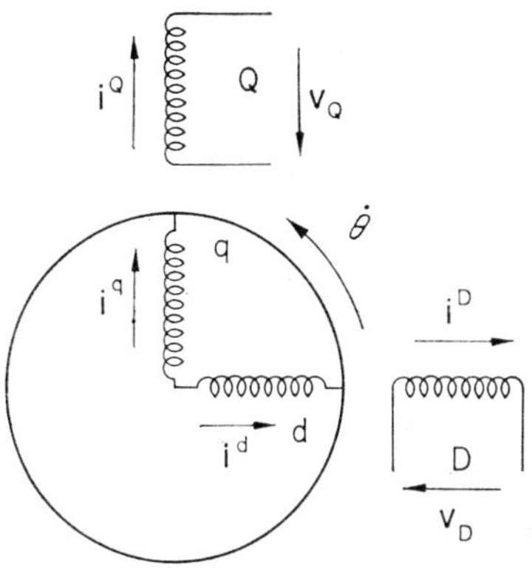

FIG. 32. Balanced induction machine.

consists wholly of zeros will always transform to another matrix consisting wholly of zeros. There are of course transformations where this would not be so, but such transformations have no place in electrical machine analysis.

If D, Q are the axes of an actual two-phase stator winding they might equally well be denoted by A′, B′ as indicated on p. 120. If the stator winding terminal voltage is at an angular frequency ω, the steady-state impedance matrix may be obtained by replacing p by $j\omega$. It is convenient also to replace $\dot{\theta}$ by $\mathrm{v}\omega$. Since ω is the synchronous angular velocity for a two-pole machine, v is the fractional, or "per unit", speed. If the products of ω and inductances are written as reactances, the steady-state impedance matrix is

	D	Q	d	q
D	R_S+jX_S		jX_M	
Q		R_S+jX_S		jX_M
d	jX_M	$\mathrm{v}X_M$	R_r+jX_r	$\mathrm{v}X_r$
q	$-\mathrm{v}X_M$	jX_M	$-\mathrm{v}X_r$	R_r+jX_r

If this matrix were partitioned between the stator and rotor axes, the sub-matrices would be either scalar or the sum of scalar and skew-symmetric matrices. This suggests transformation to symmetrical component axes.† In the case of the stator, being applied to the actual two-phase winding axes, the transformation will lead to positive- and negative-sequence axes P, N. In the case of the rotor, being applied to axes already transformed from the actual winding axes to d, q axes, the transformation will lead to forward and backward axes f, b as explained on pp. 118–19.

The transformation will have to be applied separately to the stator and rotor axes, and, since there is no transformation between stator and rotor axes, the remaining spaces of the transformation matrix will be filled with zeros. Thus

$$\mathbf{C} = \frac{1}{\sqrt{2}}$$

	P	N	f	b
D	1	1	0	0
Q	$-j$	j	0	0
d	0	0	1	1
q	0	0	$-j$	j

The transformed impedance matrix is therefore as shown on p. 165.

The order of the rows and columns may be changed provided that the same changes are made to both, so that each element has the same indices after rearrangement as before.‡ The impedance matrix may

† See Exercise 14.

‡ This is equivalent to a simple transformation which in this particular case would be defined by

$$\mathbf{C} =$$

	P	f	N	b
P	1			
N			1	
f		1		
b				1

$$\mathbf{Z}' = \mathbf{C}_t^*\mathbf{Z}\mathbf{C}$$

$$= \frac{1}{\sqrt{2}} \begin{array}{c|cccc|} & D & Q & d & q \\ \hline P & 1 & j & & \\ N & 1 & -j & & \\ f & & & 1 & j \\ b & & & 1 & -j \\ \hline \end{array} \quad \begin{array}{c|cccc|} & D & Q & d & q \\ \hline D & R_S+jX_S & & jX_M & \\ Q & & R_S+jX_S & & jX_M \\ d & jX_M & vX_M & R_r+jX_r & vX_r \\ q & -vX_M & jX_M & -vX_r & R_r+jX_r \\ \hline \end{array} \quad \frac{1}{\sqrt{2}} \begin{array}{c|cccc|} & P & N & f & b \\ \hline D & 1 & 1 & & \\ Q & -j & j & & \\ d & & & 1 & 1 \\ q & & & -j & j \\ \hline \end{array}$$

$$= \begin{array}{c|cccc|} & P & N & f & b \\ \hline P & R_S+jX_S & & jX_M & \\ N & & R_S+jX_S & & jX_M \\ f & j(1-v)X_M & & R_r+j(1-v)X_r & \\ b & & j(1+v)X_M & & R_r+j(1+v)X_r \\ \hline \end{array}$$

therefore be written

	P	f	N	b
P	R_S+jX_S	jX_M		
f	$j(1-v)X_M$	$R_r+j(1-v)X_r$		
N			R_S+jX_S	jX_M
b			$j(1+v)X_M$	$R_r+j(1+v)X_r$

In this form† the matrix shows that the axes form two independent groups, namely P, f and N, b, which have absolutely no interaction on one another and which can, therefore, be considered quite separately.

Since V_f will be zero for reasons given on pp. 162–63, the voltage equation of the positive-sequence system, P, f, can be written

$$\begin{array}{c|c} & \\ P & V_P \\ f & 0 \end{array} = \begin{array}{c|cc} & P & f \\ P & R_S+jX_S & jX_M \\ f & j(1-v)X_M & R_r+j(1-v)X_r \end{array} \; \begin{array}{c|c} & \\ P & I^P \\ f & I^f \end{array}$$

† An alternative approach which leads to this same matrix is as follows: if the order of the indices in the original D, Q, d, q matrix is changed to D, d, Q, q and the matrix is then partitioned between the d and Q axes, the resulting compound matrix is of the form $\begin{bmatrix} \mathbf{A} & \mathbf{B} \\ -\mathbf{B} & \mathbf{A} \end{bmatrix}$. The results of Exercise 14 here suggest a transformation defined by $\frac{1}{\sqrt{2}}\begin{bmatrix} \mathrm{U} & \mathrm{U} \\ -j\mathrm{U} & j\mathrm{U} \end{bmatrix}$ and this leads directly to the P, f, N, b matrix above.

Currents

By inverting the impedance matrix, we get the currents as

$$\begin{array}{c|c|}\mathrm{P} & I^{\mathrm{P}}\\ \hline \mathrm{f} & I^{\mathrm{f}}\end{array} = \frac{1}{\Delta}\;\begin{array}{c|c|c|}& \mathrm{P} & \mathrm{f}\\ \hline \mathrm{P} & R_r+j(1-\mathrm{v})X_r & -jX_{\mathrm{M}}\\ \hline \mathrm{f} & -j(1-\mathrm{v})X_{\mathrm{M}} & R_{\mathrm{S}}+jX_{\mathrm{S}}\end{array}\;\begin{array}{c|c|}\mathrm{P} & V_{\mathrm{P}}\\ \hline \mathrm{f} & 0\end{array}$$

$$= \frac{V_{\mathrm{P}}}{\Delta}\;\begin{array}{c|c|}\mathrm{P} & R_r+j\mathrm{s}X_r\\ \hline \mathrm{f} & -j\mathrm{s}X_{\mathrm{M}}\end{array}$$

where $\mathrm{s} = 1-\mathrm{v}$ is the fractional or per unit slip and

$$\begin{aligned}\Delta &= \{R_{\mathrm{S}}+jX_{\mathrm{S}}\}\{R_r+j(1-\mathrm{v})X_r\}+(1-\mathrm{v})X_{\mathrm{M}}^2\\ &= \{R_{\mathrm{S}}+jX_{\mathrm{S}}\}\{R_r+j\mathrm{s}X_r\}+\mathrm{s}X_{\mathrm{M}}^2\end{aligned}$$

Since the elements of the N, b submatrix are the same as those of the P, f submatrix except that $(1-\mathrm{v}) = \mathrm{s}$ is replaced by $(1+\mathrm{v}) = (2-\mathrm{s})$, the expressions for the N, b currents can be obtained from those for the P, f currents by replacing P, f by N, b and s by $(2-\mathrm{s})$. Hence

$$\begin{array}{c|c|}\mathrm{N} & I^{\mathrm{N}}\\ \hline \mathrm{b} & I^{\mathrm{b}}\end{array} = \frac{V_{\mathrm{N}}}{\Delta}\;\begin{array}{c|c|}\mathrm{N} & R_r+j(2-\mathrm{s})X_r\\ \hline \mathrm{b} & -j(2-\mathrm{s})X_{\mathrm{M}}\end{array}$$

where now $\Delta = \{R_{\mathrm{S}}+jX_{\mathrm{S}}\}\{R_r+j(2-\mathrm{s})X_r\}+(2-\mathrm{s})X_{\mathrm{M}}^2$. The currents in the D, Q, d, q, axes are then

$$\begin{array}{c|c|}\mathrm{D} & I^{\mathrm{D}}\\ \hline \mathrm{Q} & I^{\mathrm{Q}}\\ \hline \mathrm{d} & I^{\mathrm{d}}\\ \hline \mathrm{q} & I^{\mathrm{q}}\end{array} = \frac{1}{\sqrt{2}}\;\begin{array}{c|c|c|c|c|}& \mathrm{P} & \mathrm{N} & \mathrm{f} & \mathrm{b}\\ \hline \mathrm{D} & 1 & 1 & & \\ \hline \mathrm{Q} & -j & j & & \\ \hline \mathrm{d} & & & 1 & 1\\ \hline \mathrm{q} & & & -j & j\end{array}\;\begin{array}{c|c|}\mathrm{P} & I^{\mathrm{P}}\\ \hline \mathrm{N} & I^{\mathrm{N}}\\ \hline \mathrm{f} & I^{\mathrm{f}}\\ \hline \mathrm{b} & I^{\mathrm{b}}\end{array} = \frac{1}{\sqrt{2}}\;\begin{array}{c|c|}\mathrm{D} & I^{\mathrm{P}}+I^{\mathrm{N}}\\ \hline \mathrm{Q} & -jI^{\mathrm{P}}+jI^{\mathrm{N}}\\ \hline \mathrm{d} & I^{\mathrm{f}}+I^{\mathrm{b}}\\ \hline \mathrm{q} & -jI^{\mathrm{f}}+jI^{\mathrm{b}}\end{array}$$

The D, Q currents are the two-phase stator currents $I^{A'}$, $I^{B'}$ in complex form. The three-phase stator currents I^{A}, I^{B}, I^{C} can be obtained from them by means of the two-phase to three-phase transformation.

The d, q currents are also in complex form, but can be transformed to the two-phase rotor axes α, β or the three-phase rotor axes a, b, c, if they are first rewritten in instantaneous form. This intermediate step is necessary because the d, q currents are of stator frequency, whereas the actual rotor winding currents are of slip frequency. Because of the way in which the actual stator and rotor winding axes have been defined, which is shown in Fig. 33, at sub-synchronous speeds i^{β} leads

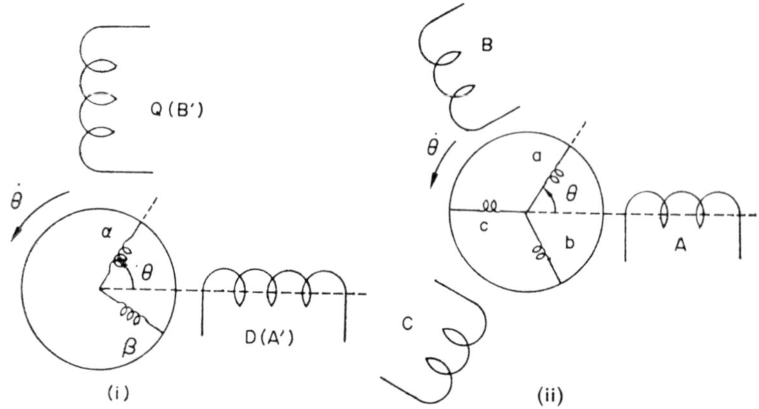

FIG. 33. (i) Balanced two-phase induction machine. (ii) Balanced three-phase induction machine.

i^{α} and i^{b} leads i^{a}, while at super-synchronous speeds i^{α} leads i^{β} and i^{a} leads i^{b}. The stator voltages have, of course, been defined with the voltage of the D or A′ phase leading that of the Q or B′ phase and, correspondingly, the three-phase A-phase voltage leading the B-phase voltage.

Equivalent Circuit

From the expression for I^P given on p. 166, it can be seen that, viewed from the primary terminals, in terms of P, f axes, the machine appears as an impedance

$$\frac{(R_S+jX_S)(R_r+jsX_r)+sX_M^2}{R_r+jsX_r} = \frac{(R_S+jX_S)(R_r/s+jX_r)+X_M^2}{R_r/s+jX_r}$$

If this is expressed in terms of x_S and x_r,† defined by the equations $X_S = X_M+x_S$ and $X_r = X_M+x_r$, it becomes

$$\frac{(R_S+jx_S+jX_M)(R_r/s+jx_r+jX_M)+X_M^2}{R_r/s+jx_r+jX_M}$$

$$= R_S+jx_S+\frac{jX_M(R_r/s+jx_r+jX_M)+X_M^2}{R_r/s+jx_r+jX_M}$$

$$= R_S+jx_S+\frac{(jX_M)(R_r/s+jx_r)}{(jX_M)+(R_r/s+jx_r)}$$

The latter part of this expression, being the product of two impedances divided by their sum, is the impedance of these two in parallel. The total impedance is, therefore, R_S+jx_S in series with the combination of $(R_r/s+jx_r)$ and jX_M in parallel. The impedance can, therefore, be represented by an equivalent circuit as shown in Fig. 34.‡

The N, b system has a similar impedance except that $(1-v)$ is replaced by $(1+v)$, i.e. s is replaced by $(2-s)$. It can, therefore, be represented by an equivalent circuit which is identical with Fig. 34 except that R_r/s is replaced by $R_r/(2-s)$. The complete equivalent circuit can be drawn as in Fig. 35 in which, although independent, the positive-sequence and negative-sequence parts are connected together

† If all impedances have been referred to the same number of turns, x_S and x_r are the leakage reactances referred to that number of turns.

‡ Compare ref. 3, p. 376.

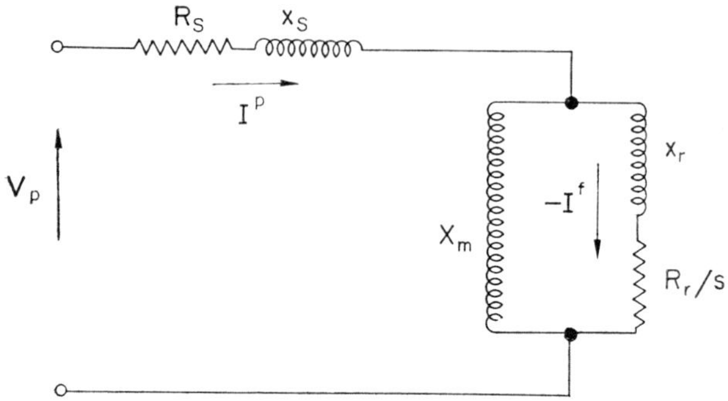

FIG. 34. Equivalent circuit.

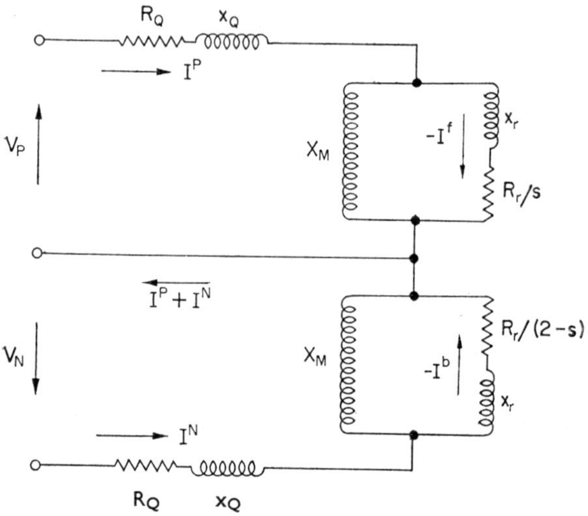

FIG. 35. Equivalent circuit of balanced induction machine.

by analogy with the circuit of the unbalanced and single-phase machine circuits.†

The current in the $(R_r/s+jx_r)$ branch of the positive-sequence circuit is $jX_M/(R_r/s+jx_r+jX_M) = jX_M/(R_r/s+jX_r)$ of the current in the R_S+jx_S branch, i.e. of I^P. It is, therefore,

$$\{jX_M/(R_r/s+jX_r)\}I^P = \{jsX_M/(R_r+jsX_r)\}I^P.$$

From the current equation on p. 167, it can be seen that this branch current is equal in magnitude and opposite in sign to I^f. The negative sign arises because of our original assumption that i^D and i^d produce m.m.f.s in the same direction when both are positive. We have thus also defined I^P and I^f as magnetizing in the same direction when they are of the same sign. In fact they magnetize in opposite directions and the analysis leads to their having opposite signs. In the equivalent circuit, however, the branch current which we have been considering is a component of the current I^P. In other words, it is the component of the primary current required to balance out the secondary current I^f and it is naturally of opposite sign to I^f.

Torque

Since $v = \dot{\theta}/\omega$, the **G′** matrix may be obtained from the impedance matrix of p. 166 by dividing the coefficients of v by ω. Therefore

$\mathbf{G}' = \dfrac{1}{\omega}$

	P	f	N	b
P				
f	$-jX_M$	$-jX_r$		
N				
b			jX_M	jX_r

† See Figs. 37 and 38.

The two-pole torque is therefore

$$T = \text{Re}\,\frac{1}{\omega}\;\begin{array}{cccc} \text{P} & \text{f} & \text{N} & \text{b} \\ \hline \multicolumn{1}{|c|}{I^{\text{P}*}} & \multicolumn{1}{c|}{I^{\text{f}*}} & \multicolumn{1}{c|}{I^{\text{N}*}} & \multicolumn{1}{c|}{I^{\text{b}*}} \\ \hline \end{array}\;\begin{array}{c|c|c|c|c|} & \text{P} & \text{f} & \text{N} & \text{b} \\ \hline \text{P} & & & & \\ \hline \text{f} & -jX_{\text{M}} & -jX_{\text{r}} & & \\ \hline \text{N} & & & & \\ \hline \text{b} & & & jX_{\text{M}} & jX_{\text{r}} \\ \hline \end{array}\;\begin{array}{c|c|} \hline \text{P} & I^{\text{P}} \\ \hline \text{f} & I^{\text{f}} \\ \hline \text{N} & I^{\text{N}} \\ \hline \text{b} & I^{\text{b}} \\ \hline \end{array}$$

$$= \text{Re}\,\frac{1}{\omega}\,[-jX_{\text{M}}I^{\text{P}}I^{\text{f}*} - jX_{\text{r}}I^{\text{f}}I^{\text{f}*} + jX_{\text{M}}I^{\text{N}}I^{\text{b}*} + jX_{\text{r}}I^{\text{b}}I^{\text{b}*}]$$

Since $I^{\text{f}}I^{\text{f}*}$ and $I^{\text{b}}I^{\text{b}*}$ are necessarily real, the terms in which they appear are imaginary and the torque is

$$T = \text{Re}\,\frac{1}{\omega}\,[-jX_{\text{M}}I^{\text{P}}I^{\text{f}*} + jX_{\text{M}}I^{\text{N}}I^{\text{b}*}]$$

Now from the equation for the currents on p. 166 it can be seen that

$$I^{\text{P}} = \frac{R_{\text{r}} + j(1-\text{v})X_{\text{r}}}{-j(1-\text{v})X_{\text{M}}}\,I^{\text{f}} = \frac{R_{\text{r}} + j\text{s}X_{\text{r}}}{-j\text{s}X_{\text{M}}}\,I^{\text{f}}$$

and it can be similarly shown that

$$I^{\text{N}} = \frac{R_{\text{r}} + j(1+\text{v})X_{\text{r}}}{-j(1+\text{v})X_{\text{M}}}\,I^{\text{b}} = \frac{R_{\text{r}} + j(2-\text{s})X_{\text{r}}}{-j(2-\text{s})X_{\text{M}}}\,I^{\text{b}}$$

If these values for I^{P} and I^{N} are substituted in the torque expression above,

$$T = \text{Re}\,\frac{1}{\omega}\left[jX_{\text{M}}\,\frac{R_{\text{r}} + j\text{s}X_{\text{r}}}{j\text{s}X_{\text{M}}}\,I^{\text{f}}I^{\text{f}*} - jX_{\text{M}}\,\frac{R_{\text{r}} + j(2-\text{s})X_{\text{r}}}{j(2-\text{s})X_{\text{M}}}\,I^{\text{b}}I^{\text{b}*}\right]$$

$$= \text{Re}\,\frac{1}{\omega}\,[\{R_{\text{r}}/\text{s} + jX_{\text{r}}\}I^{\text{f}}I^{\text{f}*} - \{R_{\text{r}}/(2-\text{s}) + jX_{\text{r}}\}I^{\text{b}}I^{\text{b}*}]$$

$$= \frac{1}{\omega}\,[|I^{\text{f}}|^2\,R/_{\text{r}}\text{s} - |I^{\text{b}}|^2\,R_{\text{r}}/(2-\text{s})]$$

The torque thus consists of a forward component in the direction D (A′) to Q (B′), i.e. counterclockwise, of which the magnitude is $1/\omega$ times the power "dissipated" in the resistance R_r/s of the positive-sequence equivalent circuit, together with a backward component in a clockwise direction of magnitude $1/\omega$ times the power "dissipated" in the resistance $R_r/(2-\mathrm{s})$ of the negative-sequence equivalent circuit.

From p. 167

$$I^f = \frac{V_P}{\Delta}(-j\mathrm{s}X_M) = V_P \frac{-j\mathrm{s}X_M}{(R_S+jX_S)(R_r+j\mathrm{s}X_r)+\mathrm{s}X_M^2}$$

$$= V_P \frac{-j\mathrm{s}X_M}{(R_S R_r-\mathrm{s}X_S X_r+\mathrm{s}X_M^2)+j(X_S R_r+\mathrm{s}R_S X_r)}$$

$$|I^f|^2 = V_P^2 \frac{\mathrm{s}^2 X_M^2}{\{R_S R_r-\mathrm{s}(X_S X_r-X_M^2)\}^2+\{X_S R_r+\mathrm{s}R_S X_r\}^2}$$

The forward component of the torque is therefore

$$\frac{1}{\omega}|I^f|^2\frac{R_r}{\mathrm{s}} = \frac{1}{\omega}V_P^2 \frac{\mathrm{s}X_M^2 R_r}{\{R_S R_r-\mathrm{s}(X_S X_r-X_M^2)\}^2+\{X_S R_r+\mathrm{s}R_S X_r\}^2}$$

$$= \frac{1}{\omega}V_P^2 \frac{X_M^2 R_r/\mathrm{s}}{\{R_S R_r/\mathrm{s}-(X_S X_r-X_M^2)\}^2+\{X_S R_r/\mathrm{s}+R_S X_r\}^2}$$

Similarly, the backward component of the torque is

$$\frac{1}{\omega}|I^b|^2\frac{R_r}{(2-\mathrm{s})}$$

$$= \frac{1}{\omega}V_N^2 \frac{X_M^2 R_r/(2-\mathrm{s})}{\{R_S R_r/(2-\mathrm{s})-(X_S X_r-X_M^2)\}^2+\{X_S R_r/(2-\mathrm{s})+R_S X_r\}^2}$$

Balanced Terminal Voltage

So far the expressions for current, and hence also for torque, have been expressed in terms of V_P and V_N. It is now necessary to consider the effect of specific terminal conditions. We will consider first the two-phase machine.

If the voltages applied to the D (A′), Q (B′) axes are balanced, they may be represented as V and $-jV$ respectively. It was shown on p. 113 that V_P is then equal to $\sqrt{2}V$ and V_N is zero. As a result only the positive-sequence axes P, f need be considered and the currents are as given on p. 167.

Since

$$\begin{array}{c|c|} & \\ \text{D (A}') & I^D \\ \text{Q (B}') & I^Q \end{array} = \frac{1}{\sqrt{2}} \begin{array}{c|c|c|} & \text{P} & \text{N} \\ \text{D (A}') & 1 & 1 \\ \text{Q (B}') & -j & j \end{array} \begin{array}{c|c|} & \\ \text{P} & I^P \\ \text{N} & I^N \end{array}$$

$$= \frac{1}{\sqrt{2}} \begin{array}{c|c|} \text{D (A}') & I^P + I^N \\ \text{Q (B}') & -jI^P + jI^N \end{array}$$

the currents under balanced conditions are

$$I^{A'} = I^D = (1/\sqrt{2})I^P, \quad I^{B'} = I^Q = -j(1/\sqrt{2})I^P$$

and correspondingly

$$I^d = (1/\sqrt{2})I^f, \quad I^q = -j(1/\sqrt{2})I^f$$

For the three-phase machine the terminal voltages may be defined as V, h^2V, hV and from p. 112 we know that V_P is then $\sqrt{3}V$ and V_N is again zero. The currents are

$$\begin{array}{c|c|} & \\ \text{A} & I^A \\ \text{B} & I^B \\ \text{C} & I^C \end{array} = \frac{1}{\sqrt{3}} \begin{array}{c|c|c|c|} & \text{P} & \text{N} & \text{O} \\ \text{A} & 1 & 1 & 1 \\ \text{B} & h^2 & h & 1 \\ \text{C} & h & h^2 & 1 \end{array} \begin{array}{c|c|} & \\ \text{P} & I^P \\ \text{N} & I^N \\ \text{O} & I^O \end{array} = \frac{1}{\sqrt{3}} \begin{array}{c|c|} \text{A} & I^P + I^N + I^O \\ \text{B} & h^2 I^P + hI^N + I^O \\ \text{C} & hI^P + h^2 I^N + I^O \end{array}$$

With balanced voltage, I^N and I^O being zero, the currents are

$$I^A = \frac{1}{\sqrt{3}} I^P, \quad I^B = \frac{1}{\sqrt{3}} h^2 I^P, \quad I^C = \frac{1}{\sqrt{3}} hI^P$$

Since with balanced voltage the D (A′), d currents and voltages are all $1/\sqrt{2}$ of the P, f currents and voltages, the equivalent circuit Fig. 34 can be used to express the relationships between the D (A′), d axes currents and voltages directly. Similarly, the same circuit can be used to represent directly the relationship between the phase voltages and currents of a three-phase machine. This is the manner in which it is used in classical theory, but it should be noted that invariance of power has not then been maintained and that consequently powers and torques calculated in this manner are per phase and not total as in the present work.

Unbalanced Terminal Voltage

If the terminal voltages of either two-phase or three-phase machines are unbalanced, neither V_P nor V_N will be zero and consequently I^P, I^f, I^N, I^b have all to be calculated and transformed back to two-phase or three-phase axes. Both components of torque will be present. It is assumed here that there is no zero-sequence current present. Although it can be taken into account,† zero-sequence performance is complicated and cannot be considered here.

Parameters

It would be possible to perform a test analogous to the open-circuit test on a transformer by supplying a polyphase slip-ring induction machine with balanced voltage with the secondary circuit open. It is not desirable to do this, however, because it subjects the rotor core to magnetization at supply frequency, whereas in operation it is subject only to slip frequency. With a squirrel-cage machine no such test is possible. The alternative is to drive the machine at synchronous speed, which eliminates the rotor core loss and also reduces the rotor current to zero, so that it is then immaterial whether the rotor winding is open circuit or not.

From p. 167 for balanced voltage,

$$I^P = \frac{V_P(R_r + jsX_r)}{(R_S + jX_S)(R_r + jsX_r) + sX_M^2}$$

† See ref. 17.

At synchronous speed s = 0 and

$$I^{\mathrm{P}} = \frac{V_{\mathrm{P}} R_{\mathrm{r}}}{(R_{\mathrm{S}}+jX_{\mathrm{S}})R_{\mathrm{r}}} = \frac{V_{\mathrm{P}}}{R_{\mathrm{S}}+jX_{\mathrm{S}}}$$

Hence, $\dfrac{V_{\mathrm{P}}}{I^{\mathrm{P}}} = R_{\mathrm{S}}+jX_{\mathrm{S}}$, but $\dfrac{V_{\mathrm{P}}}{I^{\mathrm{P}}} = \dfrac{\sqrt{3}V_{\mathrm{A}}}{\sqrt{3}I^{\mathrm{A}}} = \dfrac{V_{\mathrm{A}}}{I^{\mathrm{A}}}$ and also $\dfrac{V_{\mathrm{P}}}{I^{\mathrm{P}}} = \dfrac{\sqrt{2}V_{\mathrm{D}}}{\sqrt{2}I^{\mathrm{D}}} = \dfrac{V_{\mathrm{D}}}{I^{\mathrm{D}}} = \dfrac{V_{\mathrm{A'}}}{I^{\mathrm{A'}}}$. The ratio of phase voltage to phase current in both three-phase and two-phase machines thus gives the value of $R_{\mathrm{S}}+jX_{\mathrm{S}}$.

If, on the other hand, the rotor is locked in a stationary position, s = 1 and

$$I^{\mathrm{P}} = \frac{V_{\mathrm{P}}(R_{\mathrm{r}}+jX_{\mathrm{r}})}{(R_{\mathrm{S}}+jX_{\mathrm{S}})(R_{\mathrm{r}}+jX_{\mathrm{r}})+X_{\mathrm{M}}^2}$$

Then

$$\frac{V_{\mathrm{P}}}{I^{\mathrm{P}}} = R_{\mathrm{S}}+jX_{\mathrm{S}}+\frac{X_{\mathrm{M}}^2}{R_{\mathrm{r}}+jX_{\mathrm{r}}}$$

which is again identical with the transformer short-circuit test expression on p. 43 apart from the differences of notation. Here again, $\dfrac{V_{\mathrm{P}}}{I^{\mathrm{P}}} = \dfrac{V_{\mathrm{A}}}{I^{\mathrm{A}}} = \dfrac{V_{\mathrm{A'}}}{I^{\mathrm{A'}}}$, so that this locked rotor test measures $(X_{\mathrm{S}}-X_{\mathrm{M}}^2/X_{\mathrm{r}})$, which is approximately equal to the sum of the leakage reactances $(x_{\mathrm{S}}+x_{\mathrm{r}})$ referred to the stator number of turns, for both three-phase and two-phase machines.†

It is usually assumed that the stator and rotor winding leakage reactances are equal, unless knowledge of the design details suggests some other ratio. On this basis these two reactances can be determined from their measured sum. The mutual (magnetizing) reactance can then be determined as $(X_{\mathrm{S}}-x_{\mathrm{S}})$.

With a squirrel-cage rotor it is not possible to measure the winding resistance directly, and the only obvious solution is to subtract the stator winding resistance from the measured effective sum of the resistances.

† Compare ref. 3, p. 408.

The Unbalanced Two-phase Induction Machine

Three-phase induction motors with unbalanced windings are abnormal, but two-phase motors with unbalanced windings are common in the form of split-phase and capacitor motors† operating from a single-phase supply. Since the balanced two-phase motor can be regarded as a particular case of the unbalanced motor, and since, moreover, the performance of a balanced three-phase motor can be determined in terms of a balanced two-phase motor, the unbalanced two-phase machine is sufficiently general for all practical requirements. Unbalanced in this context means that the two stator windings have different distributions, and/or different numbers of turns and/or different resistances. The rotor windings, however, must be balanced in the form of three- or two-phase windings or squirrel-cage. For the purpose of this analysis the rotor will be represented by the stationary reference axes d, q. The stator windings are assumed to be at right angles to each other and are represented by D, Q.

It will be assumed that the D winding has the lesser L/R ratio so that the rotation will be in the positive direction of $\dot{\theta}$ when the machine is operated as a split-phase motor. For generality it will also be assumed that an external impedance Z_X is connected in series with the D winding.

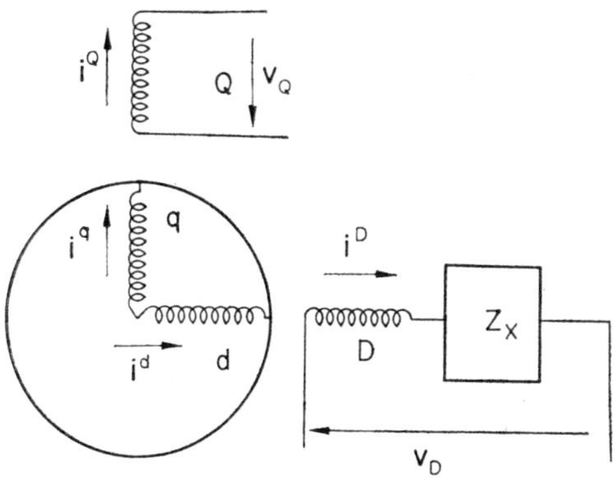

FIG. 36. Unbalanced two-phase induction machine.

† See ref. 3, pp. 265 and 453.

In most particular cases this impedance will be zero, but in the capacitor motor it will be the impedance of the capacitor.

In stationary axes the machine has the configuration shown in Fig. 36.

From this it is a routine procedure to write down the transient impedance matrix according to the rules given on pp. 74–76 as

$\mathbf{Z} =$

	D	Q	d	q
D	$R_D+L_Dp+Z_X$		$M_{Dd}p$	
Q		R_Q+L_Qp		$M_{Qq}p$
d	$M_{Dd}p$	$M_{Qq}\dot{\theta}$	R_r+L_rp	$L_r\dot{\theta}$
q	$-M_{Dd}\dot{\theta}$	$M_{Qq}p$	$-L_r\dot{\theta}$	R_r+L_rp

Let $M_{Dd}/M_{Qq} = n$. Since the rotor windings are balanced, the d and q windings are identical. In effect, therefore, we are saying that the effective turns ratio, D to Q, is n. "Effective" here means taking into account not only the numbers of turns but also the distribution of the windings.

The impedance matrix is then

$\mathbf{Z} =$

	D	Q	d	q
D	$R_D+L_Dp+Z_X$		$nM_{Qq}p$	
Q		R_Q+L_Qp		$M_{Qq}p$
d	$nM_{Qq}p$	$M_{Qq}\dot{\theta}$	R_r+L_rp	$L_r\dot{\theta}$
q	$-nM_{Qq}\dot{\theta}$	$M_{Qq}p$	$-L_r\dot{\theta}$	R_r+L_rp

This matrix leads to the "cross-field" equations of the unbalanced two-phase machine, which, however, will not be considered further here.

One objective is to obtain the equations in a form leading to an equivalent circuit. For this purpose it is necessary first to apply a transformation which is equivalent to referring the D winding to the same number of effective turns as the Q winding. This transformation is

C =

	D′	Q′	d′	q′
D	$1/n$			
Q		1		
d			1	
q				1

which leads to

$\mathbf{Z}' = \mathbf{C}_t\mathbf{Z}\mathbf{C}$ =

	D′	Q′	d′	q′
D′	$(R_D+L_Dp+Z_X)/n^2$		$M_{Qq}p$	
Q′		R_Q+L_Qp		$M_{Qq}p$
d′	$M_{Qq}p$	$M_{Qq}\dot{\theta}$	R_r+L_rp	$L_r\dot{\theta}$
q′	$-M_{Qq}\dot{\theta}$	$M_{Qq}p$	$-L_r\dot{\theta}$	R_r+L_rp

Since the axes are stationary, and the only exciting functions are the terminal voltages which are sinusoidal and of angular frequency ω, the steady-state impedance matrix may be obtained by replacing p by $j\omega$. If, at the same time, we write X_D, X_Q, X_r and X_M for ωL_D, ωL_Q, ωL_r and ωM_{Qq} respectively, and also $\mathrm{v}\omega$ for $\dot{\theta}$, i.e. v is the per-unit speed, the impedance matrix for steady-state is obtained as

$\mathbf{Z}'$ =

	D′	Q′	d′	q′
D′	$(R_D+jX_D+Z_X)/n^2$		jX_M	
Q′		R_Q+jX_Q		jX_M
d′	jX_M	$\mathrm{v}X_M$	R_r+jX_r	$\mathrm{v}X_r$
q′	$-\mathrm{v}X_M$	jX_M	$-\mathrm{v}X_r$	R_r+jX_r

where Z_X may be complex.

$\mathbf{Z''} =$

$\frac{1}{\sqrt{2}}$

	D′	Q′	d′	q′
P	1	j		
N	1	$-j$		
f			1	j
b			1	$-j$

	D′	Q′	d′	q′
D′	$(R_D+jX_D+Z_X)/n^2$		jX_M	
Q′		R_Q+jX_Q		jX_M
d′	jX_M	vX_M	R_r+jX_r	vX_r
q′	$-vX_M$	jX_M	$-vX_r$	R_r+jX_r

$\frac{1}{\sqrt{2}}$

	P	N	f	b
D′	1	1		
Q′	$-j$	j		
d′			1	1
q′			$-j$	j

	P	N	f	b
P	$R_Q+jX_Q+Z_U$	Z_U	jX_M	
N	Z_U	$R_Q+jX_Q+Z_U$		jX_M
f	jsX_M		R_r+jsX_r	
b		$j(2-s)X_M$		$R_r+j(2-s)X_r$

If this matrix is transformed in the same way as that for the balanced machine on p. 164 it becomes $\mathbf{Z}''$ as given on p. 180 where Z_U is defined as

$$\tfrac{1}{2}\{R_D+jX_D+Z_X)/n^2-(R_Q+jX_Q)\},$$

and where, as before, $s = (1-v)$ is the fractional slip. Since Z_U appears as the element designated P, N and also as that designated N, P, the P, f and N, b sets of axes are not independent, and there is no particular advantage in rearranging the order of the rows and columns as was done in the case of the balanced machine.

If, however, this matrix is partitioned between the stator and rotor axes, three of the sub-matrices are diagonal and this makes elimination of the short-circuited rotor axes f, b attractive. The resulting reduced impedance matrix is derived on p. 182.

Equivalent Circuit

This matrix is symmetric and is therefore suitable for devising an equivalent circuit. The element designated P, P is identical with the impedance of the positive-sequence system of the balanced machine given on p. 168, except that there is an additional term Z_U and the index S has been replaced by Q. If follows that, in respect of this element an equivalent circuit similar to that of the positive-sequence circuit of the balanced machine, but with an additional series component Z_U, is appropriate. The negative-sequence circuit is the same, except that s is replaced by $(2-s)$. There are, however, Z_U elements in the P, N, and N, P positions in the impedance matrix. This implies that there is a component Z_U common to both the positive- and negative-sequence circuits. The complete equivalent circuit is therefore as shown in Fig. 37 in which $x_Q = X_Q - X_M$ and $x_r = X_r - X_M$ are the leakage reactances of the stator Q winding and the rotor winding respectively if all values are referred to the Q winding number of turns. Z_U can also be defined in terms of leakage reactances, if the leakage reactance of the stator D winding is $x_D = X_D - n^2 X_M$ in which x_D and X_D are

$$
\begin{array}{c|cc|}
 & \mathrm{P} & \mathrm{N} \\ \hline
\mathrm{P} & R_Q+jX_Q+Z_U & Z_U \\
\mathrm{N} & Z_U & R_Q+jX_Q+Z_U \\ \hline
\end{array}
-
\begin{array}{c|cc|}
 & \mathrm{f} & \mathrm{b} \\ \hline
\mathrm{P} & jX_M & \\
\mathrm{N} & & jX_M \\ \hline
\end{array}
\;
\begin{array}{c|cc|}
 & \mathrm{f} & \mathrm{b} \\ \hline
\mathrm{f} & 1/\{R_r+jsX_r\} & \\
\mathrm{b} & & 1/\{R_r+j(2-s)X_r\} \\ \hline
\end{array}
\;
\begin{array}{c|cc|}
 & \mathrm{P} & \mathrm{N} \\ \hline
\mathrm{f} & jsX_M & \\
\mathrm{b} & & j(2-s)X_M \\ \hline
\end{array}
$$

$$
=
\begin{array}{c|cc|}
 & \mathrm{P} & \mathrm{N} \\ \hline
\mathrm{P} & R_Q+jX_Q+Z_U & \\
\mathrm{N} & & R_Q+jX_Q+Z_U \\ \hline
\end{array}
+
\begin{array}{c|cc|}
 & \mathrm{P} & \mathrm{N} \\ \hline
\mathrm{P} & sX_M^2/\{R_r+jsX_r\} & \\
\mathrm{N} & & (2-s)X_M^2/\{R_r+j(2-s)X_r\} \\ \hline
\end{array}
$$

$$
=
\begin{array}{c|cc|}
 & \mathrm{P} & \mathrm{N} \\ \hline
\mathrm{P} & R_Q+jX_Q+Z_U+sX_M^2/\{R_r+jsX_r\} & Z_U \\
\mathrm{N} & Z_U & R_Q+jX_Q+Z_U+(2-s)X_M^2/\{R_r+j(2-s)X_r\} \\ \hline
\end{array}
$$

$$
=
\begin{array}{c|cc|}
 & \mathrm{P} & \mathrm{N} \\ \hline
\mathrm{P} & R_Q+jX_Q+Z_U+X_M^2/\{R_r/s+jX_r\} & Z_U \\
\mathrm{N} & Z_U & R_Q+jX_Q+Z_U+X_M^2/\{R_r/(2-s)+jX_r\} \\ \hline
\end{array}
$$

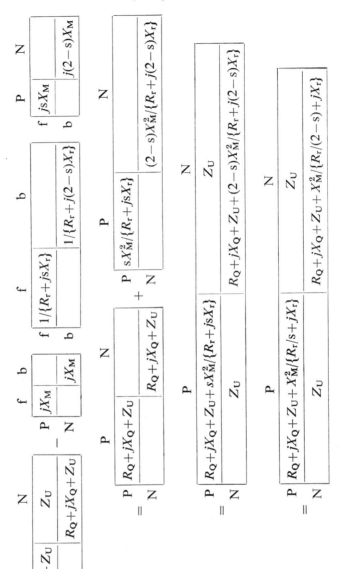

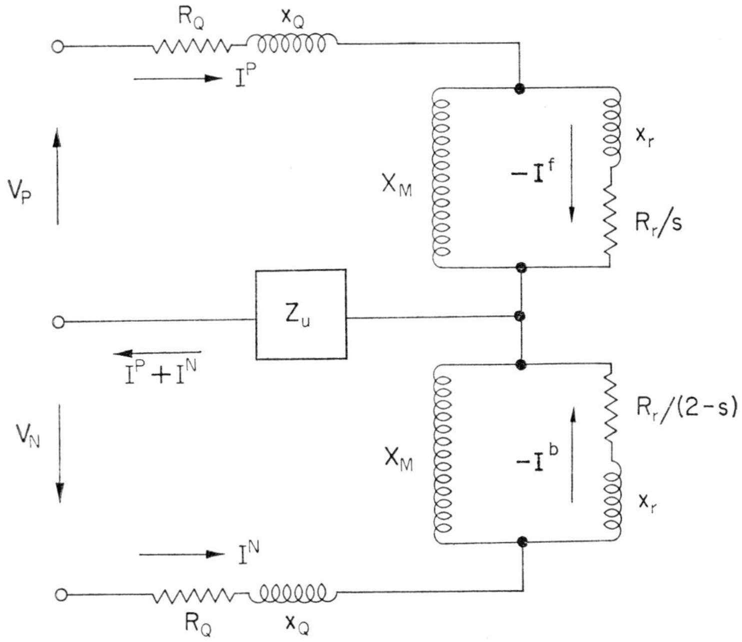

FIG. 37. Equivalent circuit of unbalanced two-phase induction machine.

referred to the D winding and X_M is referred to the Q winding. In these terms then

$$Z_U = \tfrac{1}{2}\{(R_D + jx_D + Z_X)/n^2 - (R_Q + jx_Q)\}$$

In the balanced machine $n = 1$, $R_D = R_Q$, $x_D = x_Q$, and $Z_X = 0$, so that $Z_U = 0$, in which case the impedance matrix reduces to a diagonal matrix.

Currents

If the terminal voltages are V_D, V_Q in the original reference axes, the symmetrical component terminal voltages are

$$\begin{array}{c|c|}\text{P} & V_{\text{P}} \\ \hline \text{N} & V_{\text{N}}\end{array} = \frac{1}{\sqrt{2}} \begin{array}{c|c|c|}& \text{D}' & \text{Q}' \\ \text{P} & 1 & j \\ \hline \text{N} & 1 & -j\end{array} \begin{array}{c|c|}\text{D}' & V_{\text{D}'} \\ \hline \text{Q}' & V_{\text{Q}'}\end{array}$$

$$= \frac{1}{\sqrt{2}} \begin{array}{c|c|c|}& \text{D}' & \text{Q}' \\ \text{P} & 1 & j \\ \hline \text{N} & 1 & -j\end{array} \begin{array}{c|c|c|}& \text{D} & \text{Q} \\ \text{D}' & 1/n & \\ \hline \text{Q}' & & 1\end{array} \begin{array}{c|c|}\text{D} & V_{\text{D}} \\ \hline \text{Q} & V_{\text{Q}}\end{array}$$

$$= \frac{1}{\sqrt{2}} \begin{array}{c|c|}\text{P} & V_{\text{D}}/n+jV_{\text{Q}} \\ \hline \text{N} & V_{\text{D}}/n-jV_{\text{Q}}\end{array}$$

If the currents I^{P}, I^{N} are found in terms of these voltages, the actual currents I^{D}, I^{Q} are found by

$$\begin{array}{c|c|}\text{D} & I^{\text{D}} \\ \hline \text{Q} & I^{\text{Q}}\end{array} = \frac{1}{\sqrt{2}} \begin{array}{c|c|c|}& \text{D}' & \text{Q}' \\ \text{D} & 1/n & \\ \hline \text{Q} & & 1\end{array} \begin{array}{c|c|c|}& \text{P} & \text{N} \\ \text{D}' & 1 & 1 \\ \hline \text{Q}' & -j & j\end{array} \begin{array}{c|c|}\text{P} & I^{\text{P}} \\ \hline \text{N} & I^{\text{N}}\end{array}$$

$$= \frac{1}{\sqrt{2}} \begin{array}{|c|}(I^{\text{P}}+I^{\text{N}})/n \\ \hline -j(I^{\text{P}}-I^{\text{N}})\end{array}$$

If the unbalanced two-phase motor is in fact a single-phase motor of the split-phase or capacitor type, the voltages V_{D} and V_{Q} are equal, and I^{Q} is the current of the main winding and I^{D} the current of the auxiliary winding.

Torque

The $\mathbf{G}''$ matrix of the unbalanced machine in P, N, f, b axes is the same as that of the balanced machine already considered. Moreover, the ratios I^P/I^f and I^N/I^b are the same as in the balanced machine. Consequently the expressions for torque in terms of I^P, I^f, I^N, I^b, already derived on p. 172 for the balanced machine, apply also to the unbalanced machine.

The expression for torque in terms of terminal voltage is, of course, more complicated and dependent on the actual terminal conditions.

Single-phase Operation of Induction Machine

Single-phase operation is an unbalanced condition requiring specific consideration.

Three-phase Machine

There are two ways in which a three-phase machine can be operated single-phase. In one case the supply is connected between one terminal (preferably A) and the other two (B and C) connected together. All line voltages are then known and their symmetrical components can be determined. The symmetrical components of the phase voltages, omitting zero-sequence, can be determined from those of the line voltages and then the currents determined in the manner discussed in pp. 163–75. In this connection the machine is not satisfactory as a motor but can be used as a brake.†

The alternative, which is satisfactory as a motoring condition, is to connect the supply between two terminals (preferably B and C), leaving the third terminal (A) isolated. Since the voltage of the third terminal is then unknown, it is not possible to determine the symmetrical components of the terminal voltages. On the other hand, it is known that

† See ref. 16, pp. 245–60.

the A-phase current is zero and that $I^C = -I^B$, hence

$$\begin{array}{c|c|}\mathrm{P} & I^P \\ \mathrm{N} & I^N \\ \mathrm{O} & I^O\end{array} = \frac{1}{\sqrt{3}} \begin{array}{c|c|c|c|} & \mathrm{A} & \mathrm{B} & \mathrm{C} \\ \mathrm{P} & 1 & h & h^2 \\ \mathrm{N} & 1 & h^2 & h \\ \mathrm{O} & 1 & 1 & 1\end{array} \begin{array}{c|c|}\mathrm{A} & I^A \\ \mathrm{B} & I^B \\ \mathrm{C} & I^C\end{array}$$

$$= \frac{1}{\sqrt{3}} \begin{array}{c|c|}\mathrm{P} & hI^B + h^2I^C \\ \mathrm{N} & h^2I^B + hI^C \\ \mathrm{O} & I^B + I^C\end{array} = \frac{1}{\sqrt{3}} \begin{array}{c|c|}\mathrm{P} & (h-h^2)I^B \\ \mathrm{N} & (h^2-h)I^B \\ \mathrm{O} & \mathrm{O}\end{array}$$

so that $I^P = jI^B = -I^N$. Hence $I^P + I^N = 0$ and there is no current in the middle conductor of the circuit shown in Fig. 35, which leads to the equivalent circuit of Fig. 38. Now

$$\begin{array}{c|c|}\mathrm{P} & V_P \\ \mathrm{N} & V_N \\ \mathrm{O} & V_O\end{array} = \frac{1}{\sqrt{3}} \begin{array}{c|c|c|c|} & \mathrm{A} & \mathrm{B} & \mathrm{C} \\ \mathrm{P} & 1 & h & h^2 \\ \mathrm{N} & 1 & h^2 & h \\ \mathrm{O} & 1 & 1 & 1\end{array} \begin{array}{c|c|}\mathrm{A} & V_A \\ \mathrm{B} & V_B \\ \mathrm{C} & V_C\end{array}$$

$$= \frac{1}{\sqrt{3}} \begin{array}{c|c|}\mathrm{P} & V_A + hV_B + h^2V_C \\ \mathrm{N} & V_A + h^2V_B + hV_C \\ \mathrm{O} & V_A + V_B + V_C\end{array}$$

and the terminal voltage of the equivalent circuit is

$$(V_P - V_N) = (1/\sqrt{3})\{(h-h^2)V_B + (h^2-h)V_C\} = (1/\sqrt{3})j\sqrt{3}\,(V_B - V_C$$
$$= jV_{BC}\,.$$

Since V_{BC} is the known line voltage, there are sufficient data to determine the current $I^B = -jI^P$, either from the equivalent circuit or from

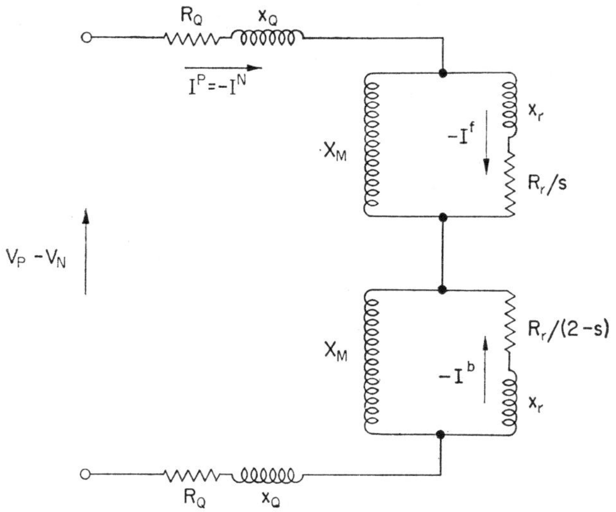

FIG. 38. Equivalent circuit of single-phase induction machine.

the impedance matrix of p. 164 by eliminating the f and b axes and applying a further transformation

$$\begin{array}{c|c|} P & I^P \\ N & I^N \end{array} = \begin{array}{c|c|} P & 1 \\ N & -1 \end{array} \boxed{I^{P'}} .$$

Two-phase Machine

The single-phase performance of the two-phase machine can be determined from the analysis of either the balanced machine (pp. 162–76) or of the unbalanced machine (pp. 177–850), by open-circuiting the D winding. Then $I^D = 0$, and since $I^D = (1/\sqrt{2})(I^P + I^N)/n$, $I^P + I^N = 0$, leading to the same equivalent circuit, Fig. 38. Here, $V_P - V_N = j\sqrt{2}V_Q$ and is consequently known. $I^Q = -j(1/\sqrt{2}) \times (I^P - I^N) = -j\sqrt{2}I^P$.

Single-phase Machine

Although the single-phase induction machine is, by definition, not polyphase, it is convenient to mention it at the end of this chapter.

It may be regarded as either a three-phase machine from which the A phase has been removed or as a two-phase machine from which the D phase has been removed.

Parameters

The parameters of the three-phase and two-phase machines operating single-phase are, of course, the same parameters as those involved in polyphase operation. In the three-phase machine operating single-phase line-to-line, the ratio of voltage to current is

$$\frac{V_{BC}}{I^{B}} = \frac{-j(V_{P}-V_{N})}{-jI^{P}} = \frac{V_{P}-V_{N}}{I^{P}}$$

whereas in the two-phase machine it is

$$\frac{V_{Q}}{I^{Q}} = \frac{-j(1/\sqrt{2})(V_{P}-V_{N})}{-j\sqrt{2}I^{P}} = \frac{1}{2}\,\frac{V_{P}-V_{N}}{I^{P}}$$

This difference was to be expected, because in the three-phase machine two winding phases are connected in series, whereas in the two-phase machine only one phase is in circuit.†

When the single-phase machine parameters are considered, there is no such simple reference to polyphase conditions. However, if it is regarded as equivalent to the three-phase machine operating line-to-line, its winding is equivalent to two phases in series. The actual values of resistance and reactance of its one winding must therefore be halved to obtain the parameters used here. If it is regarded as equivalent to one phase of a two-phase machine, the presence of the $\frac{1}{2}$ in the expression for V_Q/I^Q again shows that the elements of the equivalent circuit are half those of the single winding. In classical theory, the equivalent circuit of the single-phase machine is usually expressed in terms of the resistances and reactances of the actual winding, so that the $\frac{1}{2}$ appears explicitly in the equivalent circuit.‡

† It was shown on pp. 137 and 138 that the effective phase impedance, with the transformations used, is the same for the three-phase and two-phase machines.

‡ See ref. 3, pp. 451–4.

From the equivalent circuit, Fig. 38, the ratio of voltage to current with the rotor stationary can be seen to be

$$2\left\{R_Q+jx_Q+\frac{jX_M(R_r+jx_r)}{R_r+jx_r+jX_M}\right\}$$

This is of the same form as the corresponding expressions for the transformer and polyphase induction machines and leads to an approximate value for the sum of the leakage reactances. With the rotor running at synchronous speed, the ratio of voltage to current is

$$2(R_Q+jx_Q)+jX_M+\frac{jX_M(R_r/2+jx_r)}{R_r/2+jx_r+jX_M}$$

If the resistances are known and it is assumed, as is customary, that the sum of the leakage reactances can be divided equally or in some other known ratio, X_M can be determined and all the parameters are known.

The Polyphase Synchronous Machine with a Uniform Air-gap and no Damper Windings

The synchronous machine with a uniform air-gap will be briefly examined before considering in detail the more general salient-pole machine.

The arrangement of the machine in terms of stationary axes is shown diagrammatically in Fig. 39.

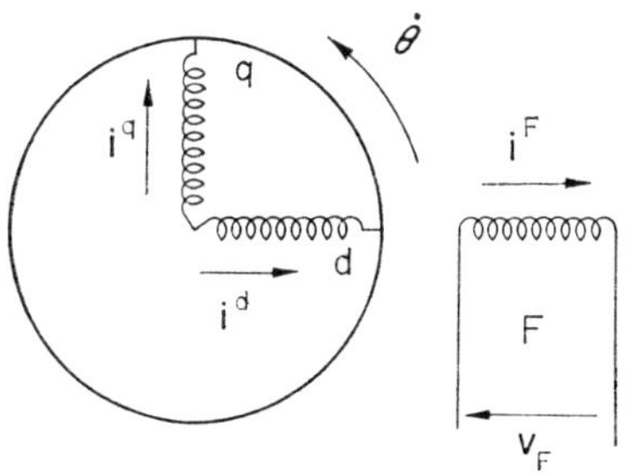

FIG. 39. Synchronous machine with uniform air-gap.

The impedance matrix in d, q, D terms can be obtained from that on p. 63 by omitting the Q axis. Here F is used in place of D to indicate that this is the field winding; d, q and a are used for the armature. The impedance matrix is then

	d	q	F
d	$R_a+L_d p$	$L_d\dot{\theta}$	$L_{Fd}p$
q	$-L_d\dot{\theta}$	$R_a+L_d p$	$-L_{Fd}\dot{\theta}$
F	$L_{Fd}p$		$R_F+L_F p$

Under balanced steady-state conditions, the exciting function is the voltage v_F applied to the field winding and, being a d.c. voltage, may be represented by the constant voltage V_F. Since all the elements of the circuit are constants in terms of the stationary reference axes, the response functions will also be constant and the operator p may be replaced by zero, leading to the voltage equation

				d	q	F			
d	V_d		d	R_a	$L_d\dot{\theta}$			d	I^d
q	V_q	=	q	$-L_d\dot{\theta}$	R_a	$-L_{Fd}\dot{\theta}$		q	I^q
F	V_F		F			R_F		F	I^F

in which capitals have been used to represent the constant voltages and currents.

Since the machine is "synchronous", its angular velocity $\dot{\theta}$ will define the angular frequency ω and it is convenient to replace the $\dot{\theta}$ of the impedance matrix by ω and the resulting products ωL can be regarded as reactances X. The final steady-state voltage equation is thus

				d	q	F			
d	V_d		d	R_a	X_d			d	I^d
q	V_q	=	q	$-X_d$	R_a	$-X_{Fd}$		q	I^q
F	V_F		F			R_F		F	I^F

Since we are investigating balanced conditions, there are no zero-sequence currents and

$$\begin{array}{c|c|}\hline a & i^{a} \\ \hline b & i^{b} \\ \hline c & i^{c} \\ \hline \end{array} = \sqrt{\tfrac{2}{3}}\; \begin{array}{c|c|c|c|} & \text{d} & \text{q} & \text{o} \\ \hline a & \cos\theta & \sin\theta & 1/\sqrt{2} \\ \hline b & \cos(\theta-2\pi/3) & \sin(\theta-2\pi/3) & 1/\sqrt{2} \\ \hline c & \cos(\theta-4\pi/3) & \sin(\theta-4\pi/3) & 1/\sqrt{2} \\ \hline \end{array} \; \begin{array}{c|c|}\hline d & I^{d} \\ \hline q & I^{q} \\ \hline o & 0 \\ \hline \end{array}$$

$$= \sqrt{\tfrac{2}{3}}\; \begin{array}{c|c|}\hline a & \cos\theta I^{d} + \sin\theta I^{q} \\ \hline b & \cos(\theta-2\pi/3)I^{d}+\sin(\theta-2\pi/3)I^{q} \\ \hline c & \cos(\theta-4\pi/3)I^{d}+\sin(\theta-4\pi/3)I^{q} \\ \hline \end{array}$$

$$= \sqrt{\tfrac{2}{3}}\; \sqrt{\{(I^{d})^{2}+(I^{q})^{2}\}}\; \begin{array}{c|c|}\hline a & \cos\theta\sin\psi + \sin\theta\cos\psi \\ \hline b & \cos(\theta-2\pi/3)\sin\psi+\sin(\theta-2\pi/3)\cos\psi \\ \hline c & \cos(\theta-4\pi/3)\sin\psi+\sin(\theta-4\pi/3)\cos\psi \\ \hline \end{array}$$

$$= \sqrt{\tfrac{2}{3}}\; \sqrt{\{(I^{d})^{2}+(I^{q})^{2}\}}\; \begin{array}{c|c|}\hline a & \sin(\theta+\psi) \\ \hline b & \sin(\theta+\psi-2\pi/3) \\ \hline c & \sin(\theta+\psi-4\pi/3) \\ \hline \end{array} = \hat{I}\; \begin{array}{c|c|}\hline a & \sin(\theta+\psi) \\ \hline b & \sin(\theta+\psi-2\pi/3) \\ \hline c & \sin(\theta+\psi-4\pi/3) \\ \hline \end{array}$$

where $\cos\psi = I^{q}/\sqrt{\{(I^{d})^{2}+(I^{q})^{2}\}}$ and $\hat{I} = \sqrt{\tfrac{2}{3}}\;\sqrt{\{(I^{d})^{2}+(I^{q})^{2}\}}$
Thus i^{a}, i^{b}, and i^{c} are a system of balanced three-phase currents of peak magnitude $\hat{I} = \sqrt{\tfrac{2}{3}}\cdot\sqrt{\{(I^{d})^{2}+(I^{q})^{2}\}}$.

From the matrix voltage equation,

$$\begin{array}{c|c|}\hline d & V_{d} \\ \hline q & V_{q} \\ \hline F & V_{F} \\ \hline \end{array} = \begin{array}{c|c|}\hline d & R_{a}I^{d}+X_{d}I^{q} \\ \hline q & -X_{d}I^{d}+R_{a}I^{q}-X_{Fd}I^{F} \\ \hline F & R_{F}I^{F} \\ \hline \end{array}$$

Hence, zero-sequence being absent,

$$\begin{aligned} v_a &= \sqrt{\tfrac{2}{3}}\{\cos\theta V_d + \sin\theta\, V_q\} \\ &= \sqrt{\tfrac{2}{3}}\{R_a\cos\theta I^d + X_d\cos\theta I^q - X_d\sin\theta I^d \\ &\quad + R_a\sin\theta I^q - X_{Fd}\sin\theta I^F\} \\ &= R_a\sqrt{\tfrac{2}{3}}(\cos\theta I^d + \sin\theta I^q) - X_d\sqrt{\tfrac{2}{3}}(\sin\theta I^d - \cos\theta I^q) \\ &\quad - \sqrt{\tfrac{2}{3}}X_{FD}I^F\sin\theta \\ &= R_a\sqrt{\tfrac{2}{3}}\sqrt{\{(I^d)^2+(I^q)^2\}}\sin(\theta+\psi) \\ &\quad + X_d\sqrt{\tfrac{2}{3}}\sqrt{\{(I^d)^2+(I^q)^2\}}\cos(\theta+\psi) - \sqrt{\tfrac{2}{3}}X_{Fd}I^F\sin\theta \end{aligned}$$

If the armature were open circuit, so that i^a, i^b, and i^c were all zero, so also would I^d and I^q be zero, and v_a would be

$$-\sqrt{\tfrac{2}{3}}X_{Fd}I^F\sin\theta = \hat{E}\sin\theta, \quad \text{where} \quad \hat{E} = -\sqrt{\tfrac{2}{3}}X_{Fd}I^F$$

The voltage equation can then be written

$$v_a = R_a\hat{I}\sin(\theta+\psi) + X_d\hat{I}\cos(\theta+\psi) + \hat{E}\sin\theta$$

Now ψ is the "internal" phase angle by which $i^a = \hat{I}\sin(\theta+\psi)$ leads the e.m.f. $E\sin\theta$, and if θ is taken as $(\omega t-\psi)$, so that the current i^a is thereby defined as $\hat{I}\sin(\theta+\psi) = \hat{I}\sin\omega t$, and thus determines the time zero,

$$v_a = R_a\hat{I}\sin\omega t + X_d\hat{I}\cos\omega t + \hat{E}\sin(\omega t-\psi)$$

which in phasor terms is

$$V = R_a I + jX_d I + E$$

Being the reactance which the machine appears to have when operating synchronously, X_d is called the "synchronous reactance".

It is important to remember in the above that in the initial equations all circuits were regarded as sinks, conforming to "motor" conventions of positive directions of current and power flow.

The e.m.f. E above is not the actual e.m.f. on load, but is the e.m.f. which the same field current I^F would produce on no-load if there were no changes of saturation. It is sometimes called the e.m.f. behind synchronous reactance. The true e.m.f. is equal to the difference (for motor conventions) of the terminal voltage and the voltage drop across the armature resistance and leakage reactance, i.e. $V-(R_a+jx_a)I$, where x_a† is the armature leakage reactance and is equal to ωl_a.

Phasor Diagram

From this voltage equation a phasor diagram‡ can be constructed as shown in Fig. 40.

Since our original conventions assumed that power was supplied to all windings of the machine, this diagram relates to a synchronous

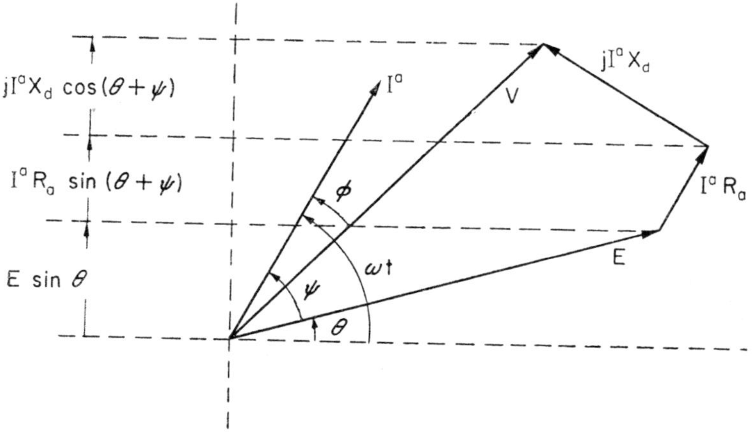

Fig. 40. Phasor diagram of synchronous machine with uniform air-gap.

motor. It will be noted that the diagram is drawn for leading current. To obtain the diagram for a generator with the usual conventions, we should have to write $-I^a$ for I^a.

† It was shown on p. 68 that in the salient-pole machine, $L_d = L_{Dd}+l_a$, when all are referred to a common number of turns. This relationship applies equally to the uniform air-gap machine, so that here $X_d = X_{Fd}+x_a$, if these reactances are referred to a common base.

‡ Compare ref. 3, p. 474.

Field Current required for Given Armature Terminal Conditions

E in the preceding analysis is the armature voltage which the field current I^F would produce on open-circuit if X_{Fd} were constant and there were no saturation. Because of saturation, the actual field current required to produce specified balanced terminal conditions would be greater than the value of I^F obtained by putting into the equation $\hat{E} = -\sqrt{\frac{2}{3}}\, X_{Fd} I^F$ the value of E obtained from the equation $V = R_a I + jX_d I + E$ by an amount which depends on the magnitude of the true e.m.f. on load, namely $V_a - (R_a + jx_a)I^a$, where x_a is the armature leakage reactance. This extra component of the field current is most conveniently obtained from the graph of open-circuit voltage against field current as the difference in the field current required to produce the true e.m.f. and the field current that would be required for this e.m.f. if there were no saturation.† This is shown in Fig. 41(i).

The same quantity can be found from the graph of X_{Fd} against field current by plotting (true e.m.f. divided by field current) and noting the values of field current where this curve cuts the actual graph of X_{Fd} and where it cuts the straight constant part of this graph extended. This is shown in Fig. 41(ii). This latter method is not to be recommended because the angle between the curves at the point of intersection is small, so that the point itself is ill-defined.

As an alternative, a lower "saturated" value of synchronous reactance X_d could be used to determine E from the equation $V = R_a I + jX_d I + E$, and a corresponding lower value of X_{Fd} used to determine I^F. This is equivalent to replacing the open-circuit graph by a straight line passing through the origin and the actual operating point on the graph, as shown in Fig. 41(i). The accuracy of this method is obviously dependent on the accuracy of the prediction of the position of the operating point. The accuracy may be improved by iteration.

Torque

The torque of the uniform air-gap machine is considered on p. 215.

† Compare ref. 3, p. 474.

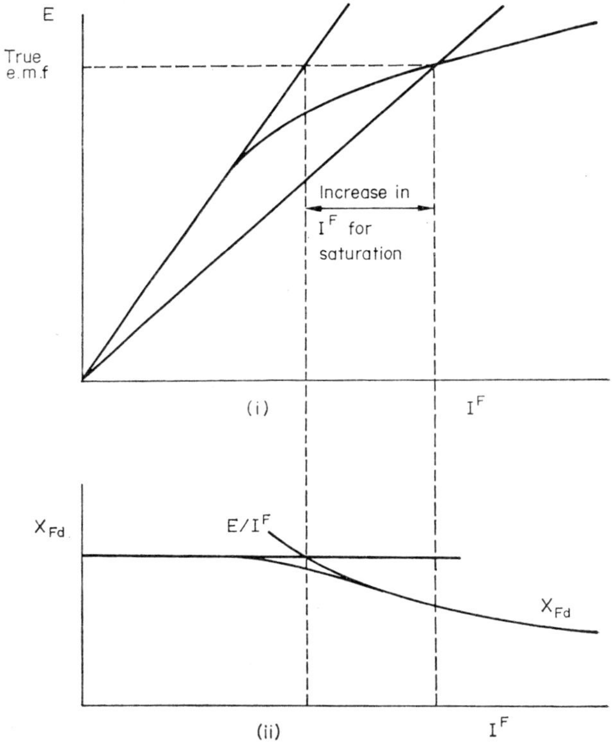

FIG. 41. Allowance for saturation.

Parameters

The steady-state voltage equation of the uniform air-gap synchronous machine is given on p. 190 in d, q, F axes. If it is driven with the armature windings open circuit, so that both I^d and I^q are zero, $V_d = 0$, and $V_q = -X_{Fd}I^F$. Hence, from the equations on p. 137, in the two-phase machine $v_\alpha = -X_{Fd}I^F \sin \omega t$, and in the three-phase machine $v_a = -(\sqrt{2}/\sqrt{3})\, X_{Fd}I^F \sin \omega t$. In r.m.s. terms, therefore, $V_\alpha = (1/\sqrt{2})\, X_{Fd}I^F$ and $V_a = (1/\sqrt{3})\, X_{Fd}I^F$. The test therefore leads directly to the value of X_{Fd}, which, of course, varies widely with saturation.

If the machine is driven with the armature windings short-circuited, V_d and V_q are both zero. The current equation, obtained by inverting the matrix, is

$$\begin{array}{c|c|}
 & \\
\hline
d & I^d \\
\hline
q & I^q \\
\hline
F & I^F \\
\hline
\end{array} = \frac{1}{\Delta}\;
\begin{array}{c|c|c|c|}
 & d & q & F \\
\hline
d & - & - & -X_d X_{Fd} \\
\hline
q & - & - & R_a X_{Fd} \\
\hline
F & - & - & R_a^2 + X_d^2 \\
\hline
\end{array}\;
\begin{array}{c|c|}
 & \\
\hline
d & 0 \\
\hline
q & 0 \\
\hline
F & V_F \\
\hline
\end{array}$$

where $\Delta = R_F(R_a^2 + X_d^2)$.

In this last admittance matrix the first two columns multiply zeros of the voltage matrix and consequently they have been omitted to show that they need not be calculated.

Hence

$$\begin{array}{c|c|}
\hline
d & I^d \\
\hline
q & I^q \\
\hline
F & I^F \\
\hline
\end{array} =
\begin{array}{c|c|}
\hline
d & -X_d X_{Fd} V_F / \{(R_a^2 + X_d^2) R_F\} \\
\hline
q & R_a X_{Fd} V_F / \{(R_a^2 + X_d^2) R_F\} \\
\hline
F & V_F / R_F \\
\hline
\end{array}$$

In the three-phase machine, therefore,

$$i^a = (\sqrt{2}/\sqrt{3}) \left\{ \frac{-X_d \cos \omega t + R_a \sin \omega t}{R_a^2 + X_d^2} \right\} \frac{X_{Fd} V_F}{R_F}$$

If R_a is negligible compared to X_d, this equation simplifies to

$$i^a = -(\sqrt{2}/\sqrt{3})\,(1/X_d)\,(X_{Fd} V_F / R_F) \cos \omega t$$

and the r.m.s. value of the current is

$$(1/\sqrt{3})\,(1/X_d)\,(X_{Fd} V_F / R_F) = (1/\sqrt{3})\,(1/X_d)\,(X_{Fd} I^F).$$

Hence the ratio of open-circuit voltage to short-circuit current for the same field current I^F is X_d.† From this it is obvious that X_d also varies widely with saturation.

In the two-phase machine

$$i^\alpha = \left\{ \frac{-X_d \cos \omega t + R_a \sin \omega t}{R_a^2 + X_d^2} \right\} \frac{X_{Fd} V_F}{R_F}$$

† Compare ref. 3, p. 470.

If R_a is again neglected, the r.m.s. value of the current I^α is $(1/\sqrt{2})$ $(1/X_d)(X_{Fd}I_F)$. Here again, therefore, the ratio of open-circuit voltage to short-circuit current for the same field current is X_d.

The Polyphase Synchronous Machine with Salient Poles and no Damper Windings

The synchronous machines of practical importance are salient-pole three-phase machines with damping windings which are effective on both the axis of the poles (the direct axis) and the axis between the poles (the quadrature axis). Even if no damping winding is fitted to the machine, there will be damping circuits in a machine of normal construction. These are provided by the yoke (or rotor body in a revolving-field machine) and the poles. Even if the poles are laminated, a circuit is provided, partly because the laminations are not effectively insulated and partly through the rivets holding the laminations together.

The ideal salient-pole machine without any damping circuits is, however, so much simpler that it will be considered first. The machine with damping circuits will be treated in the next chapter.

With the armature winding on the rotor and the field system salient with only one winding, the balanced three-phase machine is structurally that shown diagrammatically in Fig. 23 (p. 132). Here, however, the field winding will be designated F in place of the more general D.

The transient impedance matrix in terms of stationary reference axes can be written down in accordance with the rules stated on p. 74 and, of course, is the matrix of p. 72 from which those rules were derived. Hence

Z =		d	q	F
	d	$R_a+L_d p$	$L_q\dot{\theta}$	$L_{Fd}p$
	q	$-L_d\dot{\theta}$	$R_a+L_q p$	$-L_{Fd}\dot{\theta}$
	F	$L_{Fd}p$		$R_F+L_F p$

In this chapter we shall consider only balanced steady-state conditions and there is no need to include the zero-sequence axis.

The voltage equation is therefore

$$\begin{array}{c|c|} & \\ \text{d} & v_d \\ \text{q} & v_q \\ \text{F} & v_F \end{array} = \begin{array}{c|c|c|c|} & \text{d} & \text{q} & \text{F} \\ \text{d} & R_a+L_d p & L_q\dot{\theta} & L_{Fd}p \\ \text{q} & -L_d\dot{\theta} & R_a+L_q p & -L_{Fd}\dot{\theta} \\ \text{F} & L_{Fd}p & & R_F+L_F p \end{array} \begin{array}{c|c|} & \\ \text{d} & i^d \\ \text{q} & i^q \\ \text{F} & i^F \end{array}$$

The interpretation to be placed upon this equation depends upon the particular condition which is the subject of investigation.

The exciting function is the constant d.c. voltage V_F applied to the field winding, for steady-state performance therefore the p of the transient impedance matrix may be replaced by zero. Consequently i^d, i^q, i^F, v_d, and v_q will be constant d.c. currents and voltages and will be represented by capital letters. If at the same time $\dot{\theta}$ is replaced by ω, and $\omega L_d, \omega L_q, \omega L_{Fd}$ by X_d, X_q, X_{Fd} respectively, the voltage equation becomes

$$\begin{array}{c|c|} & \\ \text{d} & V_d \\ \text{q} & V_q \\ \text{F} & V_F \end{array} = \begin{array}{c|c|c|c|} & \text{d} & \text{q} & \text{F} \\ \text{d} & R_a & X_q & \\ \text{q} & -X_d & R_a & -X_{Fd} \\ \text{F} & & & R_F \end{array} \begin{array}{c|c|} & \\ \text{d} & I^d \\ \text{q} & I^q \\ \text{F} & I^F \end{array}$$

The determinant of the steady-state impedance matrix $\mathbf{Z}$ is

$$\Delta = R_F(R_a^2+X_d X_q)$$

The transpose of $\mathbf{Z}$ is

$$\mathbf{Z}_t = \begin{array}{c|c|c|c|} & \text{d} & \text{q} & \text{F} \\ \text{d} & R_a & -X_d & \\ \text{q} & X_q & R_a & \\ \text{F} & & -X_{Fd} & R_F \end{array}$$

Hence the inverse of **Z** is

$$\mathbf{Z}^{-1} = \frac{1}{\Delta}\;\begin{array}{c|c|c|c|}
 & \mathrm{d} & \mathrm{q} & \mathrm{F} \\ \hline
\mathrm{d} & R_a R_F & -X_q R_F & -X_q X_{Fd} \\ \hline
\mathrm{q} & R_F X_d & R_a R_F & R_a X_{Fd} \\ \hline
\mathrm{F} & & & (R_a^2 + X_d X_q) \\ \hline
\end{array}$$

We can now consider some particular cases of balanced steady-state operation.

Open-circuit Condition

On open circuit the armature currents i^a, i^b, i^c are zero. Consequently I^d and I^q are zero and the voltage equation reduces to

$$\begin{array}{c|c|}
\mathrm{d} & V_d \\ \hline
\mathrm{q} & V_q \\ \hline
\mathrm{F} & V_F \\
\end{array}
=
\begin{array}{c|c|c|c|}
 & \mathrm{d} & \mathrm{q} & \mathrm{F} \\ \hline
\mathrm{d} & R_a & X_q & \\ \hline
\mathrm{q} & -X_d & R_a & -X_{Fd} \\ \hline
\mathrm{F} & & & R_F \\ \hline
\end{array}
\;
\begin{array}{c|c|}
\mathrm{d} & 0 \\ \hline
\mathrm{q} & 0 \\ \hline
\mathrm{F} & I_F \\
\end{array}$$

$$=\begin{array}{c|c|}
\mathrm{d} & 0 \\ \hline
\mathrm{q} & -X_{Fd} I^F \\ \hline
\mathrm{F} & R_F I^F \\
\end{array}$$

from which it follows that $I^F = V_F/R_F$, $V_d = 0$ and $V_q = -X_{Fd} V_F/R_F$.

The actual open-circuit armature voltages are therefore given by

$$\begin{array}{c|c|}
\mathrm{a} & v_a \\ \hline
\mathrm{b} & v_b \\ \hline
\mathrm{c} & v_c \\
\end{array}
= \sqrt{\tfrac{2}{3}}\;
\begin{array}{c|c|c|c|}
 & \mathrm{d} & \mathrm{q} & \mathrm{o} \\ \hline
\mathrm{a} & \cos\theta & \sin\theta & 1/\sqrt{2} \\ \hline
\mathrm{b} & \cos(\theta - 2\pi/3) & \sin(\theta - 2\pi/3) & 1/\sqrt{2} \\ \hline
\mathrm{c} & \cos(\theta - 4\pi/3) & \sin(\theta - 4\pi/3) & 1/\sqrt{2} \\ \hline
\end{array}
\;
\begin{array}{c|c|}
\mathrm{d} & V_d \\ \hline
\mathrm{q} & V_q \\ \hline
\mathrm{o} & 0 \\
\end{array}$$

$$
= \sqrt{\tfrac{2}{3}}\;
\begin{array}{c|c|c|c|}
 & \text{d} & \text{q} & \text{o} \\ \hline
\text{a} & \cos\theta & \sin\theta & 1/\sqrt{2} \\ \hline
\text{b} & \cos(\theta-2\pi/3) & \sin(\theta-2\pi/3) & 1/\sqrt{2} \\ \hline
\text{c} & \cos(\theta-4\pi/3) & \sin(\theta-4\pi/3) & 1/\sqrt{2} \\ \hline
\end{array}
\;
\begin{array}{c|c|}
\hline
\text{d} & 0 \\ \hline
\text{q} & -X_{Fd}V_F/R_F \\ \hline
\text{o} & 0 \\ \hline
\end{array}
$$

$$
= \sqrt{\tfrac{2}{3}}\;
\begin{array}{c|c|}
\hline
\text{a} & -(X_{Fd}V_F/R_F)\sin\theta \\ \hline
\text{b} & -(X_{Fd}V_F/R_F)\sin(\theta-2\pi/3) \\ \hline
\text{c} & -(X_{Fd}V_F/R_F)\sin(\theta-4\pi/3) \\ \hline
\end{array}
$$

But since $\dot{\theta} = \omega$, integrating gives $\theta = (\omega t+\psi)$, where ψ is the angle between the a and the F axes at the independently chosen time zero.†

The open-circuit armature voltages are thus

$$
\begin{array}{c|c|}
\hline
\text{a} & v_a \\ \hline
\text{b} & v_b \\ \hline
\text{c} & v_c \\ \hline
\end{array}
= \sqrt{\tfrac{2}{3}}\;
\begin{array}{c|c|}
\hline
\text{a} & -(X_{Fd}V_F/R_F)\sin(\omega t+\psi) \\ \hline
\text{b} & -(X_{Fd}V_F/R_F)\sin(\omega t+\psi-2\pi/3) \\ \hline
\text{c} & -(X_{Fd}V_F/R_F)\sin(\omega t+\psi-4\pi/3) \\ \hline
\end{array}
$$

Short-circuit Condition

On short circuit $v_a = 0$, $v_b = 0$, and $v_c = 0$.

The transformed voltages V_d, V_q are therefore zero.

Since we have already determined the inverse of the steady-state impedance matrix, the currents are given by

$$
\begin{array}{c|c|}
\hline
\text{d} & I^d \\ \hline
\text{q} & I^q \\ \hline
\text{F} & I^F \\ \hline
\end{array}
= \frac{1}{\Delta}\;
\begin{array}{c|c|c|c|}
 & \text{d} & \text{q} & \text{F} \\ \hline
\text{d} & R_a R_F & -X_q R_F & -X_q X_{Fd} \\ \hline
\text{q} & R_F X_d & R_a R_F & R_a X_{Fd} \\ \hline
\text{F} & & & (R_a^2+X_d X_q) \\ \hline
\end{array}
\;
\begin{array}{c|c|}
\hline
\text{d} & V_d \\ \hline
\text{q} & V_q \\ \hline
\text{F} & V_F \\ \hline
\end{array}
$$

† Since ψ is here $(\theta-\omega t)$, it is opposite in sign to the ψ of p. 192. The choice may be made arbitrarily.

$$
= \frac{1}{\Delta}
\begin{array}{c|c|c|c|}
 & d & q & F \\
\hline
d & - & - & -X_q X_{Fd} \\
\hline
q & - & - & R_a X_{Fd} \\
\hline
F & - & - & (R_a^2 + X_d X_q) \\
\hline
\end{array}
\quad
\begin{array}{c|c|}
\hline
d & 0 \\
\hline
q & 0 \\
\hline
F & V_F \\
\hline
\end{array}
$$

In this last admittance matrix the first two columns multiply zeros of the voltage matrix and consequently they have been omitted to show that had they not been previously calculated, they would not have been calculated now.

The currents are therefore

$$
\begin{array}{c|c|}
\hline
d & I^d \\
\hline
q & I^q \\
\hline
F & I^F \\
\hline
\end{array}
= \frac{1}{\Delta}
\begin{array}{c|c|}
\hline
d & -X_q X_{Fd} V_F \\
\hline
q & R_a X_{Fd} V_F \\
\hline
F & (R_a^2 + X_d X_q) V_F \\
\hline
\end{array}
=
\begin{array}{c|c|}
\hline
d & -X_q X_{Fd} V_F / R_F (R_a^2 + X_d X_q) \\
\hline
q & R_a X_{Fd} V_F / R_F (R_a^2 + X_d X_q) \\
\hline
F & V_F / R_F \\
\hline
\end{array}
$$

The armature winding phase currents are then given by

$$
\begin{array}{c|c|}
\hline
a & i^a \\
\hline
b & i^b \\
\hline
c & i^c \\
\hline
\end{array}
= \sqrt{\tfrac{2}{3}}
\begin{array}{c|c|c|c|}
 & d & q & o \\
\hline
a & \cos\theta & \sin\theta & 1/\sqrt{2} \\
\hline
b & \cos(\theta - 2\pi/3) & \sin(\theta - 2\pi/3) & 1/\sqrt{2} \\
\hline
c & \cos(\theta - 4\pi/3) & \sin(\theta - 4\pi/3) & 1/\sqrt{2} \\
\hline
\end{array}
\quad
\begin{array}{c|c|}
\hline
a & I^d \\
\hline
b & I^q \\
\hline
c & I^o \\
\hline
\end{array}
$$

$$
= \sqrt{\tfrac{2}{3}}
\begin{array}{c|c|c|c|}
 & d & q & o \\
\hline
a & \cos\theta & \sin\theta & 1/\sqrt{2} \\
\hline
b & \cos(\theta - 2\pi/3) & \sin(\theta - 2\pi/3) & 1/\sqrt{2} \\
\hline
c & \cos(\theta - 4\pi/3) & \sin(\theta - 4\pi/3) & 1/\sqrt{2} \\
\hline
\end{array}
\quad
\begin{array}{c|c|}
\hline
d & -X_q X_{Fd} V_F / R_F (R_a^2 + X_d X_q) \\
\hline
q & R_a X_{Fd} V_F / R_F (R_a^2 + X_d X_q) \\
\hline
o & 0 \\
\hline
\end{array}
$$

$$
= \sqrt{\frac{2}{3}} \, \frac{X_{Fd} V_F}{R_F (R_a^2 + X_d X_q)}
\begin{array}{c|c|}
\hline
a & -X_q \cos\theta + R_a \sin\theta \\
\hline
b & -X_q \cos(\theta - 2\pi/3) + R_a \sin(\theta - 2\pi/3) \\
\hline
c & -X_q \cos(\theta - 4\pi/3) + R_a \sin(\theta - 4\pi/3) \\
\hline
\end{array}
$$

$$= \sqrt{\frac{2}{3}} \frac{X_{Fd}V_F\sqrt{(R_a^2+X_q^2)}}{R_F(R_a^2+X_dX_q)} \begin{array}{c|l|} & \\ a & \sin(\theta-\gamma) \\ b & \sin(\theta-\gamma-2\pi/3) \\ c & \sin(\theta-\gamma-4\pi/3) \end{array}$$

$$= \sqrt{\frac{2}{3}} \frac{X_{Fd}V_F\sqrt{(R_a^2+X_q^2)}}{R_F(R_a^2+X_dX_q)} \begin{array}{c|l|} & \\ a & \sin(\omega t+\psi-\gamma) \\ b & \sin(\omega t+\psi-\gamma-2\pi/3) \\ c & \sin(\omega t+\psi-\gamma-4\pi/3) \end{array}$$

where $\sin\gamma = X_q/\sqrt{(R_a^2+X_q^2)}$ and $\cos\gamma = R_a/\sqrt{(R_a^2+X_q^2)}$ and ψ is the angle which the a axis makes with the F axis at time zero.

If R_a is sufficiently small for its square to be negligible compared to X_q^2 and X_dX_q, which is usually so, the peak magnitude of the armature phase currents is $\sqrt{\frac{2}{3}}X_{Fd}V_F/(X_dR_F)$. The peak magnitude $\hat{E}$ of the open-circuit phase voltage for the same field winding voltage V_F is $\sqrt{\frac{2}{3}}X_{Fd}V_F/R_F$. The r.m.s. short-circuit current is thus E/X_d and, regarded as a generated current, i.e. changed in sign, it lags behind E by the angle γ which is $\pi/2$ when R_a is wholly negligible.

The reactances X_d and X_q of the d and q armature circuits are known as the "direct-axis synchronous reactance" and the "quadrature-axis synchronous reactance" respectively.

On Balanced Load as a Generator

Let the load be represented by an impedance Z_L consisting of resistance R_L and inductance L_L in series with each phase winding. The process of interconnection must be performed in terms of the actual winding axes. Let the impedance matrix of the machine, given on p. 136, be expressed as

	am	bm	cm	Fm
am	Z_{aa}	Z_{ab}	Z_{ac}	Z_{aF}
bm	Z_{ba}	Z_{bb}	Z_{bc}	Z_{bF}
cm	Z_{ca}	Z_{cb}	Z_{cc}	Z_{cF}
Fm	Z_{Fa}	Z_{Fb}	Z_{Fc}	Z_{FF}

The load impedance matrix may be written

	aL	bL	cL
aL	Z_L		
bL		Z_L	
cL			Z_L

The combined impedance matrix before connection is therefore

$\mathbf{Z} =$

	am	bm	cm	Fm	aL	bL	cL
am	Z_{aa}	Z_{ab}	Z_{ac}	Z_{aF}			
bm	Z_{ba}	Z_{bb}	Z_{bc}	Z_{bF}			
cm	Z_{ca}	Z_{cb}	Z_{cc}	Z_{cF}			
Fm	Z_{Fa}	Z_{Fb}	Z_{Fc}	Z_{FF}			
aL					Z_L		
bL						Z_L	
cL							Z_L

Since the am, bm, cm windings are connected to, and have the same currents as, the aL, bL, cL loads respectively, the connection matrix is

$\mathbf{C}_1 =$

	a	b	c	F
am	1			
bm		1		
cm			1	
Fm				1
aL	1			
bL		1		
cL			1	

The transformation $\mathbf{C}_{1t}\mathbf{Z}\mathbf{C}_1$ leads to

$\mathbf{Z}' =$

	a	b	c	F
a	$Z_{aa}+Z_L$	Z_{ab}	Z_{ac}	Z_{aF}
b	Z_{ba}	$Z_{bb}+Z_L$	Z_{bc}	Z_{bF}
c	Z_{ca}	Z_{cb}	$Z_{cc}+Z_L$	Z_{cF}
F	Z_{Fa}	Z_{Fb}	Z_{Fc}	Z_{FF}

$$= \mathbf{Z}_m + \mathbf{Z}_L$$

where $\mathbf{Z}_m$ is the machine impedance matrix and

$\mathbf{Z}_L =$

	a	b	c	F
a	Z_L			
b		Z_L		
c			Z_L	
F				

$=$

	a	b	c	F
a	R_L+L_Lp			
b		R_L+L_Lp		
c			R_L+L_Lp	
F				

If now the a, b, c axes are transformed to d, q, o axes by

$\mathbf{C}_2 = \sqrt{\frac{2}{3}}$

	d	q	o	F
a	$\cos\theta$	$\sin\theta$	$1/\sqrt{2}$	
b	$\cos(\theta-2\pi/3)$	$\sin(\theta-2\pi/3)$	$1/\sqrt{2}$	
c	$\cos(\theta-4\pi/3)$	$\sin(\theta-4\pi/3)$	$1/\sqrt{2}$	
F				$\sqrt{\frac{3}{2}}$

$$\mathbf{Z}'' = \mathbf{C}_{2t}\mathbf{Z}'\mathbf{C}_2 = \mathbf{C}_{2t}\mathbf{Z}_m\mathbf{C}_2 + \mathbf{C}_{2t}\mathbf{Z}_L\mathbf{C}_2$$

Now $\mathbf{C}_{2t}\mathbf{Z}_m\mathbf{C}_2$ is the usual machine impedance matrix in d, q, F axes with the addition of the zero-sequence row and column, namely

	d	q	o	F
d	$R_a+L_d p$	$L_q\dot{\theta}$		$M_{Fd}p$
q	$-L_d\dot{\theta}$	$R_a+L_q p$		$-M_{Fd}\dot{\theta}$
o			$R_a+L_o p$	
F	$M_{Fd}p$			$R_F+L_F p$

$\mathbf{C}_{2t}\mathbf{Z}_L\mathbf{C}_2$ has to be evaluated by first forming the product $\mathbf{Z}_L\mathbf{C}_2$, performing the differentiations in a manner similar to that on p. 72, and, lastly, pre-multiplying by $\mathbf{C}_{2t}$. It is

	d	q	o	F
d	$R_L+L_L p$	$L_L\dot{\theta}$		
q	$-L_L\dot{\theta}$	$R_L+L_L p$		
o				
F				

Hence the matrix representing machine and connected load in d, q, o, F axes is

	d	q	o	F
d	$(R_a+R_L)+(L_d+L_L)p$	$(L_q+L_L)\dot{\theta}$		$M_{Fd}p$
q	$-(L_d+L_L)\dot{\theta}$	$(R_a+R_L)+(L_q+L_L)p$		$-M_{Fd}\dot{\theta}$
o			$R_a+L_o p$	
F	$M_{Fd}p$			$R_F+L_F p$

If the field voltage v_F is a constant (d.c.) voltage V_F, the steady-state impedance matrix is obtained by replacing p by zero, and if, at the same time, we write ω for $\dot{\theta}$ and X for ωL and ωM and also omit the zero-sequence row because the system is balanced and there will be no zero-sequence currents,

$$
\mathbf{Z}'' = \begin{array}{c|c|c|c|}
 & \text{d} & \text{q} & \text{F} \\
\hline
\text{d} & R_a+R_L & X_q+X_L & \\
\hline
\text{q} & -X_d-X_L & R_a+R_L & -X_{Fd} \\
\hline
\text{F} & & & R_F \\
\hline
\end{array}
$$

The voltages of the a, b, c circuits, which are closed, are zero. If this is not obvious, it should be remembered that the terminal voltages of the machine windings and the loads connected to them are equal and opposite when summed around the mesh. If it is still not obvious, the voltage transformations must be done in their correct order. The result is, of course, that the voltages v_d and v_q are both zero. The form of the voltage equation is thus the same as in the short-circuit case already considered except that

$$
\begin{aligned}
R_a &\text{ is replaced by } (R_a+R_L)\\
X_d &\text{ is replaced by } (X_d+X_L)\\
X_q &\text{ is replaced by } (X_q+X_L)
\end{aligned}
$$

The currents can therefore be written down without more ado as

$$
\begin{array}{c|c|}
\hline
\text{d} & I^d \\
\hline
\text{q} & I^q \\
\hline
\text{F} & I^F \\
\hline
\end{array}
=
\begin{array}{c|c|}
\hline
\text{d} & -(X_q+X_L)X_{Fd}V_F/R_F\{(R_a+R_L)^2+(X_d+X_L)(X_q+X_L)\} \\
\hline
\text{q} & (R_a+R_L)X_{Fd}V_F/R_F\{(R_a+R_L)^2+(X_d+X_L)(X_d+X_L)\} \\
\hline
\text{F} & V_F/R_F \\
\hline
\end{array}
$$

Then

$$
\begin{aligned}
i^a &= \sqrt{\tfrac{2}{3}}\{\cos\theta I^d + \sin\theta I^q + (1/\sqrt{2})I^o\}\\
&= \sqrt{\frac{2}{3}}\,\frac{-(X_q+X_L)\cos\theta+(R_a+R_L)\sin\theta}{(R_a+R_L)^2+(X_d+X_L)(X_q+X_L)}\,\frac{X_{Fd}V_F}{R_F}\\
&= \sqrt{\frac{2}{3}}\,\frac{\sqrt{\{(R_a+R_L)^2+(X_q+X_L)^2\}}}{(R_a+R_L)^2+(X_d+X_L)(X_q+X_L)}\,\frac{X_{Fd}V_F}{R_F}\sin(\theta-\chi)
\end{aligned}
$$

where $\tan\chi = (X_q+X_L)/(R_a+R_L)$.

It will be noted that, again, $\sqrt{\frac{2}{3}}X_{Fd}V_F/R_F$ is the e.m.f. which would be produced on open circuit by the same excitation if there were no saturation change. Whilst this equation gives a relationship between the load current and the field current, in practice the excitation required by a specified load would be greater than that given by this equation by an amount dependent upon the degree of saturation on the field, i.e. F, axis. The relevant voltage is therefore not the true e.m.f. itself, but its component in phase with the open-circuit voltage.

The a-phase voltage on open circuit is given on p. 200 as $-\sqrt{\frac{2}{3}}(X_{Fd}V_F/R_F)\sin\theta$. The a-phase current, regarded as a generated current (i.e. reversed in sign) thus lags behind the open-circuit e.m.f. by an angle χ. The true e.m.f. is $\{(R_a+R_L)+j(x_a+X_L)\}I^a$ and thus leads the (generated) current by an angle ζ such that $\tan\zeta = (x_a+X_L)/(R_a+R_L)$. The true e.m.f. thus lags behind the open-circuit e.m.f. by an angle $(\chi-\zeta)$, the cosine of which is

$$\cos\chi\cos\zeta+\sin\chi\sin\zeta$$
$$= \frac{(X_q+X_L)(x_a+X_L)+(R_a+R_L)(R_a+R_L)}{\sqrt{\{(R_a+R_L)^2+(X_q+X_L)^2\}}\sqrt{\{(R_a+R_L)^2+(x_a+X_L)^2\}}}$$

The component of the true e.m.f. in phase with the open-circuit e.m.f. is therefore of magnitude

$$I^a\sqrt{\{(R_a+R_L)^2+(x_a+X_L)^2\}}$$
$$\times\frac{(R_a+R_L)^2+(x_a+X_L)(X_q+X_L)}{\sqrt{\{(R_a+R_L)^2+(X_q+X_L)^2\}}\sqrt{\{(R_a+R_L)^2+(x_a+X_L)^2\}}}$$
$$= \frac{(R_a+R_L)^2+(x_a+X_L)(X_q+X_L)}{\sqrt{(R_a+R_L)^2+(X_q+X_L)^2\}}}I^a$$

The component of field current required for saturation can be found from the graph of open-circuit voltage (or of X_{Fd}) against field current using this component of the true e.m.f. in a manner similar to that used for the machine with a uniform air-gap on p. 194.

Phasor Diagram

It must be remembered that the equations for V_d and V_q are algebraic equations not phasor equations. V_d, V_q, V_F, I^d, I^q, I^F, v_a, v_b, v_c, i^a, i^b, and i^c are all instantaneous values of which only the last six are not constant (i.e. d.c.) under steady-state conditions. There is no question, therefore, of phase relationships in equations expressed in terms of the reference axes d, q, F. Phase relationships arise, however in the equations for i^a, i^b, i^c, v_a, v_b, and v_c from the introduction of $\cos\theta$ and $\sin\theta$ terms by the transformation matrix, since θ is a function of time of the form $(\omega t+\psi)$.

Again, I^d and I^q, being d.c., are not identical with the I_d and I_q of conventional theory, which are a.c. components of the armature current I. These quantities are, however, related, and it is necessary to determine this relationship.

Assuming that the balanced armature currents are defined as

$$\begin{array}{c|c|} a & i^a \\ b & i^b \\ c & i^c \end{array} = \begin{array}{c|c|} a & \hat{I}\sin\omega t \\ b & \hat{I}\sin(\omega t-2\pi/3) \\ c & \hat{I}\sin(\omega t-4\pi/3) \end{array}$$

we have

$$\begin{array}{c|c|} d & I^d \\ q & I^q \\ o & I^o \end{array} = \sqrt{\tfrac{2}{3}}\; \begin{array}{c|c|c|c|} & a & b & c \\ \hline d & \cos\theta & \cos(\theta-2\pi/3) & \cos(\theta-4\pi/3) \\ q & \sin\theta & \sin(\theta-2\pi/3) & \sin(\theta-4\pi/3) \\ o & 1/\sqrt{2} & 1/\sqrt{2} & 1/\sqrt{2} \end{array}\; \begin{array}{c|c|} a & \hat{I}\sin\omega t \\ b & \hat{I}\sin(\omega t-2\pi/3) \\ c & \hat{I}\sin(\omega t-4\pi/3) \end{array}$$

$$= \sqrt{\tfrac{2}{3}}\hat{I}\; \begin{array}{c|c|} d & \sin\omega t\cos\theta+\sin(\omega t-2\pi/3)\cos(\theta-2\pi/3) \\ & +\sin(\omega t-4\pi/3)\cos(\theta-4\pi/3) \\ \hline q & \sin\omega t\sin\theta+\sin(\omega t-2\pi/3)\sin(\theta-2\pi/3) \\ & +\sin(\omega t-4\pi/3)\sin(\theta-4\pi/3) \\ \hline o & (1/\sqrt{2})\{\sin\omega t+\sin(\omega t-2\pi/3) \\ & +\sin(\omega t-4\pi/3)\} \end{array}$$

$$= \sqrt{\tfrac{2}{3}}\hat{I}\ \begin{array}{c|c|} \text{d} & \tfrac{3}{2}\sin(\omega t-\theta) \\ \hline \text{q} & \tfrac{3}{2}\cos(\omega t-\theta) \\ \hline \text{o} & 0 \end{array} = \sqrt{\tfrac{3}{2}}\hat{I}\ \begin{array}{c|c|} \text{d} & -\sin\psi \\ \hline \text{q} & \cos\psi \\ \hline \text{o} & 0 \end{array}\ \dagger$$

In conventional theory, the direct- and quadrature-axis components I_d and I_q of I are alternating currents defined by

$$I_d = I\sin\psi \quad \text{lagging } (\pi/2-\psi) \text{ behind } I$$
$$I_q = I\cos\psi \quad \text{leading } I \text{ by } \psi$$

whereas I^d and I^q are direct currents equal to $-\sqrt{\tfrac{3}{2}}\hat{I}\sin\psi$ and $\sqrt{\tfrac{3}{2}}\hat{I}\cos\psi$ respectively.

If I is the a-phase current which has the instantaneous value $\hat{I}\sin\omega t$, and i_d, i_q are the instantaneous values of I_d and I_q respectively,

$$\begin{aligned} i_d &= \hat{I}\sin\psi\sin\{\omega t-(\pi/2-\psi)\} = -\hat{I}\sin\psi\cos(\omega t+\psi) \\ &\qquad\qquad\qquad\qquad\qquad\qquad = -\hat{I}\sin\psi\cos\theta \\ i_q &= \hat{I}\cos\psi\sin(\omega t+\psi) \qquad\quad = \hat{I}\cos\psi\sin\theta \end{aligned}$$

Now $v_a = \sqrt{\tfrac{2}{3}}\{\cos\theta\, V_d+\sin\theta\, V_q\}$ and with the values of V_d and V_q given by the voltage equation on p. 198,

$$\begin{aligned} v_a &= \sqrt{\tfrac{2}{3}}\cos\theta\{R_a I^d+X_q I^q\}+\sqrt{\tfrac{2}{3}}\sin\theta\{-X_d I^d+R_a I^q-X_{Fd}I^F\} \\ &= \sqrt{\tfrac{2}{3}}\{R_a I^d\cos\theta+X_q I^q\cos\theta-X_d I^d\sin\theta+R_a I^q\sin\theta \\ &\quad -X_{Fd}I^F\sin\theta\} \\ &= -R_a\hat{I}\sin\psi\cos\theta+X_q\hat{I}\cos\psi\cos\theta+X_d\hat{I}\sin\psi\sin\theta \\ &\quad +R_a\hat{I}\cos\psi\sin\theta-\sqrt{\tfrac{2}{3}}X_{Fd}I^F\sin\theta \\ &= R_a\{-\hat{I}\sin\psi\cos\theta\}+X_q\{\hat{I}\cos\psi\sin(\theta+\pi/2)\} \\ &\quad +X_d\{-\hat{I}\sin\psi\cos(\theta+\pi/2)\}+R_a\{\hat{I}\cos\psi\sin\theta\} \\ &\quad -\sqrt{\tfrac{2}{3}}X_{Fd}I^F\sin\theta \end{aligned}$$

† The armature currents and ψ have been so defined here that the latter is equal to the angle between E and I to conform with the usual symbol of classical theory for this angle.

In complex form this is

$$V = R_a I_d + jX_q I_q + jX_d I_d + R_a I_q + E$$

where E is the e.m.f. with an instantaneous value $-\sqrt{\frac{2}{3}}X_{Fd}I^F \sin\theta$ as on open circuit. This is, of course, the "motor" equation, the "generator" equation is obtained by writing $-I$ for I, as

$$E = V + R_a I_d + R_a I_q + jX_d I_d + jX_q I_q$$

which leads to the well-known phasor diagram Fig. 42.†

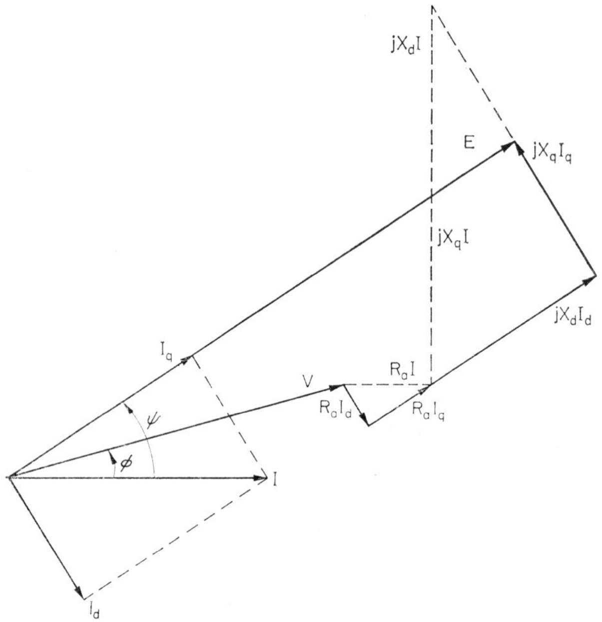

FIG. 42. Phasor diagram of salient-pole synchronous machine.

† Compare ref. 3, p. 483.

Torque

The coefficients of $\dot{\theta}$ in the impedance matrix in terms of the stationary reference axes given on p. 197 define the torque matrix as

$$\mathbf{G} = \begin{array}{c} \\ \mathrm{d} \\ \mathrm{q} \\ \mathrm{F} \end{array} \begin{array}{ccc} \mathrm{d} & \mathrm{q} & \mathrm{F} \\ \hline & L_{\mathrm{q}} & \\ -L_{\mathrm{d}} & & -L_{\mathrm{Fd}} \\ & & \end{array}$$

In terms of reactances the two-pole torque is

$$T = \frac{1}{\omega} \begin{array}{ccc} \mathrm{d} & \mathrm{q} & \mathrm{F} \\ \hline i^{\mathrm{d}} & i^{\mathrm{q}} & i^{\mathrm{F}} \end{array} \quad \begin{array}{c} \\ \mathrm{d} \\ \mathrm{q} \\ \mathrm{F} \end{array} \begin{array}{ccc} \mathrm{d} & \mathrm{q} & \mathrm{F} \\ \hline & X_{\mathrm{q}} & \\ -X_{\mathrm{d}} & & -X_{\mathrm{Fd}} \\ & & \end{array} \quad \begin{array}{c} \mathrm{d} \\ \mathrm{q} \\ \mathrm{F} \end{array} \begin{array}{c} i^{\mathrm{d}} \\ i^{\mathrm{q}} \\ i^{\mathrm{F}} \end{array}$$

$$= -\frac{1}{\omega}\{i^{\mathrm{d}} i^{\mathrm{q}}(X_{\mathrm{d}} - X_{\mathrm{q}}) + i^{\mathrm{q}} i^{\mathrm{F}} X_{\mathrm{Fd}}\}$$

If the armature terminal voltage and the excitation are specified, it is possible to determine an expression for the torque in terms of $(\theta - \omega t) = \psi$.

If the terminal voltages are defined as†

$$\begin{array}{c} \mathrm{a} \\ \mathrm{b} \\ \mathrm{c} \end{array} \begin{array}{c} v_{\mathrm{a}} \\ v_{\mathrm{b}} \\ v_{\mathrm{c}} \end{array} = \begin{array}{c} \mathrm{a} \\ \mathrm{b} \\ \mathrm{c} \end{array} \begin{array}{c} -\hat{V} \sin \omega t \\ -\hat{V} \sin(\omega t - 2\pi/3) \\ -\hat{V} \sin(\omega t - 4\pi/3) \end{array} \quad \text{then}$$

† These particular voltages have been chosen because they are in phase with the open-circuit terminal voltages (given on p. 200) if the angle ψ is zero. The significance of this will be explained later.

$$\begin{array}{c|c|}\hline \mathrm{d} & v_{\mathrm{d}} \\ \hline \mathrm{q} & v_{\mathrm{q}} \\ \hline \mathrm{o} & v_{\mathrm{o}} \\ \hline \mathrm{F} & v_{\mathrm{F}} \\ \hline \end{array} = \sqrt{\tfrac{2}{3}} \begin{array}{c|c|c|c|c|} & \mathrm{a} & \mathrm{b} & \mathrm{c} & \mathrm{F} \\ \hline \mathrm{d} & \cos\theta & \cos(\theta-2\pi/3) & \cos(\theta-4\pi/3) & \\ \hline \mathrm{q} & \sin\theta & \sin(\theta-2\pi/3) & \sin(\theta-4\pi/3) & \\ \hline \mathrm{o} & 1/\sqrt{2} & 1/\sqrt{2} & 1/\sqrt{2} & \\ \hline \mathrm{F} & & & & \sqrt{\tfrac{3}{2}} \\ \hline \end{array} \begin{array}{c|c|}\hline \mathrm{a} & -\hat{V}\sin\omega t \\ \hline \mathrm{b} & -\hat{V}\sin(\omega t-2\pi/3) \\ \hline \mathrm{c} & -\hat{V}\sin(\omega t-4\pi/3) \\ \hline \mathrm{F} & V_{\mathrm{F}} \\ \hline \end{array}$$

$$= \sqrt{\tfrac{2}{3}} \begin{array}{c|c|}\hline \mathrm{d} & -\hat{V}\{\cos\theta\sin\omega t+\cos(\theta-2\pi/3)\sin(\omega t-2\pi/3)+\cos(\theta-4\pi/3)\sin(\omega t-4\pi/3)\} \\ \hline \mathrm{q} & -\hat{V}\{\sin\theta\sin\omega t+\sin(\theta-2\pi/3)\sin(\omega t-2\pi/3)+\sin(\theta-4\pi/3)\sin(\omega t-4\pi/3)\} \\ \hline \mathrm{o} & -(1/\sqrt{2})\,\hat{V}\{\sin\omega t+\sin(\omega t-2\pi/3)+\sin(\omega t-4\pi/3)\} \\ \hline \mathrm{F} & \sqrt{\tfrac{3}{2}}V_{\mathrm{F}} \\ \hline \end{array}$$

$$= \sqrt{\tfrac{2}{3}} \begin{array}{c|c|}\hline \mathrm{d} & -\hat{V}\tfrac{3}{2}\sin(\omega t-\theta) \\ \hline \mathrm{q} & -\hat{V}\tfrac{3}{2}\cos(\omega t-\theta) \\ \hline \mathrm{o} & \\ \hline \mathrm{F} & \sqrt{\tfrac{3}{2}}V_{\mathrm{F}} \\ \hline \end{array}$$

$$
= \begin{array}{c|c|}
\mathrm{d} & \sqrt{\tfrac{3}{2}}\hat{V}\sin\psi \\
\mathrm{q} & -\sqrt{\tfrac{3}{2}}\hat{V}\cos\psi \\
\mathrm{o} & \\
\mathrm{F} & V_{\mathrm{F}}
\end{array}
$$

The currents are therefore given by

$$
\begin{array}{c|c|}
\mathrm{d} & I^{\mathrm{d}} \\
\mathrm{q} & I^{\mathrm{q}} \\
\mathrm{F} & I^{\mathrm{F}}
\end{array}
$$

$$
= \frac{1}{R_{\mathrm{F}}(R_{\mathrm{a}}^{2}+X_{\mathrm{d}}X_{\mathrm{q}})}
\begin{array}{c|c|c|c|}
 & \mathrm{d} & \mathrm{q} & \mathrm{F} \\
\mathrm{d} & R_{\mathrm{a}}R_{\mathrm{F}} & -X_{\mathrm{q}}R_{\mathrm{F}} & -X_{\mathrm{q}}X_{\mathrm{Fd}} \\
\mathrm{q} & R_{\mathrm{F}}X_{\mathrm{d}} & R_{\mathrm{a}}R_{\mathrm{F}} & R_{\mathrm{a}}X_{\mathrm{Fd}} \\
\mathrm{F} & & & (R_{\mathrm{a}}^{2}+X_{\mathrm{d}}X_{\mathrm{q}})
\end{array}
\;
\begin{array}{c|c|}
\mathrm{d} & \sqrt{\tfrac{3}{2}}\hat{V}\sin\psi \\
\mathrm{q} & -\sqrt{\tfrac{3}{2}}\hat{V}\cos\psi \\
\mathrm{F} & V_{\mathrm{F}}
\end{array}
$$

$$
= \frac{1}{R_{\mathrm{F}}(R_{\mathrm{a}}^{2}+X_{\mathrm{d}}X_{\mathrm{q}})}
\begin{array}{c|c|}
\mathrm{d} & \sqrt{\tfrac{3}{2}}\,\hat{V}R_{\mathrm{F}}(R_{\mathrm{a}}\sin\psi+X_{\mathrm{q}}\cos\psi)-V_{\mathrm{F}}X_{\mathrm{q}}X_{\mathrm{Fd}} \\
\mathrm{q} & \sqrt{\tfrac{3}{2}}\,\hat{V}R_{\mathrm{F}}(X_{\mathrm{d}}\sin\psi-R_{\mathrm{a}}\cos\psi)+V_{\mathrm{F}}R_{\mathrm{a}}X_{\mathrm{Fd}} \\
\mathrm{F} & V_{\mathrm{F}}(R_{\mathrm{a}}^{2}+X_{\mathrm{d}}X_{\mathrm{q}})
\end{array}
$$

This expression can be substituted in the torque expression; here, however, the usual assumption that R_a is small compared to X_d and X_q will be made for simplicity. In this case the currents are

$$
\begin{array}{c|c|}
\mathrm{d} & I^{\mathrm{d}} \\
\mathrm{q} & I^{\mathrm{q}} \\
\mathrm{F} & I^{\mathrm{F}}
\end{array}
=
\begin{array}{c|c|}
\mathrm{d} & \sqrt{\tfrac{3}{2}}\hat{V}\cos\psi/X_{\mathrm{d}}-(X_{\mathrm{Fd}}V_{\mathrm{F}}/R_{\mathrm{F}})/X_{\mathrm{d}} \\
\mathrm{q} & \sqrt{\tfrac{3}{2}}\hat{V}\sin\psi/X_{\mathrm{q}} \\
\mathrm{F} & V_{\mathrm{F}}/R_{\mathrm{F}}
\end{array}
$$

If these values be substituted in the torque expression, we get the two-pole torque as

$$T = -\frac{1}{\omega}\left[I^{d}I^{q}(X_{d}-X_{q})+I^{q}I^{F}X_{Fd}\right]$$

$$= -\frac{1}{\omega}\left[(X_{d}-X_{q})\left\{\frac{3}{2}\hat{V}^{2}\sin\psi\cos\psi/X_{d}X_{q}\right.\right.$$

$$\left.-\sqrt{\frac{3}{2}}\hat{V}\sin\psi(X_{Fd}V_{F}/R_{F})/X_{d}X_{q}\right\}$$

$$\left.+X_{Fd}\sqrt{\frac{3}{2}}\hat{V}\sin\psi(V_{F}/R_{F})/X_{q}\right]$$

$$= -\frac{1}{\omega}\left[(X_{d}-X_{q})\left\{\frac{3}{2}\hat{V}^{2}\sin\psi\cos\psi/X_{d}X_{q}\right\}\right.$$

$$\left.+\sqrt{\frac{3}{2}}\hat{V}\sin\psi(X_{Fd}V_{F}/R_{F})/X_{d}\right]$$

$$= -\frac{1}{\omega}\,\frac{3}{2}\left[\hat{V}\left\{\sqrt{\frac{2}{3}}X_{Fd}V_{F}/R_{F}\right\}\sin\psi/X_{d}+\hat{V}^{2}\frac{X_{d}-X_{q}}{2X_{d}X_{q}}\sin 2\psi\right]$$

$$= -\frac{1}{\omega}\,\frac{3}{2}\left[\frac{\hat{V}\hat{E}}{X_{d}}\sin\psi+\frac{\hat{V}^{2}}{2}\left(\frac{1}{X_{q}}-\frac{1}{X_{d}}\right)\sin 2\psi\right]$$

$$= -\frac{3}{\omega}\left[\frac{VE}{X_{d}}\sin\psi+\frac{V^{2}}{2}\left(\frac{1}{X_{q}}-\frac{1}{X_{d}}\right)\sin 2\psi\right]^{\dagger}$$

where V and E are r.m.s. values. $E = (1/\sqrt{3})X_{Fd}V_{F}/R_{F}$ is the e.m.f. which the excitation would produce on open circuit if there were no changes of saturation.

The armature terminal voltage was defined on p. 211 so that it lagged behind the open-circuit e.m.f., given on p. 200, by the angle $\psi = \theta - \omega t$. It follows that, when $\psi = 0$, the terminal voltage and the e.m.f. are in phase with one another. If resistance is neglected, any current differs in phase from the voltages by $\pi/2$, the power is accordingly zero and so also is the torque. The angle ψ is thus both the angle between the e.m.f. and the terminal voltage and also the angle by which the arma-

† Compare ref. 3, p. 506.

ture is displaced from the position it would have occupied at any given time had the machine not been loaded. The conditions have been deliberately specified so that ψ is here the load or torque angle usually denoted by δ. The torque expression above is thus identical with that of classical theory except that ψ appears in place of δ, that there is a factor 3 which occurs because this expression represents total torque and not torque per phase, and that it is negative. The torque is negative when ψ is positive because $\theta > \omega t$ is inherently a generating condition, yet the machine is here analysed as if it were a motor. For a motoring condition ψ would have to be negative if it is defined, as here, as $\theta - \omega t$.

If the machine is not salient, $X_q = X_d$ and the second term of the torque expression vanishes, leaving

$$T = -\frac{3}{\omega}\frac{VE}{X_d}\sin\psi$$

In this case, however, the result is not unduly complicated when the armature resistance is not neglected.

The currents in a machine with a uniform air-gap could be obtained from the equation on p. 190, but can now be obtained by putting $X_q = X_d$ in the expression in p. 213.

$$\begin{array}{c|c|} \mathrm{d} & I^d \\ \hline \mathrm{q} & I^q \\ \hline \mathrm{F} & I^F \end{array} = \frac{1}{R_F(R_a^2+X_d^2)} \begin{array}{c|c|} \mathrm{d} & \sqrt{\tfrac{3}{2}}\,\hat{V}R_F(R_a\sin\psi+X_d\cos\psi)-V_FX_dX_{Fd} \\ \hline \mathrm{q} & \sqrt{\tfrac{3}{2}}\,\hat{V}R_F(X_d\sin\psi-R_a\cos\psi)+V_FR_aX_{Fd} \\ \hline \mathrm{F} & V_F(R_a^2+X_d^2) \end{array}$$

If $(R_a^2+X_d^2)$ is replaced by Z_d^2, R_a/Z_d by $\sin\gamma$ and X_d/Z_d by $\cos\gamma$,

$$\begin{array}{c|c|} \mathrm{d} & I^d \\ \hline \mathrm{q} & I^q \\ \hline \mathrm{F} & I^F \end{array} = \begin{array}{c|c|} \mathrm{d} & \sqrt{\tfrac{3}{2}}\hat{V}\cos(\psi-\gamma)/Z_d-(X_{Fd}V_F/R_F)X_d/Z_d^2 \\ \hline \mathrm{q} & \sqrt{\tfrac{3}{2}}\hat{V}\sin(\psi-\gamma)/Z_d+(X_{Fd}V_F/R_F)R_a/Z_d^2 \\ \hline \mathrm{F} & V_F/R_F \end{array}$$

and since $X_q = X_d$, the two-pole torque is

$$T = -\frac{1}{\omega} I^q I^F X_{Fd}$$

$$= -\frac{1}{\omega}\left[\sqrt{\frac{3}{2}}\hat{V}(X_{Fd}V_F/R_F)\sin(\psi-\gamma)/Z_d + (X_{Fd}V_F/R_F)^2 R_a/Z_d^2\right]$$

$$= -\frac{1}{\omega}\frac{3}{2}\left[\frac{\hat{V}\hat{E}}{Z_d}\sin(\psi-\gamma)+\frac{\hat{E}^2 R_a}{Z_d^2}\right]$$

$$= -\frac{3}{\omega}\left[\frac{VE}{Z_d}\sin(\psi-\gamma)+\frac{E^2 R_a}{Z_d^2}\right]^{\dagger}$$

Parameters

The open-circuit and short-circuit tests considered for the uniform air-gap machine lead to values of the same parameters X_{Fd} and X_d of the salient-pole machine. For the calculation of steady-state conditions, however, the quadrature-axis synchronous reactance X_q is also required.

If the machine is separately driven at a speed ω with the field winding unexcited and balanced voltage of positive phase-sequence applied to the armature, irrespective of whether the field winding is open-circuited or short-circuited,

$$I^d = \frac{1}{\Delta}\{R_a V_d - X_q V_q\}$$

$$I^q = \frac{1}{\Delta}\{X_d V_d + R_a V_q\}$$

where $\Delta = (R_a^2 + X_d X_q)$

If the voltage applied to a three-phase machine is

$$\begin{array}{c|c|}
a & v_a \\ \hline
b & v_b \\ \hline
c & v_c
\end{array}
=
\begin{array}{c|c|}
a & \hat{V}\cos\omega t \\ \hline
b & \hat{V}\cos(\omega t - 2\pi/3) \\ \hline
c & \hat{V}\cos(\omega t - 4\pi/3)
\end{array}$$

† Compare ref. 3, p. 497.

$$
\begin{array}{c|c|}
 & \\
\hline
\mathrm{d} & v_d \\
\hline
\mathrm{q} & v_q \\
\hline
\end{array}
= \sqrt{\tfrac{2}{3}}
\begin{array}{c|c|c|c|}
 & \mathrm{a} & \mathrm{b} & \mathrm{c} \\
\hline
\mathrm{d} & \cos\theta & \cos(\theta-2\pi/3) & \cos(\theta-4\pi/3) \\
\hline
\mathrm{q} & \sin\theta & \sin(\theta-2\pi/3) & \sin(\theta-4\pi/3) \\
\hline
\end{array}
\begin{array}{c|c|}
\hline
\mathrm{a} & \hat{V}\cos\omega t \\
\hline
\mathrm{b} & \hat{V}\cos(\omega t-2\pi/3) \\
\hline
\mathrm{c} & \hat{V}\cos(\omega t-4\pi/3) \\
\hline
\end{array}
$$

$$
= \sqrt{\tfrac{2}{3}}
\begin{array}{c|c|}
\hline
\mathrm{d} & \hat{V}\{\cos\theta\cos\omega t+\cos(\theta-2\pi/3)\cos(\omega t-2\pi/3) \\
 & +\cos(\theta-4\pi/3)\cos(\omega t-4\pi/3)\} \\
\hline
\mathrm{q} & \hat{V}\{\sin\theta\cos\omega t+\sin(\theta-2\pi/3)\cos(\omega t-2\pi/3) \\
 & +\sin(\theta-4\pi/3)\cos(\omega t-4\pi/3)\} \\
\hline
\end{array}
$$

$$
= \sqrt{\tfrac{2}{3}}
\begin{array}{c|c|}
\hline
\mathrm{d} & \hat{V}(\tfrac{3}{2})\cos(\theta-\omega t) \\
\hline
\mathrm{q} & \hat{V}(\tfrac{3}{2})\sin(\theta-\omega t) \\
\hline
\end{array}
= \sqrt{\tfrac{3}{2}}
\begin{array}{c|c|}
\hline
\mathrm{d} & \hat{V}\cos\psi \\
\hline
\mathrm{q} & \hat{V}\sin\psi \\
\hline
\end{array}
$$

where $\psi = \theta-\omega t$.

Then
$$
\begin{aligned}
i^a &= \sqrt{\tfrac{2}{3}}\{i^d\cos\theta+i^q\sin\theta\} \\
&= \frac{1}{\Delta}\{R_a\hat{V}\cos\psi\cos\theta-X_q\hat{V}\sin\psi\cos\theta \\
&\qquad +X_d\hat{V}\cos\psi\sin\theta+R_a\hat{V}\sin\psi\sin\theta\}
\end{aligned}
$$

If the phase of the applied voltage and the position of the armature at time zero are so related that $\psi = 0$,

$$i^a = \frac{\hat{V}}{R_a^2+X_dX_q}\{R_a\cos\omega t+X_d\sin\omega t\}$$

If R_a can be neglected, this reduces to

$$i^a = (\hat{V}/X_q)\sin\omega t$$

Since $v_a = \hat{V}\cos\omega t$, we have, in r.m.s. terms,

$$V_a = (1/\sqrt{2})\hat{V} \quad \text{and} \quad I^a = -j(1/\sqrt{2})(\hat{V}/X_q)$$

Hence $X_q = |V_a/I^a|$.

If $\psi = \pi/2$, a similar analysis shows that $|V_a/I^a|$ is then equal to X_d. At intermediate values of ψ, $|V_a/I^a|$ lies between the values X_d and X_q.

As shown on p. 137, V_α/I^α for the two-phase machine is equal to V_a/I^a for the three-phase machine. Hence X_q and X_d for the two-phase machine can be measured by an identical method to that of the three-phase machine.

It is possible to control the value of ψ, as required above, only under laboratory or special test-bed conditions. In practice, therefore, the test may be performed as a "slip" test,† i.e. the machine is made to run at a speed only very slightly different from synchronous speed, so that the effects of the speed difference are negligible yet the whole range of ψ is swept. The current and voltage are recorded oscillographically and the maximum and minimum values of impedance correspond to X_d and X_q respectively. There is some practical difficulty in performing this test unless the driving machine is large compared to that under test.

In all the above X_d, X_q and X_{Fd} are the "direct-axis synchronous reactance", the "quadrature-axis synchronous reactance" and the "armature-reaction reactance" of classical theory respectively, since they are measured in an identical manner. X_d can be regarded as the arithmetic sum of the armature leakage reactance $x_a = \omega l_a$ and X_{Fd} referred to the armature number of turns.

† Compare ref. 3, p. 482.

CHAPTER 11

Transient and Negative-sequence Conditions in A.C. Machines

Balanced Induction Machine Transients

The transient impedance matrix of the balanced induction machine in terms of stationary axes is given on p. 162 as

	D	Q	d	q
D	R_S+L_Sp		Mp	
Q		R_S+L_Sp		Mp
d	Mp	$M\dot{\theta}$	R_r+L_rp	$L_r\dot{\theta}$
q	$-M\dot{\theta}$	Mp	$-L_r\dot{\theta}$	R_r+L_rp

which can be transformed to P, N, f, b axes by means of the matrix

$$\mathbf{C} = \frac{1}{\sqrt{2}}$$

	P	N	f	b
D	1	1		
Q	$-j$	j		
d			1	1
q			$-j$	j

to obtain

$$
\mathbf{Z}' = \begin{array}{c|c|c|c|c|}
 & \text{P} & \text{N} & \text{f} & \text{b} \\
\hline
\text{P} & R_S+L_S p & & Mp & \\
\hline
\text{N} & & R_S+L_S p & & Mp \\
\hline
\text{f} & M(p-j\dot{\theta}) & & R_r+L_r(p-j\dot{\theta}) & \\
\hline
\text{b} & & M(p+j\dot{\theta}) & & R_r+L_r(p+j\dot{\theta}) \\
\hline
\end{array}
$$

This can be rearranged as

$$
\mathbf{Z}' = \begin{array}{c|c|c|c|c|}
 & \text{P} & \text{f} & \text{N} & \text{b} \\
\hline
\text{P} & R_S+L_S p & Mp & & \\
\hline
\text{f} & M(p-j\dot{\theta}) & R_r+L_r(p-j\dot{\theta}) & & \\
\hline
\text{N} & & & R_S+L_S p & Mp \\
\hline
\text{b} & & & M(p+j\dot{\theta}) & R_r+L_r(p+j\dot{\theta}) \\
\hline
\end{array}
$$

From this it can be seen that the P, f axes and the N, b axes again form two completely independent systems and that, moreover, the elements of the N, b system are the complex conjugates of those of the P, f system.

Considering the P, f system first, we have the voltage equation

$$
\begin{array}{c|c|}
\text{P} & v_P \\
\text{f} & v_f
\end{array}
=
\begin{array}{c|c|c|}
 & \text{P} & \text{f} \\
\hline
\text{P} & R_S+L_S p & Mp \\
\hline
\text{f} & M(p-j\dot{\theta}) & R_r+L_r(p-j\dot{\theta}) \\
\hline
\end{array}
\begin{array}{c|c|}
\text{P} & i^P \\
\text{f} & i^f
\end{array}
$$

in which v_f will be zero because the rotor windings are short-circuited.

The operational current equation is therefore

$$
\begin{array}{c|c|}
\text{P} & i^P \\
\text{f} & i^f
\end{array}
= \frac{1}{\Delta}
\begin{array}{c|c|c|}
 & \text{P} & \text{f} \\
\hline
\text{P} & R_r+L_r(p-j\dot{\theta}) & -Mp \\
\hline
\text{f} & -M(p-j\dot{\theta}) & R_S+L_S p \\
\hline
\end{array}
\begin{array}{c|c|}
\text{P} & v_P \\
\text{f} & 0
\end{array}
$$

The Laplace transform equation for the currents is thus

$$\begin{array}{c|c|} \text{P} & \bar{i}^{\text{P}} \\ \hline \text{f} & \bar{i}^{\text{f}} \end{array} = \frac{\bar{v}_{\text{P}}}{\Delta} \begin{array}{c|c|} \text{P} & R_r + L_r(s - j\dot{\theta}) \\ \hline \text{f} & -M(s - j\dot{\theta}) \end{array}$$

where $\Delta = \{R_S + L_S s\}\{R_r + L_r(s - j\dot{\theta})\} - M^2 s(s - j\dot{\theta})$

$$= (L_S L_r - M^2)s^2 + \{(L_S R_r + L_r R_S) - j(L_S L_r + M^2)\dot{\theta}\}s + \{R_S R_r - jR_S L_r \dot{\theta}\}$$

$$= (L_S L_r - M^2)\left[s^2 + \left\{\frac{R_r/L_r + R_S/L_S}{1 - M^2/L_S L_r} - j\dot{\theta}\right\}s + \left\{\frac{(R_S/L_S)(R_r/L_r)}{1 - M^2/L_S L_r} - j\frac{R_S/L_S}{1 - M^2/L_S L_r}\dot{\theta}\right\}\right]$$

The factors of this quadratic in s are

$$s + \frac{1}{2}\left(\left\{\frac{R_S/L_S + R_r/L_r}{1 - M^2/L_S L_r} - j\dot{\theta}\right\} \pm \left[\left\{\frac{R_S/L_S + R_r/L_r}{1 - M^2/L_S L_r}\right\}^2 - j2\frac{R_S/L_S + R_r/L_r}{1 - M^2/L_S L_r}\dot{\theta} - \dot{\theta}^2 - 4\frac{(R_S/L_S)(R_r/L_r)}{1 - M^2/L_S L_r} + j4\frac{R_S/L_S}{1 - M^2/L_S L_r}\dot{\theta}\right]^{1/2}\right)$$

$$= s + \frac{1}{2}\left(\left\{\frac{R_S/L_S + R_r/L_r}{1 - M^2/L_S L_r} - j\dot{\theta}\right\} \pm \left[\frac{(R_S/L_S - R_r/L_r)^2 + 4M^2 R_S R_r/(L_S L_r)^2}{(1 - M^2/L_S L_r)^2} - \dot{\theta}^2 + j2\frac{R_S/L_S - R_r/L_r}{1 - M^2/L_S L_r}\dot{\theta}\right]^{1/2}\right)$$

From this it is clear that:

(i) any further attempt at a general solution is extremely complicated and tedious, although simple in principle; and
(ii) a numerical solution for a particular case is fairly simple, especially if it may be assumed that $R_S/L_S = R_r/L_r$.

On this assumption, the term under the square root becomes purely real and equal to

$$\frac{4M^2 R_S R_r}{(L_S L_r - M^2)^2} - \dot{\theta}^2$$

At synchronous speed with a rated frequency of 50 Hz or more, with typical values of the resistances and inductances, $\dot{\theta}^2$ will be one or two orders of magnitude greater than the other term. The square root itself is then purely imaginary and only slightly less in magnitude than $\dot{\theta}$, say $(\dot{\theta}-\delta\dot{\theta})$. The zeros of the quadratic are then

$$-\frac{R_S/L_S+R_r/L_r}{2(1-M^2/L_SL_r)}+j(\dot{\theta}-\delta\dot{\theta}/2) \quad \text{and} \quad -\frac{R_S/L_S+R_r/L_r}{2(1-M^2/L_SL_r)}+j\,\delta\dot{\theta}/2$$

Since the determinant of the N, b submatrix is the complex conjugate of that of the P, f submatrix, it follows that its zeros are the complex conjugates of those of the P, f determinant.

The determinant of the complete impedance matrix is equal to the product of the determinants of the two submatrices and thus consists of the product of these two pairs of complex conjugates. The transient response will therefore consist of two damped sine waves, one of frequency slightly less than that corresponding to the speed of rotation and the other quite small, so that the sum of the two frequencies is equal to that of the speed of rotation. Both waves have the same damping.

At speeds of rotation lower than synchronous speed, the difference between the two frequencies is less, until at a speed

$$\dot{\theta} = \frac{2M\sqrt{(R_SR_r)}}{(L_SL_r-M^2)}$$

both frequencies are $\dot{\theta}/2$. At lower speeds still, the frequencies remain at $\dot{\theta}/2$, but the dampings are different. At standstill the transients are, of course, unidirectional.

It must be remembered that there is an inherent assumption in the above analysis in terms of Laplace transforms that $\dot{\theta}$ is constant.

Sudden Application of Terminal Voltage

To take a particular case, let us consider the sudden application of balanced terminal voltage to a motor already running with zero currents at a speed $\dot{\theta}$. The instantaneous symmetrical component voltages for balanced three-phase and two-phase conditions have been deduced on p. 113 as

$$\frac{\sqrt{3}}{2}\;\begin{array}{c|c|}\cline{2-2} \mathrm{P} & \hat{V}(\cos \omega t+j \sin \omega t) \\ \cline{2-2} \mathrm{N} & \hat{V}(\cos \omega t-j \sin \omega t) \\ \cline{2-2} 0 & 0 \\ \cline{2-2}\end{array}$$

and

$$1/\sqrt{2}\;\begin{array}{c|c|}\cline{2-2} \mathrm{P} & \hat{V}(\cos \omega t+j \sin \omega t) \\ \cline{2-2} \mathrm{N} & \hat{V}(\cos \omega t-j \sin \omega t) \\ \cline{2-2} 0 & 0 \\ \cline{2-2}\end{array}$$

respectively.

The positive-sequence current Laplace transform equation is

$$\bar{i}^{\mathrm{P}} = \frac{\bar{v}_{\mathrm{P}}}{\Delta}\{R_{\mathrm{r}}+L_{\mathrm{r}}(s-j\dot{\theta})\}$$

which, for the three-phase machine leads to

$$\bar{i}^{\mathrm{P}} = \frac{\sqrt{3}\,\hat{V}(s+j\omega)\,\{R_{\mathrm{r}}+L_{\mathrm{r}}(s-j\dot{\theta})\}}{2(L_{\mathrm{S}}L_{\mathrm{r}}-M^2)\,(s^2+\omega^2)\,(s+a+jb)\,(s+a-jb)}$$

where $s+a\pm jb$ are the factors of Δ already determined. After resolving this expression into its partial fractions, we shall be able to determine i^{P} as a complex time function. Since both the impedance and the voltage of the negative-sequence system N, b are the conjugates of those of the positive-sequence system P, f, it follows that i^{N} is the conjugate of i^{P}.

The transformation to three-phase currents, in the absence of zero-sequence components, leads to

$$\begin{aligned} i^{\mathrm{A}} &= (1/\sqrt{3})\,(i^{\mathrm{P}}+i^{\mathrm{N}}) \quad = (1/\sqrt{3})\,(i^{\mathrm{P}}+i^{\mathrm{P}*}) \\ &= (2/\sqrt{3})\,(\mathrm{Re}\; i^{\mathrm{P}}) \\ i^{\mathrm{B}} &= (1/\sqrt{3})\,(h^2 i^{\mathrm{P}}+h i^{\mathrm{N}}) = (1/\sqrt{3})\,(h^2 i^{\mathrm{P}}+h i^{\mathrm{P}*}) \\ &= (1/\sqrt{3})\,(-\mathrm{Re}\; i^{\mathrm{p}}+\sqrt{3}\;\mathrm{Im}\; i^{\mathrm{P}}) \\ i^{\mathrm{C}} &= (1/\sqrt{3})\,(h i^{\mathrm{P}}+h^2 i^{\mathrm{N}}) = (1/\sqrt{3})\,(h i^{\mathrm{P}}+h^2 i^{\mathrm{P}*}) \\ &= (1/\sqrt{3})\,(-\mathrm{Re}\; i^{\mathrm{P}}-\sqrt{3}\;\mathrm{Im}\; i^{\mathrm{P}}) \end{aligned}$$

whilst the transformation to two-phase currents leads to

$$i^D = (1/\sqrt{2})(i^P + i^N) \quad = (1/\sqrt{2})(i^P + i^{P*}) \quad = \sqrt{2}\,\mathrm{Re}\, i^P$$
$$i^Q = (1/\sqrt{2})(-ji^P + ji^N) = (1/\sqrt{2})(-ji^P + ji^{P*}) = \sqrt{2}\,\mathrm{Im}\, i^P$$

Since v_P is smaller in the two-phase case than in the the three-phase case in the ratio $\sqrt{\frac{2}{3}}$, the expression for $i^D = i^{A'}$ in terms of $v_D = v_{A'}$ will be the same as that for i^A in terms of v_A.

i^f and i^b, also conjugates, can be found in a similar manner to i^P and i^N, and the torque then follows.

If the stator voltage were suddenly re-applied after an interruption, the analysis would be similar to the above, but would have to include initial rotor currents, if the period of the interruption were short.

This analysis takes no account of changes of value of M, L_S, and L_r which might be significant under some transient conditions.

The problem of induction machine transients can be solved directly in D, Q, d, q axes at the expense of heavier algebra. The determinant of the impedance matrix in these axes—which is of fourth degree in p—is the product of two complex quadratic factors, namely the determinant of the P, f impedance matrix above and its conjugate, the determinant of the N, b impedance matrix.

The Synchronous Machine with Salient Poles and no Damper Windings

The transient impedance matrix of the synchronous machine without a damping winding, omitting the zero-sequence axis, is given on p. 197 as

$$\mathbf{Z} = \begin{array}{c|c|c|c|} & \text{d} & \text{q} & \text{F} \\ \hline \text{d} & R_a + L_d p & L_q \omega & L_{Fd} p \\ \hline \text{q} & -L_d \omega & R_a + L_q p & -L_{Fd} \omega \\ \hline \text{F} & L_{Fd} p & & R_F + L_F p \\ \hline \end{array}$$

if ω is written for $\dot{\theta}$.

It is necessary to find the inverse $\mathbf{Z}^{-1}$.

$$
\mathbf{Z}_t = \begin{array}{c|c|c|c|}
 & \mathrm{d} & \mathrm{q} & \mathrm{F} \\
\hline
\mathrm{d} & R_a + L_d p & -L_d\omega & L_{Fd}p \\
\hline
\mathrm{q} & L_q\omega & R_a + L_q p & \\
\hline
\mathrm{F} & L_{Fd}p & -L_{Fd}\omega & R_F + L_F p \\
\hline
\end{array}
$$

and

$$
\mathbf{Z}^{-1} = \frac{1}{\Delta}\begin{array}{c|c|c|c|}
 & \mathrm{d} & \mathrm{q} & \mathrm{F} \\
\hline
\mathrm{d} & (R_a+L_q p)(R_F+L_F p) & -L_q\omega(R_F+L_F p) & \begin{array}{c}-L_{Fd}L_q\omega^2 \\ -L_{Fd}p(R_a+L_q p)\end{array} \\
\hline
\mathrm{q} & \begin{array}{c}L_d\omega(R_F+L_F p) \\ -L_{Fd}^2\omega p\end{array} & \begin{array}{c}(R_a+L_d p)(R_F+L_F p) \\ -L_{Fd}^2 p^2\end{array} & \begin{array}{c}-L_{Fd}L_d\omega p \\ +L_{Fd}\omega(R_a+L_d p)\end{array} \\
\hline
\mathrm{F} & -L_{Fd}p(R_a+L_q p) & L_{Fd}L_q\omega p & \begin{array}{c}(R_a+L_d p)(R_a+L_q p) \\ +L_d L_q\omega^2\end{array} \\
\hline
\end{array}
$$

where

$$
\begin{aligned}
\Delta &= (R_F+L_F p)\{(R_a+L_d p)(R_a+L_q p)+L_d L_q\omega^2\} \\
&\quad -L_{Fd}p\{L_{Fd}p(R_a+L_q p)+L_{Fd}L_q\omega^2\} \\
&= (L_F L_d L_q - L_q L_{Fd}^2)p^3+\{L_F(L_d+L_q)R_a+L_d L_q R_F - L_{Fd}^2 R_a\}p^2 \\
&\quad +\{L_F(R_a^2+L_d L_q\omega^2)+R_F R_a(L_d+L_q)-L_{Fd}^2 L_q\omega^2\}p \\
&\quad +R_F(R_a^2+L_d L_q\omega^2)
\end{aligned}
$$

If we write L'_d for $(L_d - L_{Fd}^2/L_F)$,

$$
\begin{aligned}
\Delta &= L_F L_q L'_d p^3+\{L_F L_q R_a+L_d L_q R_F+L_F L'_d R_a\}p^2 \\
&\quad +\{R_a^2 L_F+R_a R_F(L_d+L_q)+L_F L_q L'_d\omega^2\}p \\
&\quad +R_F(R_a^2+L_d L_q\omega^2) \\
&= L_F L_q L'_d[p^3+(R_F L_d/L_F L'_d+R_a/L'_d+R_a/L_q)p^2 \\
&\quad +(R_a^2/L'_d L_q+R_F R_a/L_F L'_d+R_F R_a L_d/L_F L_q L'_d+\omega^2)p \\
&\quad +(R_a^2/L_q L'_d+L_d\omega^2/L'_d)R_F/L_F]
\end{aligned}
$$

Consideration of transients involves finding the zeros of this polynomial in p, i.e. the roots of the cubic equation $\Delta = 0$. In a numerical case the roots can be found to any required degree of accuracy by the techniques described in books on numerical methods. There is, however, no exact algebraic solution and consequently only an approximate general solution can be obtained. Use is made of the empirical knowledge that the armature resistance R_a is small compared with the reactances represented by $L'_d\omega$ and $L_q\omega$. First powers of R_a can be accurately represented, but second powers only approximately. This is equivalent to saying that we must assume that the time-constants L'_d/R_a and L_q/R_a are sufficiently large for the products of their inverses to be negligible compared with ω^2.

Since the equation is a cubic, one root must be real. The others may both be real or may form a conjugate pair. From the nature of the problem it is reasonable to assume, at least at first, that they form a conjugate pair. Then Δ is of the form

$$L_F L_q L'_d(p+\alpha)\{(p+\beta)^2+\gamma^2\}$$
$$= L_F L_q L'_d\{p^3+(\alpha+2\beta)p^2+(2\alpha\beta+\beta^2+\gamma^2)p+\alpha(\beta^2+\gamma^2)\}$$

By comparing the coefficients of this expression with those of Δ we can obtain an approximate solution.

Unless the system is unstable α and β are both positive and from the coefficient of p^2 we can see that they are both necessarily composed of R/L terms, hence we have:

From the coefficients of p, neglecting products of R/L, $\gamma \approx \omega$.

From the constant terms, if $R_a^2/L_q L'_d$ is negligible compared with $L_d\omega^2/L'_d$ and β^2 is negligible compared with $\gamma^2 \approx \omega^2$,

$$\alpha \approx R_F L_d/L_F L'_d$$

Then from the coefficients of p^2,

$$\beta \approx (1/L_q+1/L'_d)R_a/2$$

Hence as a first approximation

$$\Delta = L_F L_q L'_d(p+\alpha)\{(p+\beta)^2+\omega^2\}$$

where α and β have the above values. If these values are substituted in this expression and the result compared with the actual value of Δ, it will be seen that it is a better approximation than we had supposed, since the terms we have neglected are not in fact replaced by zeros but by other expressions which although not identical are nevertheless not zero.

L_F/R_F is the time-constant of the field winding but has the title of "direct-axis transient open-circuit time-constant" and is represented by T'_{do}. The open circuit here refers to the armature circuit.

$1/\alpha$ is called the "direct-axis transient short-circuit time-constant" and is represented by T'_d.

$1/\beta$ is called the "short-circuit time-constant of the armature winding" and is represented by T_a.†

Consideration of transient conditions will here be illustrated by the most important case, namely sudden symmetrical short circuit from open circuit.

Sudden Three-phase Short Circuit from Open Circuit

If all phases of a three-phase synchronous machine are short-circuited, $v_a = 0$, $v_b = 0$, and $v_c = 0$, consequently $v_d = 0$ and $v_q = 0$. The solution to this problem may be obtained by super-position after finding the response to a sudden application to the armature terminals of a voltage equal and opposite to the open-circuit terminal voltage. Here, however, the initial values of the currents will be included to utilize to the full the Laplace transform technique. With a short circuit from open circuit the only non-zero initial current is the field current which has a value V_F/R_F. If, however, the machine had been on load the initial values of the armature currents would have been included in a similar manner to that in which the field current is included.

† In machines with damping circuits the value of T_a would be different. Actual machines always have some damping, even if no damper winding as such is fitted.

The transform equation in matrix form is of the form

$$\bar{\mathbf{v}} = \mathbf{Z}\bar{\mathbf{i}} - \mathbf{L}\mathbf{i}_o$$

where $\mathbf{i}_o$ is the initial current matrix.

After short circuit the terminal voltage is

$$\mathbf{v} = \begin{array}{c|c|} \text{d} & 0 \\ \hline \text{q} & 0 \\ \hline \text{F} & V_F \end{array} \qquad \text{hence} \qquad \bar{\mathbf{v}} = \begin{array}{c|c|} \text{d} & 0 \\ \hline \text{q} & 0 \\ \hline \text{F} & V_F/s \end{array}$$

The effective excitation function is thus

$$\begin{array}{c|c|} \text{d} & 0 \\ \hline \text{q} & 0 \\ \hline \text{F} & V_F/s \end{array} + \begin{array}{c|c|c|c|} & \text{d} & \text{q} & \text{F} \\ \hline \text{d} & L_d & & L_{Fd} \\ \hline \text{q} & & L_q & \\ \hline \text{F} & L_{Fd} & & L_F \\ \hline \end{array} \; \begin{array}{c|c|} \text{d} & 0 \\ \hline \text{q} & 0 \\ \hline \text{F} & V_F/R_F \end{array}$$

$$= \begin{array}{c|c|} \text{d} & 0 \\ \hline \text{q} & 0 \\ \hline \text{F} & V_F/s \end{array} + \begin{array}{c|c|} \text{d} & L_{Fd}V_F/R_F \\ \hline \text{q} & 0 \\ \hline \text{F} & L_F V_F/R_F \end{array}$$

$$= \begin{array}{c|c|} \text{d} & L_{Fd}V_F/R_F \\ \hline \text{q} & 0 \\ \hline \text{F} & V_F/s + L_F V_F/R_F \end{array} = \frac{V_F}{R_F} \begin{array}{c|c|} \text{d} & L_{Fd} \\ \hline \text{q} & 0 \\ \hline \text{F} & R_F/s + L_F \end{array}$$

The transform voltage equation is therefore

$$\frac{V_F}{R_F} \begin{array}{c|c|} \text{d} & L_{Fd} \\ \hline \text{q} & 0 \\ \hline \text{F} & R_F/s + L_F \end{array} = \begin{array}{c|c|c|c|} & \text{d} & \text{q} & \text{F} \\ \hline \text{d} & R_a + L_d s & L_q\omega & L_{Fd}s \\ \hline \text{q} & -L_d\omega & R_a + L_q s & -L_{Fd}\omega \\ \hline \text{F} & L_{Fd}s & & R_F + L_F s \\ \hline \end{array} \; \begin{array}{c|c|} \text{d} & \bar{i}^d \\ \hline \text{q} & \bar{i}^q \\ \hline \text{F} & \bar{i}^F \end{array}$$

Inverting the impedance matrix gives the equation

$$
\begin{array}{c|c|}
\hline
\bar{i}^{\mathrm{d}} \\
\hline
\bar{i}^{\mathrm{q}} \\
\hline
\bar{i}^{\mathrm{F}} \\
\hline
\end{array}
=\frac{1}{\Delta}
\begin{array}{c}
\\ \mathrm{d} \\ \mathrm{q} \\ \mathrm{F}
\end{array}
\begin{array}{|c|c|c|}
\multicolumn{1}{c}{\mathrm{d}} & \multicolumn{1}{c}{\mathrm{q}} & \multicolumn{1}{c}{\mathrm{F}} \\
\hline
\begin{array}{c}(R_a+L_q s)\\ \times(R_F+L_F s)\end{array} & \begin{array}{c}-L_q\omega\\ \times(R_F+L_F s)\end{array} & \begin{array}{c}-L_{Fd}L_q\omega^2\\ -L_{Fd}s\\ \times(R_a+L_q s)\end{array} \\
\hline
\begin{array}{c}L_d\omega(R_F+L_F s)\\ -L_{Fd}^2\omega s\end{array} & \begin{array}{c}(R_a+L_d s)\\ \times(R_F+L_F s)\\ -L_{Fd}^2 s^2\end{array} & \begin{array}{c}L_{Fd}\omega(R_a+L_d s)\\ -L_{Fd}L_d\omega s\end{array} \\
\hline
\begin{array}{c}-L_{Fd}s\\ \times(R_a+L_q s)\end{array} & L_{Fd}L_q\omega s & \begin{array}{c}(R_a+L_d s)\\ \times(R_a+L_q s)\\ +L_d L_q\omega^2\end{array} \\
\hline
\end{array}
\;\frac{V_F}{R_F}\;
\begin{array}{c}
\mathrm{d} \\ \mathrm{q} \\ \mathrm{F}
\end{array}
\begin{array}{|c|}
\hline
L_{Fd} \\
\hline
0 \\
\hline
R_F/s+L_F \\
\hline
\end{array}
$$

from which the currents may be determined.

Field Current

It is preferable to determine the field current first, since i^{F} is the actual field winding current, whereas for the armature currents it is first necessary to find i^{d} and i^{q} and then to apply the transformation to a, b, c axes to find the actual currents.

From the current transform equation,

$$
\begin{aligned}
\bar{i}^{\mathrm{F}} &= \frac{1}{\Delta}\frac{V_F}{R_F}[-L_{Fd}^2 s(R_a+L_q s)+\{(R_a+L_d s)(R_a+L_q s) \\
&\quad +L_d L_q\omega^2\}(R_F/s+L_F)] \\
&= \frac{1}{\Delta}\frac{1}{s}\frac{V_F}{R_F}[(R_F+L_F s)\{(R_a+L_d s)(R_a+L_q s)+L_d L_q\omega^2\} \\
&\quad -L_{Fd}s\{L_{Fd}s(R_a+L_q s)+L_{Fd}L_q\omega^2-L_{Fd}L_q\omega^2\}] \\
&= \frac{1}{\Delta}\frac{1}{s}\frac{V_F}{R_F}[\Delta+L_{Fd}^2 L_q\omega^2 s] \\
&= \frac{V_F}{R_F}\left[\frac{1}{s}+\frac{L_{Fd}^2 L_q\omega^2}{\Delta}\right]
\end{aligned}
$$

Taking $\Delta = L_F L_q L_d'(s+\alpha)\{(s+\beta)^2+\omega^2\}$ let

$$\frac{1}{(s+\alpha)\{(s+\beta)^2+\omega^2\}} = \frac{A}{s+\alpha}+\frac{B(s+\beta)+C\omega}{(s+\beta)^2+\omega^2}$$

Putting $s=-\alpha$ gives $A = 1/\{(\alpha-\beta)^2+\omega^2\}$
Multiplying out

$$A\{(s+\beta)^2+\omega^2\}+B(s+\beta)(s+\alpha)+C\omega(s+\alpha) = 1$$

or $(A+B)s^2+(2A\beta+B\alpha+B\beta+C\omega)s+(A\beta^2+A\omega^2+B\alpha\beta+C\omega\alpha) = 1$

so that $$B = -A = -1/\{(\alpha-\beta)^2+\omega^2\}$$

and $$-C\omega = 2A\beta+B\alpha+B\beta = 2A\beta-A\alpha-A\beta = -A(\alpha-\beta)$$

whence $C = A(\alpha-\beta)/\omega$.

The field current is therefore

$$\begin{aligned} i^F &= \frac{V_F}{R_F}+\frac{V_F}{R_F}\frac{L_{Fd}^2 L_q\omega^2}{L_F L_q L_d'}[A\varepsilon^{-\alpha t}+B\varepsilon^{-\beta t}\cos\omega t+C\varepsilon^{-\beta t}\sin\omega t] \\ &= \frac{V_F}{R_F}+\frac{V_F}{R_F}\frac{L_{Fd}^2\omega^2}{L_F L_d'}A\left[\varepsilon^{-\alpha t}-\varepsilon^{-\beta t}\cos\omega t+\frac{\alpha-\beta}{\omega}\varepsilon^{-\beta t}\sin\omega t\right] \end{aligned}$$

where $A = 1/\{(\alpha-\beta)^2+\omega^2\}$.

The approximate values for α and β on p. 226 may be substituted here. As a further approximation, however, we may say that $(\alpha-\beta)$, which is the difference of the inverses of the two time-constants T_d' and T_a, is sufficiently small compared with ω for its square to be neglected in the denominator of A. Then

$$i^F = \frac{V_F}{R_F}+\frac{V_F}{R_F}\frac{L_{Fd}^2}{L_F L_d'}\left[\varepsilon^{-\alpha t}-\varepsilon^{-\beta t}\cos\omega t+\frac{\alpha-\beta}{\omega}\varepsilon^{-\beta t}\sin\omega t\right]$$

The field current thus consists of three components:

(i) the original, and final, steady-state field current V_F/R_F;

(ii) an induced unidirectional transient current in the same direction as the original current and initially about $L_{Fd}^2/L_F/L_d'$ times as great as the original current and having a time-constant $1/\alpha$, which is the transient short-circuit time-constant T_d' and has the approximate value $L_F L_d'/R_F L_d$;

(iii) an alternating component of the same initial peak magnitude as the transient unidirectional component but with a time constant $1/\beta$, which is the armature short-circuit time-constant T_a and has the approximate value

$$2/R_a(1/L_q+1/L'_d) = 2L_qL'_d/R_a(L'_d+L_q)$$

The sine component of the alternating component is small compared with the cosine component and will make little difference to the magnitude but will produce a slight phase-shift.

Armature Currents

For the armature we have to find both i^d and i^q before we can obtain the phase currents. From the matrix equation of p. 229 we have

$$\begin{aligned}
\bar{i}^d &= \frac{1}{\Delta}\frac{V_F}{R_F}[(R_a+L_qs)(R_F+L_Fs)L_{Fd} \\
&\quad -\{L_qL_{Fd}\omega^2+L_{Fd}s(R_a+L_qs)\}(R_F+L_Fs)/s] \\
&= \frac{1}{\Delta}\frac{V_F}{R_F}L_{Fd}(R_F+L_Fs)[(R_a+L_qs)-\{L_q\omega^2/s+(R_a+L_qs)\}] \\
&= \frac{1}{\Delta}\frac{V_F}{R_F}L_{Fd}(R_F+L_Fs)(-L_q\omega^2/s) \\
&= -\frac{V_F}{R_F}L_{Fd}L_q\omega^2\frac{R_F+L_Fs}{s\Delta} \\
&= -\frac{V_F}{R_F}\frac{L_{Fd}\omega^2}{L_FL_qL'_q}\frac{L_q(R_F+L_Fs)}{s(s+\alpha)\{(s+\beta)^2+\omega^2\}} \\
&= -\frac{V_F}{R_F}\frac{L_{Fd}\omega^2}{L_FL_qL'_d}\left[\frac{F}{s}+\frac{G}{s+\alpha}+\frac{H(s+\beta)+J\omega}{(s+\beta)^2+\omega^2}\right] \quad \text{say,}
\end{aligned}$$

where L_q has been retained in the denominator for convenience later when i^d and i^q are combined to give i^a, i^b and i^c.

Proceeding as with i^F we find that

$$F = \frac{R_F}{\alpha(\beta^2+\omega^2)}L_q$$

$$G = -\frac{R_F-\alpha L_F}{\alpha\{(\alpha-\beta)^2+\omega^2\}}L_q$$

$$H = -\frac{(\alpha-2\beta)R_F+(\beta^2+\omega^2)L_F}{(\beta^2+\omega^2)\{(\alpha-\beta)^2+\omega^2\}}L_q$$

$$J = -\frac{(\alpha\beta-\beta^2+\omega^2)R_F-(\alpha-\beta)(\beta^2+\omega^2)L_F}{\omega(\beta^2+\omega^2)\{(\alpha-\beta)^2+\omega^2\}}L_q$$

If, as an approximation, we neglect β^2, $\beta(\alpha-\beta)$ and $(\alpha-\beta)^2$ when added to ω^2, and $(\alpha-2\beta)R_F$ when added to $L_F\omega^2$, we get from the approximate values of α and β,

$$\begin{aligned} F &= L_F L_q L'_d/L_d\omega^2 \\ G &= L_F L_q(1-L'_d/L_d)/\omega^2 \\ H &= -L_F L_q/\omega^2 \\ J &= -[R_F(1-L_d/L'_d)+L_F R_a(1/L_q+1/L'_d)/2]L_q/\omega^3 \end{aligned}$$

Hence

$$i^d = -\frac{V_F}{R_F}\frac{L_{Fd}\omega^2}{L_F L_q L'_d}[F+G\varepsilon^{-\alpha t}+H\varepsilon^{-\beta t}\cos\omega t+J\varepsilon^{-\beta t}\sin\omega t]$$

in which accurate values may be used if known for F, G, H, and J or the approximate values in default of anything better.

From the matrix equation on p. 229 we have also

$$\begin{aligned} \bar{i}^q &= \frac{1}{\Delta}\frac{V_F}{R_F}[L_{Fd}\{L_d\omega(R_F+L_F s)-L_{Fd}^2\omega s\} \\ &\quad +(R_F/s+L_F)\{L_{Fd}\omega(R_a+L_d s)-L_{Fd}L_d\omega s\}] \\ &= \frac{1}{\Delta}\frac{V_F}{R_F}\frac{L_{Fd}\omega}{s}[(R_a+L_d s)(R_F+L_F s)-L_{Fd}^2 s^2] \\ &= \frac{V_F}{R_F}L_{Fd}\omega^2\left[\frac{(R_a+L_d s)(R_F+L_F s)-L_{Fd}^2 s^2}{\omega s\Delta}\right] \\ &= \frac{V_F}{R_F}\frac{L_{Fd}\omega^2}{L_F L_q L'_d}\left[\frac{L_F L'_d s^2+(R_F L_d+R_a L_F)s+R_a R_F}{\omega s(s+\alpha)\{(s+\beta)^2+\omega^2\}}\right] \\ &= \frac{V_F}{R_F}\frac{L_{Fd}\omega^2}{L_F L_q L'_d}\left[\frac{K}{s}+\frac{N}{s+\alpha}+\frac{P(s+\beta)+Q\omega}{(s+\beta)^2+\omega^2}\right] \quad \text{say.} \end{aligned}$$

The values of K, N, P, and Q are not so easy to deal with as the numerators for i^{F} and i^{d} since, except for Q, they are smaller and the former approximations cannot be applied with the same confidence. The approximate values based on similar assumptions to before are, however,

$$K = R_a L_F L'_d / L_d \omega^3$$
$$N = R_a L_F (1 - L'_d/L_d)/\omega^3$$
$$P = -R_a L_F/\omega^3$$
$$Q = L'_d L_F/\omega^2$$

Hence

$$i^q = \frac{V_F}{R_F}\frac{L_{Fd}\omega^2}{L_F L_q L'_d}[K + N\varepsilon^{-\alpha t} + P\varepsilon^{-\beta t}\cos\omega t + Q\varepsilon^{-\beta t}\sin\omega t]$$

The phase currents can now be determined by transforming to a, b, c axes. Only the a-phase current will be considered here.

$$\begin{aligned}
i^a &= \sqrt{\tfrac{2}{3}}\{\cos\theta i^d + \sin\theta i^q\} \\
&= -\sqrt{\frac{2}{3}}\frac{V_F}{R_F}\frac{L_{Fd}\omega^2}{L_F L_q L'_d}[(F\cos\theta - K\sin\theta) \\
&\quad + (G\cos\theta - N\sin\theta)\varepsilon^{-\alpha t} \\
&\quad + \{H\cos\theta\cos\omega t - P\sin\theta\cos\omega t + J\cos\theta\sin\omega t \\
&\quad - Q\sin\theta\sin\omega t\}\varepsilon^{-\beta t}]
\end{aligned}$$

If $\theta = \omega t + \psi$,

$$\begin{aligned}
i^a = -\sqrt{\frac{2}{3}}\frac{V_F}{R_F}\frac{L_{Fd}\omega^2}{L_F L_q L'_d}\Big[&F\cos(\omega t+\psi) - K\sin(\omega t+\psi) \\
&+ \{G\cos(\omega t+\psi) - N\sin(\omega t+\psi)\}\varepsilon^{-\alpha t} \\
&+ \Big\{\frac{H}{2}[\cos(2\omega t+\psi) + \cos\psi] \\
&+ \frac{J}{2}[\sin(2\omega t+\psi) - \sin\psi] \\
&- \frac{P}{2}[\sin(2\omega t+\psi) + \sin\psi] \\
&- \frac{Q}{2}[-\cos(2\omega t+\psi) + \cos\psi]\Big\}\varepsilon^{-\beta t}\Big]
\end{aligned}$$

The phase current thus consists of:

(i) a component of angular frequency ω and of constant amplitude; this is the steady-state short-circuit armature current;
(ii) a component of angular frequency ω with a magnitude diminishing according to the time-constant $1/\alpha = T'_d$;
(iii) a component of angular frequency 2ω, i.e. a second harmonic, decaying according to the time-constant $1/\beta = T_a$;
(iv) a unidirectional component which also decays according to the time-constant $1/\beta = T_a$.

Bearing in mind that products of resistance are usually small compared with products of reactance, comparison of F and K shows that magnitude of the steady-state current is approximately

$$\sqrt{\frac{2}{3}}\frac{V_F}{R_F}\frac{L_{Fd}\omega^2}{L_F L_d L'_d}F = \sqrt{\frac{2}{3}}\frac{V_F}{R_F}\frac{L_{Fd}\omega^2}{L_F L_q L'_d}\frac{L_F L_q L'_d}{L_d\omega^2}$$

$$= \sqrt{\frac{2}{3}}\frac{V_F}{R_F}\frac{L_{Fd}}{L_d} = \sqrt{\frac{2}{3}}\frac{V_F}{R_F}\frac{X_{Fd}}{X_d}$$

But $\sqrt{\frac{2}{3}}(V_F/R_F)X_{Fd}$ is the open-circuit e.m.f., neglecting changes of saturation, for a field winding terminal voltage V_F, and is represented by $\hat{E}$. The steady-state short-circuit r.m.s. current is thus E/X_d as was shown previously on p. 202.

Similarly, neglecting N when compared with G, gives the transient fundamental frequency component as approximately

$$\sqrt{\frac{2}{3}}\frac{V_F}{R_F}\frac{L_{Fd}\omega^2}{L_F L_q L'_d}G = \sqrt{\frac{2}{3}}\frac{V_F}{R_F}\frac{L_{Fd}\omega^2}{L_F L_q L'_d}\frac{L_F L_q(1-L'_d/L_d)}{\omega^2}$$

$$= \sqrt{\frac{2}{3}}\frac{V_F}{R_F}\frac{L_{Fd}}{L'_d}(1-L'_d/L_d)$$

$$= \sqrt{\frac{2}{3}}\frac{V_F L_{Fd}}{R_F}\left\{\frac{1}{L'_d}-\frac{1}{L_d}\right\} = \frac{\hat{E}}{X'_d}-\frac{\hat{E}}{X_d}$$

The total initial current (r.m.s.) of angular frequency ω is thus $\{(E/X'_d)-(E/X_d)\}+E/X_d = E/X'_d$, of which E/X_d is the steady-state component and $\{(E/X_d')-(E/X_d)\}$ is the transient component with a time-constant T_d'. X_d', which thus gives the total a.c. component of current after short circuit in terms of the open-circuit voltage, is called the "direct-axis transient reactance".

Comparison of H, J, P, and Q shows that the principal terms are those containing H and Q. Neglecting the others as a first approximation gives the second harmonic and unidirectional components as

$$-\sqrt{\frac{2}{3}}\frac{V_F}{R_F}\frac{L_{Fd}\omega^2}{L_F L_q L_d'}\frac{1}{2}[-L_F L_q\{\cos(2\omega t+\psi)+\cos\psi\}/\omega^2$$

$$+L_F L_d'\{\cos(2\omega t+\psi)-\cos\psi\}/\omega^2]\varepsilon^{-\beta t}$$

$$=\sqrt{\frac{2}{3}}\frac{V_F}{R_F}\frac{L_{Fd}\omega^2}{L_F L_q L_d'}\frac{1}{2}[L_F(L_q-L_d')\cos(2\omega t+\psi)/\omega^2$$

$$+L_F(L_q+L_d')\cos\psi/\omega^2]\varepsilon^{-\beta t}$$

$$=\sqrt{\frac{2}{3}}\frac{V_F}{R_F}\frac{L_{Fd}}{2}\left\{\frac{1}{L_d'}-\frac{1}{L_q}\right\}\varepsilon^{-\beta t}\cos(2\omega t+\psi)$$

$$+\sqrt{\frac{2}{3}}\frac{V_F}{R_F}\frac{L_{Fd}}{2}\left\{\frac{1}{L_d'}+\frac{1}{L_q}\right\}\varepsilon^{-\beta t}\cos\psi$$

$$=\frac{\hat{E}}{2}\left\{\frac{1}{X_d'}-\frac{1}{X_q}\right\}\varepsilon^{-\beta t}\cos(2\omega t+\psi)+\frac{\hat{E}}{2}\left\{\frac{1}{X_d'}+\frac{1}{X_q}\right\}\varepsilon^{-\beta t}\cos\psi$$

The second harmonic thus tends to be small only if X_d' and X_q are of the same order of magnitude.

Both the second harmonic and the unidirectional component decay according to the time-constant $1/\beta = T_a$.

Impedance to Negative Sequence

For system studies it is necessary to have a value for the impedance of a synchronous machine to negative-sequence current and voltage. Since harmonics are introduced in the current even if the voltage is sinusoidal, and in the voltage even if the current is sinusoidal, the impedance to negative sequence is regarded for this purpose as the ratio of the fundamental component of the voltage to the fundamental component of the current. The result is slightly different, if the voltage is sinusoidal, from what it is if the current is sinusoidal. Both cases will be considered here.

For this purpose the field winding has no terminal voltage since it is regarded as closed through a negligible resistance. This is justified since the exciter armature has a very small resistance compared with that of the synchronous machine field winding if the efficiency of the exciter is reasonably high. There is no excitation under purely negative-sequence conditions. The synchronous machine field winding being short-circuited, it is convenient to eliminate it by partitioning the impedance matrix.

$$\mathbf{Z} = \begin{array}{c|cc|c} & \mathrm{d} & \mathrm{q} & \mathrm{F} \\ \hline \mathrm{d} & R_a + L_d p & L_q \omega & L_{Fd} p \\ \mathrm{q} & -L_d \omega & R_a + L_q p & -L_{Fd} \omega \\ \hline \mathrm{F} & L_{Fd} p & & R_F + L_F p \end{array}$$

$$\mathbf{Z}' = \begin{array}{c|cc} & \mathrm{d} & \mathrm{q} \\ \hline \mathrm{d} & R_a + L_d p & L_q \omega \\ \mathrm{q} & -L_d \omega & R_a + L_q p \end{array}$$

$$- \begin{array}{c|c} & \mathrm{F} \\ \hline \mathrm{d} & L_{Fd} p \\ \mathrm{q} & -L_{Fd} \omega \end{array} \quad \begin{array}{c|c} & \mathrm{F} \\ \hline \mathrm{F} & \dfrac{1}{R_F + L_F p} \end{array} \quad \begin{array}{c|cc} & \mathrm{d} & \mathrm{q} \\ \hline \mathrm{F} & L_{Fd} p & \end{array}$$

$$= \begin{array}{c|cc} & \mathrm{d} & \mathrm{q} \\ \hline \mathrm{d} & R_a + L_d p & L_q \omega \\ \mathrm{q} & -L_d \omega & R_a + L_q p \end{array} \; - \; \begin{array}{c|cc} & \mathrm{d} & \mathrm{q} \\ \hline \mathrm{d} & \dfrac{L_{Fd}^2 p^2}{R_F + L_F p} & \\ \mathrm{q} & \dfrac{-L_{Fd}^2 \omega p}{R_F + L_F p} & \end{array}$$

$$= \begin{array}{c|c|c|} & \text{d} & \text{q} \\ \hline \text{d} & R_a + L_d p - \dfrac{L_{Fd}^2 p^2}{R_F + L_F p} & L_q \omega \\ \hline \text{q} & -L_d \omega + \dfrac{L_{Fd}^2 \omega p}{R_F + L_F p} & R_a + L_q p \\ \hline \end{array}$$

This cannot be reduced further unless we neglect the resistance of the field winding. With this approximation the impedance matrix is

$$\begin{array}{c|c|c|} & \text{d} & \text{q} \\ \hline \text{d} & R_a + (L_d - L_{Fd}^2/L_F)p & L_q \omega \\ \hline \text{q} & -(L_d - L_{Fd}^2/L_F)\omega & R_a + L_q p \\ \hline \end{array} = \begin{array}{c|c|c|} & \text{d} & \text{q} \\ \hline \text{d} & R_a + L'_d p & L_q \omega \\ \hline \text{q} & -L'_d \omega & R_a + L_q p \\ \hline \end{array}$$

Impedance to Negative-sequence Current

Let us consider currents $i^a = \hat{I} \sin \omega t$, $i^b = \hat{I} \sin (\omega t + 2\pi/3)$, and $i^c = \hat{I} \sin (\omega t + 4\pi/3)$, which have the required negative sequence, since i^c leads i^b by $2\pi/3$ and i^b leads i^a by $2\pi/3$. Transforming to d, q, o reference axes, gives

$$\begin{array}{|c|} \hline i^d \\ \hline i^q \\ \hline i^o \\ \hline \end{array} = \sqrt{\tfrac{2}{3}} \begin{array}{c|c|c|c|} & \text{a} & \text{b} & \text{c} \\ \hline \text{d} & \cos\theta & \cos(\theta - 2\pi/3) & \cos(\theta - 4\pi/3) \\ \hline \text{q} & \sin\theta & \sin(\theta - 2\pi/3) & \sin(\theta - 4\pi/3) \\ \hline \text{o} & 1/\sqrt{2} & 1/\sqrt{2} & 1/\sqrt{2} \\ \hline \end{array} \begin{array}{c|c|} \hline \text{a} & \hat{I} \sin \omega t \\ \hline \text{b} & \hat{I} \sin(\omega t + 2\pi/3) \\ \hline \text{c} & \hat{I} \sin(\omega t + 4\pi/3) \\ \hline \end{array}$$

$$= \sqrt{\tfrac{2}{3}} \begin{array}{c|c|} \hline \text{d} & \begin{array}{c} \hat{I}\{\sin \omega t \cos\theta + \sin(\omega t + 2\pi/3)\cos(\theta - 2\pi/3) \\ + \sin(\omega t + 4\pi/3)\cos(\theta - 4\pi/3)\} \end{array} \\ \hline \text{q} & \begin{array}{c} \hat{I}\{\sin \omega t \sin\theta + \sin(\omega t + 2\pi/3)\sin(\theta - 2\pi/3) \\ + \sin(\omega t + 4\pi/3)\sin(\theta - 4\pi/3)\} \end{array} \\ \hline \text{o} & \\ \hline \end{array}$$

$$= \sqrt{\tfrac{2}{3}}\ \begin{array}{c|c|} \text{d} & \tfrac{3}{2}\hat{I}\sin(\omega t+\theta) \\ \hline \text{q} & -\tfrac{3}{2}\hat{I}\cos(\omega t+\theta) \\ \hline \text{o} & \end{array}$$

$$= \sqrt{\tfrac{3}{2}}\ \begin{array}{c|c|} \text{d} & \hat{I}\sin(2\omega t+\psi) \\ \hline \text{q} & -\hat{I}\cos(2\omega t+\psi) \\ \hline \text{o} & \end{array}$$

where $\theta = \omega t+\psi$.

If the zero-sequence axis is omitted, the voltage equation is now

$$\begin{array}{c|c|} \text{d} & v_{\text{d}} \\ \hline \text{q} & v_{\text{q}} \end{array} = \begin{array}{c|c|c|} & \text{d} & \text{q} \\ \text{d} & R_{\text{a}}+L'_{\text{d}}p & L_{\text{q}}\omega \\ \hline \text{q} & -L'_{\text{d}}\omega & R_{\text{a}}+L_{\text{q}}p \end{array}\ \sqrt{\tfrac{3}{2}}\ \begin{array}{c|c|} \text{d} & \hat{I}\sin(2\omega t+\psi) \\ \hline \text{q} & -\hat{I}\cos(2\omega t+\psi) \end{array}$$

$$= \sqrt{\tfrac{3}{2}}\ \begin{array}{c|c|} \text{d} & R_{\text{a}}\hat{I}\sin(2\omega t+\psi)+L'_{\text{d}}p\{\hat{I}\sin(2\omega t+\psi)\}-L_{\text{q}}\omega\hat{I}\cos(2\omega t+\psi) \\ \hline \text{q} & -L'_{\text{d}}\omega\hat{I}\sin(2\omega t+\psi)-R_{\text{a}}\hat{I}\cos(2\omega t+\psi)+L_{\text{q}}p\{-\hat{I}\cos(2\omega t+\psi)\} \end{array}$$

Re-arranging and performing the differentiations, gives

$$\begin{array}{c|c|} \text{d} & v_{\text{d}} \\ \hline \text{q} & v_{\text{q}} \end{array} = \sqrt{\tfrac{3}{2}}\ \begin{array}{c|c|} \text{d} & R_{\text{a}}\hat{I}\sin(2\omega t+\psi)+2\omega L'_{\text{d}}\hat{I}\cos(2\omega t+\psi)-L_{\text{q}}\omega\hat{I}\cos(2\omega t+\psi) \\ \hline \text{q} & -R_{\text{a}}\hat{I}\cos(2\omega t+\psi)-L'_{\text{d}}\omega\hat{I}\sin(2\omega t+\psi)+2\omega L_{\text{q}}\hat{I}\sin(2\omega t+\psi) \end{array}$$

$$= \sqrt{\tfrac{3}{2}}\ \begin{array}{c|c|} \text{d} & R_{\text{a}}\hat{I}\sin(2\omega t+\psi)+(2L'_{\text{d}}-L_{\text{q}})\omega\hat{I}\cos(2\omega t+\psi) \\ \hline \text{q} & -R_{\text{a}}\hat{I}\cos(2\omega t+\psi)+(2L_{\text{q}}-L'_{\text{d}})\omega\hat{I}\sin(2\omega t+\psi) \end{array}$$

from which

$$\begin{aligned} v_{\text{a}} &= \sqrt{\tfrac{2}{3}}\,[\cos\theta\, v_{\text{d}}+\sin\theta\, v_{\text{q}}+v_{\text{o}}/\sqrt{2}] \\ &= [R_{\text{a}}\hat{I}\sin(2\omega t+\psi)\cos\theta+(2L'_{\text{d}}-L_{\text{q}})\omega\hat{I}\cos(2\omega t+\psi)\cos\theta \\ &\quad -R_{\text{a}}\hat{I}\cos(2\omega t+\psi)\sin\theta+(2L_{\text{q}}-L'_{\text{d}})\omega\hat{I}\sin(2\omega t+\psi)\sin\theta] \end{aligned}$$

$$= [R_a \hat{I} \sin(2\omega t+\psi-\theta)$$
$$+(2L'_d - L_q)(\omega/2)\hat{I}\{\cos(2\omega t+\psi+\theta)+\cos(2\omega t+\psi-\theta)\}$$
$$+(2L_q - L'_d)(\omega/2)\hat{I}\{-\cos(2\omega t+\psi+\theta)+\cos(2\omega t+\psi-\theta)\}]$$
$$= [R_a \hat{I} \sin \omega t + (L'_d - L_q/2)\,\omega \hat{I}\{\cos(3\omega t+2\psi)+\cos \omega t\}$$
$$+(L'_d/2 - L_q)\,\omega \hat{I}\{\cos(3\omega t+2\psi)-\cos \omega t\}]$$
$$= R_a \hat{I} \sin \omega t + \frac{X'_d + X_q}{2}\hat{I}\cos \omega t + 3\frac{X'_d - X_q}{2}\hat{I}\cos(3\omega t+2\psi)$$

Since the current in the a phase is $i^a = \hat{I}\sin \omega t$, the fundamental component of the voltage includes the resistance drop $R_a i^a$ in phase with the current and $(X'_d+X_q)i^a/2$ leading the current by $\pi/2$. The reactance to negative-sequence current is therefore $(X'_d+X_q)/2$ which is the arithmetic mean of the direct-axis transient reactance and the quadrature-axis synchronous reactance.

There is also a large third harmonic voltage unless X'_d and X_q are approximately equal.

Impedance to Negative-sequence Voltage

Since the negative-sequence voltage is given and we require to find the negative-sequence current, it is necessary to invert the impedance matrix. We can, however, start from the matrix on p. 237 from which the field winding row and column have been eliminated. Furthermore we will neglect the armature resistance during the calculation of the reactance in order to reduce the algebra although this is not essential.

The impedance matrix is then

	d	q
d	$L'_d p$	$L_q \omega$
q	$-L'_d \omega$	$L_q p$

of which the determinant is $L_q L'_d(p^2+\omega^2)$ and the inverse is

$$\frac{1}{L_q L'_d(p^2+\omega^2)}$$

	d	q
d	$L_q p$	$-L_q \omega$
q	$L'_d \omega$	$L'_d p$

and the current equation is

$$\begin{array}{c|c|} d & i^{d} \\ \hline q & i^{q} \end{array} = \frac{1}{L_q L_d'(p^2+\omega^2)} \begin{array}{c|c|c|} & d & q \\ \hline d & L_q p & -L_q\omega \\ \hline q & L_d'\omega & L_d' p \\ \hline \end{array} \begin{array}{c|c|} d & v_d \\ \hline q & v_q \end{array}$$

The negative-sequence terminal voltage may be taken as

$$\begin{array}{c|c|} a & v_a \\ \hline b & v_b \\ \hline c & v_c \end{array} = \begin{array}{c|c|} a & \hat{V}\sin\omega t \\ \hline b & \hat{V}\sin(\omega t+2\pi/3) \\ \hline c & \hat{V}\sin(\omega t+4\pi/3) \end{array}$$

and transforming to d, q, o axes gives

$$\begin{array}{c|c|} d & v_d \\ \hline q & v_q \\ \hline o & v_o \end{array} = \sqrt{\tfrac{3}{2}} \begin{array}{c|c|} d & \hat{V}\sin(2\omega t+\psi) \\ \hline q & -\hat{V}\cos(2\omega t+\psi) \\ \hline o & \end{array}$$

from which the zero-sequence axis may be omitted, and where, as before, $\theta = \omega t+\psi$.

The Laplace transform equation is therefore

$$\begin{array}{c|c|} d & \bar{\imath}^{d} \\ \hline q & \bar{\imath}^{q} \end{array} = \sqrt{\frac{3}{2}} \frac{\hat{V}}{L_q L_d'(s^2+\omega^2)} \begin{array}{c|c|c|} & d & q \\ \hline d & L_q s & -L_q\omega \\ \hline q & L_d'\omega & L_d' s \\ \hline \end{array} \begin{array}{c|c|} d & \dfrac{2\omega\cos\psi+s\sin\psi}{s^2+(2\omega)^2} \\ \hline q & \dfrac{2\omega\sin\psi-s\cos\psi}{s^2+(2\omega)^2} \end{array}$$

$$= \sqrt{\frac{3}{2}} \frac{\hat{V}}{L_q L_d'(s^2+\omega^2)\{s^2+(2\omega)^2\}} \begin{array}{c|c|} d & L_q(s^2\sin\psi+3\omega s\cos\psi-2\omega^2\sin\psi) \\ \hline q & L_d'(-s^2\cos\psi+3\omega s\sin\psi+2\omega^2\cos\psi) \end{array}$$

Taking the two current transforms separately,

$$\bar{i}^{\mathrm{d}} = \sqrt{\frac{3}{2}}\frac{\hat{V}}{L'_{\mathrm{d}}}\left[\frac{s^2 \sin \psi + 3\omega s \cos \psi - 2\omega^2 \sin \psi}{(s^2+\omega^2)\,\{s^2+(2\omega)^2\}}\right]$$

$$= \sqrt{\frac{3}{2}}\frac{\hat{V}}{L'_{\mathrm{d}}}\left[\frac{As+B}{s^2+\omega^2}+\frac{Cs+D}{s^2+(2\omega)^2}\right] \quad \text{say.}$$

If we put $s = j\omega$,

$$(j\omega A+B)\,(3\omega^2) = -3\omega^2 \sin \psi + j3\omega^2 \cos \psi$$

from which,

$$A = \cos \psi/\omega \quad \text{and} \quad B = -\sin \psi$$

If we put $s = j2\omega$,

$$(j2\omega C+D)\,(-3\omega^2) = -6\omega^2 \sin \psi + j6\omega^2 \cos \psi$$

from which

$$C = -\cos \psi/\omega \quad \text{and} \quad D = 2 \sin \psi$$

Hence

$$\bar{i}^{\mathrm{d}} = \sqrt{\frac{3}{2}}\frac{\hat{V}}{\omega L'_{\mathrm{d}}}\left[\frac{s \cos \psi - \omega \sin \psi}{s^2+\omega^2}+\frac{-s \cos \psi + 2\omega \sin \psi}{s^2+(2\omega)^2}\right]$$

and

$$i^{\mathrm{d}} = \sqrt{\frac{3}{2}}\frac{\hat{V}}{X'_{\mathrm{d}}}\{(\cos \omega t \cos \psi - \sin \omega t \sin \psi)$$

$$+(-\cos 2\omega t \cos \psi + \sin 2\omega t \sin \psi)\}$$

$$= \sqrt{\frac{3}{2}}\frac{\hat{V}}{X'_{\mathrm{d}}}\{\cos (\omega t+\psi) - \cos (2\omega t+\psi)\}$$

We now have to find i^{q} by a similar process:

$$\bar{i}^{\mathrm{q}} = \sqrt{\frac{3}{2}}\frac{\hat{V}}{L_{\mathrm{q}}}\left[\frac{-s^2 \cos \psi + 3\omega s \sin \psi + 2\omega^2 \cos \psi}{(s^2+\omega^2)\,\{s^2+(2\omega)^2\}}\right]$$

$$= \sqrt{\frac{3}{2}}\frac{\hat{V}}{L_{\mathrm{q}}}\left[\frac{Es+F}{s^2+\omega^2}+\frac{Gs+H}{s^2+(2\omega)^2}\right] \quad \text{say.}$$

If we put $s = j\omega$,

$$(j\omega E+F)(3\omega^2) = 3\omega^2 \cos\psi + j3\omega^2 \sin\psi$$

from which

$$E = \sin\psi/\omega \quad \text{and} \quad F = \cos\psi$$

If we put $s = j2\omega$,

$$(j2\omega G+H)(-3\omega^2) = 6\omega^2 \cos\psi + j6\omega^2 \sin\psi$$

from which

$$G = -\sin\psi/\omega \quad \text{and} \quad H = -2\cos\psi$$

Hence

$$\bar{i}^{\mathrm{q}} = \sqrt{\frac{3}{2}}\,\frac{\hat{V}}{\omega L_{\mathrm{q}}}\left[\frac{s\sin\psi+\omega\cos\psi}{s^2+\omega^2}+\frac{-s\sin\psi-2\omega\cos\psi}{s^2+(2\omega)^2}\right]$$

and

$$i^{\mathrm{q}} = \sqrt{\frac{3}{2}}\,\frac{\hat{V}}{X_{\mathrm{q}}}\{(\cos\omega t\sin\psi+\sin\omega t\cos\psi)$$

$$-(\cos 2\omega t\sin\psi+\sin 2\omega t\cos\psi)\}$$

$$= \sqrt{\frac{3}{2}}\,\frac{\hat{V}}{X_{\mathrm{q}}}\{\sin(\omega t+\psi)-\sin(2\omega t+\psi)\}$$

We can find the a-phase current as

$$i^{\mathrm{a}} = \sqrt{\tfrac{2}{3}}\,\{i^{\mathrm{d}}\cos\theta+i^{\mathrm{q}}\sin\theta\}$$

$$= \frac{\hat{V}}{X'_{\mathrm{d}}}\{\cos\theta\cos(\omega t+\psi)-\cos\theta\cos(2\omega t+\psi)\}$$

$$+\frac{\hat{V}}{X_{\mathrm{q}}}\{\sin\theta\sin(\omega t+\psi)-\sin\theta\sin(2\omega t+\psi)\}$$

$$= \frac{\hat{V}}{2X'_{\mathrm{d}}}\{1+\cos 2(\omega t+\psi)-\cos\omega t-\cos(3\omega t+2\psi)\}$$

$$+\frac{\hat{V}}{2X_{\mathrm{q}}}\{1-\cos 2(\omega t+\psi)-\cos\omega t+\cos(3\omega t+2\psi)\}^{\dagger}$$

† Note: the d.c. component of i^{a} arises from the initial conditions. It is in fact a transient, but by neglecting the resistances we have lost the exponential factor. If we had taken v_{a} as $\hat{V}\cos\omega t$ instead of as $\hat{V}\sin\omega t$, this unidirectional component would not have appeared.

The magnitude of the fundamental component of i^a is thus $\hat{V}\{1/X_d'+1/X_q\}/2$, and the reactance is

$$1/[\tfrac{1}{2}\{1/X_d'+1/X_q\}] = 2X_d'X_q/(X_d'+X_q)$$

It will be noted that the value of the impedance differs according to whether the current or the voltage is assumed to be sinusoidal. When the current is sinusoidal, the effective reactance is the arithmetic mean of X_d' and X_q; when the voltage is sinusoidal, it is the harmonic mean of X_d' and X_q. The value appropriate in practice depends upon the external circuit, and since neither the current nor the voltage is likely to be entirely free of harmonic, the effective value may be expected to lie between these two limiting values.

It may also be noted that the β of p. 226 and consequently the armature short-circuit time-constant T_a, can be expressed in terms ofthe armature winding resistance and the reactance to negative-sequence voltage. There is, of course, an explanation for this in physical terms.

The Synchronous Machine with Salient Poles and Damper Windings

The preceding analysis of synchronous machines is idealized to the extent that no damping circuits have been included. We must now consider, as far as practicable, the effects of damper windings on the balanced steady-state, the transient and the negative-sequence performance.

The damper windings of synchronous machines are mechanically simple but electrically complicated devices. It is a considerable undertaking to analyse the performance of a machine with dampers taking into account their actual circuit arrangement. As an approximation therefore the damper windings may be treated as two separate windings, one each on the direct and quadrature axes. The diagrammatic arrangement of the machine is then as shown in Fig. 43 in which F is the field winding, D the direct-axis damper winding and Q the quadrature-axis damper winding.

In stationary reference axes the impedance matrix may be written down according to the rules on p. 74, although the number of windings is greater than has been considered hitherto. It is sometimes assumed that the mutual inductances are all equal when referred to the same number of turns. This assumption does not help in the

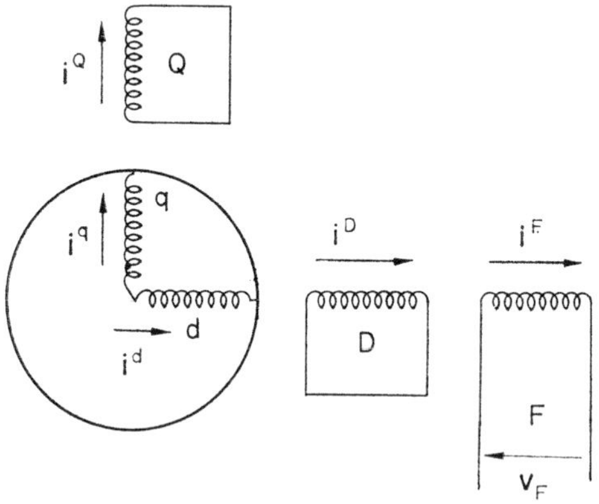

FIG. 43. Salient-pole synchronous machine with damper windings.

analysis but simplifies the visualization of the self-inductances as combinations of leakage and a common mutual flux. This assumption has not been made here.

	d	q	F	D	Q
d	$R_a+L_d p$	$L_q\dot{\theta}$	$M_{Fd}p$	$M_{Dd}p$	$M_{Qq}\dot{\theta}$
q	$-L_d\dot{\theta}$	$R_a+L_q p$	$-M_{Fd}\dot{\theta}$	$-M_{Dd}\dot{\theta}$	$M_{Qq}p$
F	$M_{Fd}p$		$R_F+L_F p$	$M_{DF}p$	
D	$M_{Dd}p$		$M_{DF}p$	$R_D+L_D p$	
Q		$M_{Qq}p$			$R_Q+L_Q p$

When the machine is operating synchronously with its terminal voltage, $\dot{\theta} = \omega$, at other speeds, e.g. during acceleration of a synchronous motor, $\dot{\theta} \neq \omega$.

Since the damper windings are closed windings, they have zero terminal voltage and it is advantageous to eliminate the D and Q axes. But before proceeding to do so, we will note that the determinant of this matrix is of fifth degree in p and, remembering the difficulty and the necessity for approximation in the case of the cubic equation of the machine without a damper, it is obviously impracticable to attempt an algebraic solution of this fifth degree equation without approximating. A numerical case can, however, be solved with any required degree of accuracy, so that any particular machine could be investigated without making the sweeping assumptions which we shall need to make in considering the general case.

Eliminating the D and Q axes we get an impedance

d	q	F
$R_a+L_d p$	$L_q\dot{\theta}$	$M_{Fd}p$
$-L_d\dot{\theta}$	$R_a+L_q p$	$-M_{Fd}\dot{\theta}$
$M_{Fd}p$		$R_F+L_F p$

	D	Q
d	$M_{Dd}p$	$M_{Qq}\dot{\theta}$
q	$-M_{Dd}\dot{\theta}$	$M_{Qq}p$
F	$M_{DF}p$	

	D	Q
D	$\dfrac{1}{R_D+L_D p}$	
Q		$\dfrac{1}{R_Q+L_Q p}$

	d	q	F
D	$M_{Dd}p$		$M_{DF}p$
Q		$M_{Qq}p$	

	d	q	F
d	$R_a+L_d p$	$L_q\dot{\theta}$	$M_{Fd}p$
q	$-L_d\dot{\theta}$	$R_a+L_q p$	$-M_{Fd}\dot{\theta}$
F	$M_{Fd}p$		$R_F+L_F p$

$$
\begin{array}{c|c|c|c|}
 & \text{d} & \text{q} & \text{F} \\
\hline
\text{d} & \dfrac{M_{\text{Dd}}^2 p^2}{R_{\text{D}}+L_{\text{D}}p} & \dfrac{M_{\text{Qq}}^2 \dot{\theta} p}{R_{\text{Q}}+L_{\text{Q}}p} & \dfrac{M_{\text{Dd}} M_{\text{DF}} p^2}{R_{\text{D}}+L_{\text{D}}p} \\
\hline
-\ \text{q} & -\dfrac{M_{\text{Dd}}^2 \dot{\theta} p}{R_{\text{D}}+L_{\text{D}}p} & \dfrac{M_{\text{Qq}}^2 p^2}{R_{\text{Q}}+L_{\text{Q}}p} & -\dfrac{M_{\text{Dd}} M_{\text{DF}} \dot{\theta} p}{R_{\text{D}}+L_{\text{D}}p} \\
\hline
\text{F} & \dfrac{M_{\text{DF}} M_{\text{Dd}} p^2}{R_{\text{D}}+L_{\text{D}}p} & & \dfrac{M_{\text{DF}}^2 p^2}{R_{\text{D}}+L_{\text{D}}p} \\
\hline
\end{array}
$$

$$
\begin{array}{c|c|c|c}
 & \text{d} & \text{q} & \text{F} \\
\hline
\text{d} & R_a+\left[L_d-\dfrac{M_{\text{Dd}}^2 p}{R_{\text{D}}+L_{\text{D}}p}\right]p & \left[L_q-\dfrac{M_{\text{Qq}}^2 p}{R_{\text{Q}}+L_{\text{Q}}p}\right]\dot{\theta} & \left[M_{\text{Fd}}-\dfrac{M_{\text{DF}} M_{\text{Dd}} p}{R_{\text{D}}+L_{\text{D}}p}\right] \\
\hline
=\ \text{q} & -\left[L_d-\dfrac{M_{\text{Dd}}^2 p}{R_{\text{D}}+L_{\text{D}}p}\right]\dot{\theta} & R_a+\left[L_q-\dfrac{M_{\text{Qq}}^2 p}{R_{\text{Q}}+L_{\text{Q}}p}\right]p & -\left[M_{\text{Fd}}-\dfrac{M_{\text{DF}} M_{\text{Dd}} p}{R_{\text{D}}+L_{\text{D}}p}\right. \\
\hline
\text{F} & \left[M_{\text{Fd}}-\dfrac{M_{\text{DF}} M_{\text{Dd}} p}{R_{\text{D}}+L_{\text{D}}p}\right]p & & R_{\text{F}}+\left[L_{\text{F}}-\dfrac{M_{\text{DF}}^2 p}{R_{\text{D}}+L_{\text{D}}p}\right. \\
\hline
\end{array}
$$

Balanced Steady-state Conditions

Under balanced steady-state conditions with a d.c. voltage applied to the field winding, the impedance is obtained by replacing p by zero, and is identical with that of the machine without damper windings.

Transient Conditions

We need, however, to consider further the transient and unbalanced conditions.

The determinant of the above matrix will still lead to a fifth degree equation and to effect any reduction in the degree we have to neglect the resistances of the damper windings. This may be justified for generators for which low resistance dampers are ideal, but for synchronous motors it may be a wholly false assumption, since high resistance dampers may be used to obtain adequate starting torque. We have, however, no alternative, and the resulting impedance matrix is

	d	q	F
d	$R_a+\left[L_d-\dfrac{M_{Dd}^2}{L_D}\right]p$	$\left[L_q-\dfrac{M_{Qq}^2}{L_Q}\right]\dot{\theta}$	$\left[M_{Fd}-\dfrac{M_{DF}M_{Dd}}{L_D}\right]p$
q	$-\left[L_d-\dfrac{M_{Dd}^2}{L_D}\right]\dot{\theta}$	$R_a+\left[L_q-\dfrac{M_{Qq}^2}{L_Q}\right]p$	$-\left[M_{Fd}-\dfrac{M_{DF}M_{Dd}}{L_D}\right]\dot{\theta}$
F	$\left[M_{Fd}-\dfrac{M_{DF}M_{Dd}}{L_D}\right]p$		$R_F+\left[L_F-\dfrac{M_{DF}^2}{L_D}\right]p$

This impedance matrix is of exactly the same form as that of the machine without a damper given on p. 197 except that the inductances are more complicated. The determinant is now a cubic in p, since each expression in brackets reduces to a single inductance.

Sudden Three-phase Short Circuit from Open Circuit

The results of the earlier analysis of the sudden three-phase short circuit of the machine without a damper can be used here to save effort, by replacing L_d by $\{L_d-(M_{Dd}^2/L_D)\}$ and so on. The components of current which flow in the armature and field windings will be exactly as before except that their magnitudes and decrements may be different. We must remember, however, that we have neglected the damper resistances and hence we have lost two time-constants which we must endeavour to recover later.

Without a damper winding the effective direct-axis reactance was X'_d,† corresponding to a transient inductance L'_d having the value $\{L_d-(M_{Fd}^2/L_F)\}$. In the present impedance matrix, however, where L_d formerly appeared, there is now $\{L_d-(M_{Dd}^2/L_D)\}$, $\{M_{Fd}-(M_{DF}M_{Dd}/L_D)\}$ appears in place of M_{Fd} and $\{L_F-(M_{DF}^2/L_D)\}$ appears in place of L_F. Hence we can immediately say that the effective

† See p. 234.

inductance on the direct axis in place of L'_d is

$$L''_d = \left\{L_d - \frac{M^2_{Dd}}{L_D}\right\} - \frac{\left\{M_{Fd} - \dfrac{M_{DF}M_{Dd}}{L_D}\right\}^2}{\left\{L_F - \dfrac{M^2_{DF}}{L_D}\right\}}$$

$$= L_d - \frac{M^2_{Dd}}{L_D} - \frac{L^2_D M^2_{Fd} - 2L_D M_{Fd} M_{DF} M_{Dd} + M^2_{DF} M^2_{Dd}}{L_D(L_D L_F - M^2_{DF})}$$

$$= L_d - \frac{\begin{bmatrix} L_D L_F M^2_{Dd} - M^2_{Dd} M^2_{DF} + L^2_D M^2_{Fd} \\ -2L_D M_{Fd} M_{DF} M_{Dd} + M^2_{DF} M^2_{Dd} \end{bmatrix}}{L_D(L_D L_F - M^2_{DF})}$$

$$= L_d - \frac{L_F M^2_{Dd} - 2M_{Fd} M_{DF} M_{Dd} + L_D M^2_{Fd}}{L_D L_F - M^2_{DF}}$$

L''_d is called the direct-axis sub-transient inductance, and the corresponding reactance $\omega L''_d = X''_d$ is the direct-axis sub-transient reactance.

The magnitude of the a.c. component of the phase current immediately after sudden short circuit will now be E/X''_d. This can be written

$$\left\{\frac{E}{X''_d} - \frac{E}{X'_d}\right\} + \left\{\frac{E}{X'_d} - \frac{E}{X_d}\right\} + \frac{E}{X_d}$$

where E/X_d is the steady-state component, as in the machine without dampers; $(E/X'_d - E/X_d)$ is the "transient" component as in the machine without dampers; and $(E/X''_d - E/X'_d)$ is the "sub-transient" component due to the presence of the dampers and which accordingly has no counterpart in the machine without dampers.

It will have been noticed that in the case without a damper, the time-constant of the a.c. transient armature current is the same as that of the induced unidirectional current in the field winding. Here a similar analysis would show that the decrements of the a.c. transient and sub-transient components of the armature current are the same as those of the induced undirectional components in the field and damper windings respectively. We have, of course, no means of discovering the time-constants of the sub-transient component if we neglect the damper winding resistances. Other than in a numerical case we can proceed only by using the empirical knowledge that the time-constants of the damper windings are much less than those of the armature and field windings and that consequently the decrement due to the resistances of the latter two windings is negligible compared with that of the damper winding itself. We therefore neglect the resistances of the armature and field circuits whilst calculating the time-constants of the damper windings. With this approximation the impedance matrix has the form

	d	q	F	D	Q
d	$L_d p$	$L_q\dot{\theta}$	$M_{Fd}p$	$M_{Dd}p$	$M_{Qq}\dot{\theta}$
q	$-L_d\dot{\theta}$	$L_q p$	$-M_{Fd}\dot{\theta}$	$-M_{Dd}\dot{\theta}$	$M_{Qq}p$
F	$M_{Fd}p$		$L_F p$	$M_{DF}p$	
D	$M_{Dd}p$		$M_{DF}p$	$R_D+L_D p$	
Q		$M_{Qq}p$			$R_Q+L_Q p$

The determinant of this matrix is

$$\left\{R_Q+\left(L_Q-\frac{M_{Qq}^2}{L_q}\right)p\right\}L_q p(p^2+\dot{\theta}^2)$$
$$\times\{(L_D L_d L_F-L_D M_{Fd}^2-L_d M_{DF}^2+2M_{DF}M_{Fd}M_{Dd}-L_F M_{Dd}^2)p+(L_d L_F-M_{Fd}^2)R_D\}$$

We thus have a "direct-axis sub-transient short-circuit time constant"

$$T_d''=\frac{L_D L_d L_F-L_D M_{Fd}^2-L_d M_{DF}^2+2M_{DF}M_{Fd}M_{Dd}-L_F M_{Dd}^2}{R_D(L_d L_F-M_{Fd}^2)}$$

$$=\frac{L_D L_F-M_{DF}^2}{L_d L_F-M_{Fd}^2}\times\frac{L_D L_d L_F-L_D M_{Fd}^2-L_d M_{DF}^2+2M_{DF}M_{Fd}M_{Dd}-L_F M_{Dd}^2}{R_D(L_D L_F-M_{DF}^2)}$$

$$= \frac{L_D - M_{DF}^2/L_F}{L_d - M_{Fd}^2/L_F} \frac{1}{R_D} \left(L_d - \frac{L_F M_{Dd}^2 - 2M_{Fd} M_{DF} M_{Dd} + L_D M_{Fd}^2}{L_F L_D - M_{DF}^2} \right)$$

$$= \frac{L_D - M_{DF}^2/L_F}{L'_d} \frac{L''_d}{R_D}$$

$$= \frac{L_D - M_{DF}^2/L_F}{R_D} \frac{L''_d}{L'_d}$$

which relates to the unidirectional current in the direct-axis damper winding and hence also to the corresponding sub-transient component of alternating current in the armature windings.

We also have a "quadrature-axis sub-transient short-circuit time constant"

$$T''_q = \frac{L_Q - M_{Qq}^2/L_q}{R_Q} = \frac{L_Q}{R_Q} \frac{L_q - M_{Qq}^2/L_Q}{L_q}$$

$$= \frac{L_Q}{R_Q} \frac{L''_q}{L_q} \quad \text{where} \quad L''_q = L_q - M_{Qq}^2/L_Q$$

T''_q relates to any unidirectional induced current in the quadrature-axis damper winding, and is not relevant to our present case.

The reactance $X''_q = \omega L''_q$ is called the "quadrature-axis sub-transient reactance".

The value of the field winding time-constant being much greater than that of the direct-axis damper winding, there will be but little decrease in the induced unidirectional current in the field winding by the time that that of the damper winding has become negligibly small and the damper will play a negligible part in determining the rate at which this component of the field current decays. The corresponding alternating component in the armature winding, which is of initial magnitude $\{E/X'_d - E/X_d\}$, will thus have effectively the same time-constant as in the machine without a damper winding, namely T'_d, the direct-axis transient short-circuit time-constant.

The magnitude of the alternating component of the armature current after sudden three-phase short-circuit is thus approximately

$$E\left[\left\{\frac{1}{X''_d} - \frac{1}{X'_d}\right\}\varepsilon^{-t/T''_d} + \left\{\frac{1}{X'_d} - \frac{1}{X_d}\right\}\varepsilon^{-t/T'_d} + \frac{1}{X_d}\right]$$

The armature time-constant T_a which will determine the decrement of the unidirectional component of the armature current and the alternating component of the field and damper currents is different from the value without a damper, but can be found by modifying the inductances in accordance with the changes from the impedance matrix of the machine without a damper to the impedance matrix of p. 247. In the machine without a damper it is

$$\frac{2}{R_a\left(\frac{1}{L'_d}+\frac{1}{L_q}\right)} = \frac{2L'_d L_q}{R_a(L'_d+L_q)}$$

hence in the machine with a damper it will be

$$\frac{2L''_d L''_q}{R_a(L''_d+L''_q)}$$

To summarize therefore, in a machine with damper windings under transient conditions in general, the current components and their time-constants are as follows:

The armature carries:

(i) a unidirectional transient current with a time-constant T_a;
(ii) an alternating component corresponding to unidirectional current in the direct-axis damper with a time-constant T''_d;
(iii) an alternating component corresponding to unidirectional transient current in the field winding with a time-constant T'_d;
(iv) a steady-state alternating component;
(v) an alternating component corresponding to unidirectional current in the quadrature-axis damper with a time-constant T''_q;
(vi) a second harmonic alternating component with the same time-constant as the unidirectional component, namely T_a.

The field winding carries:

(i) an alternating component corresponding to the unidirectional component of the armature current with a time-constant T_a;
(ii) an induced unidirectional current with a time-constant T'_d;
(iii) the steady-state direct current.

The direct-axis damper winding carries:

(i) an alternating component corresponding to the unidirectional component of the armature current with a time-constant T_a;
(ii) an induced unidirectional current with a time-constant T''_d.

The quadrature-axis damper winding carries:

(i) an alternating component corresponding to the unidirectional component of the armature current with a time-constant T_a;
(ii) an induced unidirectional current with a time-constant T''_q.

Impedance to Negative Sequence

The reactance to negative-sequence current of a machine without damper windings was deduced† as $\omega(L'_d+L_q)/2$, i.e. as $\omega(L_d-M^2_{Fd}/L_F+L_q)/2$.

For a machine with damper windings this will become

$$\omega\left[\left(L_d-\frac{M^2_{Dd}}{L_D}\right)-\frac{\left(M_{Fd}-\dfrac{M_{DF}M_{Dd}}{L_D}\right)^2}{L_F-\dfrac{M^2_{DF}}{L_D}}+\left(L_q-\frac{M^2_{Qq}}{L_Q}\right)\right]\cdot\frac{1}{2}$$

$$=\omega(L''_d+L''_q)/2=(X''_d+X''_q)/2$$

Similarly, the reactance to negative-sequence voltage of the machine without damper circuits was $2\omega L'_dL_q/(L'_d+L_q)$.‡ The corresponding value for the machine with damper windings will be $2\omega L''_dL''_q/(L''_d+L''_q)=2X''_dX''_q/(X''_d+X''_q)$.

It can now be seen that the armature short-circuit time-constant T_a of a machine with damper windings can be expressed simply in terms of the armature winding resistance and the reactance to negative-sequence voltage as for a machine without damper windings in the form $X_n/(\omega R_a)$. Since this is a more general expression than the others, it is to be preferred.

† See p. 237.
‡ See p. 239.

Parameters

The values of X'_d and X''_d can be determined from the oscillographic records of a sudden symmetrical short-circuit test using the analysis of this condition given in pp. 247–51. The measurement of X''_q is, however, not so simple a matter.

X'_d, X''_d, and X''_q are the "direct-axis transient reactance", the "direct-axis sub-transient reactance", and the "quadrature-axis sub-transient reactance" of classical theory.

The time-constants can be determined by plotting the logarithms of the relevant components of the currents, when the curves should be effectively straight and the slope is the inverse of the time-constant.

The negative-sequence impedances can be calculated from the values of X''_d and X''_q (or X'_d and X_q in the absence of any damping circuits) or, where practicable, measured directly as the ratio of fundamental voltage to fundamental current with a negative-sequence voltage applied to the armature and the field winding short-circuited.

These various tests are described in ref. 18.

CHAPTER 12

Small Oscillations

IF A MACHINE operating at a constant speed under steady load conditions is subject to a step function or impulse function change of load or terminal voltage, it will normally return to steady-state constant speed conditions with or without oscillations of speed and current about the steady-state values. Again, if a machine is subject to periodic variations of voltage or torque about a mean value, there will be associated periodic variations of speed and current. These phenomena are investigated under the title of "small oscillations". The significance of the word "small" is that the variations about the mean value are assumed to be linear, which would not in general be justifiable if the oscillations were of large amplitude.

The mechanical time-constants of electrical machines and their mechanically coupled equipment are usually so great compared with the electrical time-constants that it is sufficiently accurate to use the steady-state electrical equations to determine the currents and electrodynamic torque during acceleration and during oscillation. This is, however, an approximation which is not justified in the case of such machines as low-inertia servo-motors.

In terms of stationary reference axes, we have the voltage equation

$$\mathbf{v} = \mathbf{Ri} + \mathbf{L}p\mathbf{i} + \mathbf{G}\dot{\theta}\mathbf{i}$$

and an analogous mechanical equation

$$\begin{aligned}\text{Applied torque} &= \text{friction torque} + \text{inertia torque}\\ &\quad - \text{electrodynamic (motoring) torque}\end{aligned}$$

in which it is assumed for convenience that the applied torque is conventionally in the same direction as the electrodynamic torque. For a two-pole machine this equation may be written†

$$T_s = F\dot{\theta} + Jp\dot{\theta} - \mathbf{i}_t\mathbf{G}\mathbf{i}$$

in which T_s is the shaft torque, $F\dot{\theta}$ is the friction torque, assumed to be proportional to speed, i.e. "viscous", $Jp\dot{\theta} = J\,\mathrm{d}^2\theta/\mathrm{d}t^2$ is the inertia torque in which J is the moment of inertia, and $\mathbf{i}_t\mathbf{G}\mathbf{i}$ is the electrodynamic torque in the expression for which, for the present, the currents are presumed to be real, i.e. not complex.

Suppose that the system is running under steady-state constant speed conditions represented by the equations above, when there is a small displacement from these steady-state conditions, so that the equations may be written

$$(\mathbf{v}+\Delta\mathbf{v}) = \mathbf{R}(\mathbf{i}+\Delta\mathbf{i}) + \mathbf{L}p(\mathbf{i}+\Delta\mathbf{i}) + \mathbf{G}(\dot{\theta}+\Delta\dot{\theta})\,(\mathbf{i}+\Delta\mathbf{i})$$
$$(T_s+\Delta T_s) = F(\dot{\theta}+\Delta\dot{\theta}) + Jp(\dot{\theta}+\Delta\dot{\theta}) - (\mathbf{i}+\Delta\mathbf{i})_t\,\mathbf{G}(\mathbf{i}+\Delta\mathbf{i})$$

Now $(\mathbf{i}+\Delta\mathbf{i})_t = (\mathbf{i}_t+\Delta\mathbf{i}_t)$, and multiplying thes eequations out, gives

$$\mathbf{v}+\Delta\mathbf{v} = \mathbf{R}\mathbf{i} + \mathbf{R}\,\Delta\mathbf{i} + \mathbf{L}p\mathbf{i} + \mathbf{L}p\,\Delta\mathbf{i} + \mathbf{G}\dot{\theta}\mathbf{i} + \mathbf{G}\,\Delta\dot{\theta}\mathbf{i} + \mathbf{G}\dot{\theta}\,\Delta\mathbf{i} + \mathbf{G}\,\Delta\dot{\theta}\,\Delta\mathbf{i}$$
$$T_s+\Delta T_s = F\dot{\theta} + F\,\Delta\dot{\theta} + Jp\dot{\theta} + Jp\,\Delta\dot{\theta} - \mathbf{i}_t\mathbf{G}\mathbf{i} - \Delta\mathbf{i}_t\mathbf{G}\mathbf{i} - \mathbf{i}_t\mathbf{G}\,\Delta\mathbf{i} - \Delta\mathbf{i}_t\mathbf{G}\,\Delta\mathbf{i}$$

In both cases the last terms are second-order small quantities and may be neglected. Then subtracting the steady-state equations from these equations gives the equations between the small displacements as

$$\Delta\mathbf{v} = \mathbf{R}\,\Delta\mathbf{i} + \mathbf{L}p\,\Delta\mathbf{i} + \mathbf{G}\Delta\dot{\theta}\mathbf{i} + \mathbf{G}\dot{\theta}\,\Delta\mathbf{i}$$
$$\Delta T_s = F\,\Delta\dot{\theta} + Jp\,\Delta\dot{\theta} - \Delta\mathbf{i}_t\mathbf{G}\mathbf{i} - \mathbf{i}_t\mathbf{G}\,\Delta\mathbf{i}$$

Now $\Delta\dot{\theta}$ is a scalar so that $\mathbf{G}\,\Delta\dot{\theta}\mathbf{i} = \mathbf{G}\mathbf{i}\,\Delta\dot{\theta}$ and $\Delta\mathbf{i}_t\mathbf{G}\mathbf{i}$ is a scalar and is therefore equal to its own transpose $\mathbf{i}_t\mathbf{G}_t\,\Delta\mathbf{i}$.

The equations may therefore be written

$$\Delta\mathbf{v} = (\mathbf{R}+\mathbf{L}p+\mathbf{G}\dot{\theta})\,\Delta\mathbf{i} + \mathbf{G}\mathbf{i}\,\Delta\dot{\theta}$$
$$\Delta T_s = (F+Jp)\,\Delta\dot{\theta} - \mathbf{i}_t\mathbf{G}_t\,\Delta\mathbf{i} - \mathbf{i}_t\mathbf{G}\,\Delta\mathbf{i}$$
$$= -\mathbf{i}_t(\mathbf{G}+\mathbf{G}_t)\,\Delta\mathbf{i} + (F+Jp)\,\Delta\dot{\theta}$$

† For a machine with p pole-pairs, $\dot{\theta}$ will be replaced by $\dot{\theta}/\mathrm{p}$, and $\mathbf{i}_t\mathbf{G}\mathbf{i}$ by $\mathrm{p}\mathbf{i}_t\mathbf{G}\mathbf{i}$.

or as a compound matrix equation

$$\begin{bmatrix} \Delta \mathbf{v} \\ \Delta T_s \end{bmatrix} = \begin{bmatrix} \mathbf{R}+\mathbf{L}p+\mathbf{G}\dot{\theta} & \mathbf{Gi} \\ -\mathbf{i}_t(\mathbf{G}+\mathbf{G}_t) & F+Jp \end{bmatrix} \begin{bmatrix} \Delta \mathbf{i} \\ \Delta \dot{\theta} \end{bmatrix}$$

in which $\dot{\theta}$, $\mathbf{i}$, and $\mathbf{i}_t$ are the steady-state or equilibrium values.

This equation thus gives the relationship between small changes of applied voltage and torque and the corresponding small changes of current and speed from the steady-state values.

$\Delta\mathbf{v}$, $\Delta\mathbf{i}$, and $\mathbf{Gi}$ are column matrices; $-\mathbf{i}_t(\mathbf{G}+\mathbf{G}_t)$ is a row matrix, $(\mathbf{R}+\mathbf{L}p+\mathbf{G}\dot{\theta})$ is a square matrix and is the normal electrical transient impedance matrix $\mathbf{Z}$ for a constant speed $\dot{\theta}$. ΔT_s, $F+Jp$ and $\Delta\dot{\theta}$ are one-by-one matrices, i.e. scalars. The complete square compound matrix is called the motional impedance matrix and is commonly given the symbol $\mathbf{Ƶ}$.

In the application of the above analysis, it must be remembered that it is expressed in terms of, and applies only to, stationary reference axes, i.e. axes that are all at rest relative to one another. It is not necessary for such axes to be at rest relative to either stator or rotor.

It is apparent that two cases arise:

(i) when $\mathbf{i}$ and $\mathbf{i}_t$ are constant, i.e. d.c. and hence real in the mathematical sense; and

(ii) when $\mathbf{i}$ and $\mathbf{i}_t$ are a.c. so that in the electrical equations, and hence here also, they are represented by complex numbers.

Only the first of these is adequately covered by the above analysis. All cases of d.c. machines and balanced a.c. machines on balanced voltage can be treated in these terms, however, by choosing reference axes which are at rest relative to the main flux. This has already been seen to be the case in the analysis of the synchronous machine on pp. 190 and 198, and will be shown later for the balanced induction machine on pp. 260–5.

The second case necessarily arises when the system is in any way unbalanced and a technique for its treatment has been given by Kron.†

† See ref. 2, vol. 39, June 1936, p. 302. See also ref. 19, pp. 61–62.

Separately Excited D.C. Machine

The transient impedance matrix of a separately excited d.c. machine is

$$\mathbf{Z} = \begin{array}{c|c|c|} & \mathbf{a} & \mathbf{F} \\ \hline \mathrm{a} & R_a + L_a p & -M\dot{\theta} \\ \hline \mathrm{F} & & R_F + L_F p \\ \hline \end{array}$$

from which

$$\mathbf{G} = \begin{array}{c|c|c|} & \mathrm{a} & \mathrm{F} \\ \hline \mathrm{a} & & -M \\ \hline \mathrm{F} & & \\ \hline \end{array} \quad \text{and} \quad \mathbf{G}_t = \begin{array}{c|c|c|} & \mathrm{a} & \mathrm{F} \\ \hline \mathrm{a} & & \\ \hline \mathrm{F} & -M & \\ \hline \end{array}$$

Therefore

$$\mathbf{G}+\mathbf{G}_t = \begin{array}{c|c|c|} & \mathrm{a} & \mathrm{F} \\ \hline \mathrm{a} & & -M \\ \hline \mathrm{F} & -M & \\ \hline \end{array}$$

so that

$$\mathbf{Gi} = \begin{array}{c|c|c|} & \mathrm{a} & \mathrm{F} \\ \hline \mathrm{a} & & -M \\ \hline \mathrm{F} & & \\ \hline \end{array} \; \begin{array}{c|c|} \hline \mathrm{a} & I^a \\ \hline \mathrm{F} & I^F \\ \hline \end{array} = \begin{array}{c|c|} \hline \mathrm{a} & -MI^F \\ \hline \mathrm{F} & \\ \hline \end{array}$$

and

$$-\mathbf{i}_t(\mathbf{G}+\mathbf{G}_t) = - \begin{array}{|c|c|} \mathrm{a} & \mathrm{F} \\ \hline I^a & I^F \\ \hline \end{array} \; \begin{array}{c|c|c|} & \mathrm{a} & \mathrm{F} \\ \hline \mathrm{a} & & -M \\ \hline \mathrm{F} & -M & \\ \hline \end{array}$$

$$= \begin{array}{|c|c|} \mathrm{a} & \mathrm{F} \\ \hline MI^F & MI^a \\ \hline \end{array}$$

The steady-state impedance matrix is

$$\mathbf{Z} = \begin{array}{c|c|c|} & \mathrm{a} & \mathrm{F} \\ \hline \mathrm{a} & R_a & -M\dot{\theta} \\ \hline \mathrm{F} & & R_F \\ \hline \end{array}$$

of which the inverse is

$$\mathbf{Z}^{-1} = \frac{1}{R_a R_F} \begin{array}{c|c|c|} & \text{a} & \text{F} \\ \hline \text{a} & R_F & M\dot{\theta} \\ \hline \text{F} & & R_a \\ \hline \end{array}$$

so that if the terminal voltages V_a and V_F are constant, the currents I^a and I^F above are given by

$$\mathbf{Z}^{-1}\mathbf{v} = \frac{1}{R_a R_F} \begin{array}{c|c|c|} & \text{a} & \text{F} \\ \hline \text{a} & R_F & M\dot{\theta} \\ \hline \text{F} & & R_a \\ \hline \end{array} \begin{array}{c|c|} \hline \text{a} & V_a \\ \hline \text{F} & V_F \\ \hline \end{array}$$

$$= \begin{array}{c|c|} \hline \text{a} & V_a/R_a + (M\dot{\theta}/R_a)(V_F/R_F) \\ \hline \text{F} & V_F/R_F \\ \hline \end{array}$$

The motional impedance matrix is therefore

$$\mathbf{Z} = \begin{array}{c|c|c|c|} & \text{a} & \text{F} & \text{s} \\ \hline \text{a} & R_a + L_a p & -M\dot{\theta} & -MI^F \\ \hline \text{F} & & R_F + L_F p & \\ \hline \text{s} & MI^F & MI^a & F + Jp \\ \hline \end{array}$$

where s is used as the index for the mechanical axis.

The determinant of this matrix is

$$= (R_F + L_F p)\{(R_a + L_a p)(F + Jp) + M^2 (I^F)^2\}$$
$$= (R_F + L_F p)\{L_a J p^2 + (L_a F + R_a J)p + R_a F + M^2 (I^F)^2\}$$

The factors of this are thus

$$L_F L_a J (p + R_F/L_F) \left\{ p + \frac{L_a F + JR_a + \sqrt{\{(L_a F - JR_a)^2 - 4L_a J M^2 (I^F)^2\}}}{2L_a J} \right\}$$
$$\times \left\{ p + \frac{L_a F + JR_a - \sqrt{\{(L_a F - JR_a)^2 - 4L_a J M^2 (I^F)^2\}}}{2_a LJ} \right\}$$

The factor $(p+R_F/L_F)$ will lead to a decaying unidirectional component of the response. The last pair of factors will lead to an oscillatory response if $4L_aJM^2(I^F)^2 > (L_aF-JR_a)^2$, which is probable, but again this component will decay since the real part of the factor is then positive. If $(L_aF-JR_a)^2 > 4L_aJM^2(I^F)^2$, all factors will have positive real parts and the transient response will consist of three decaying unidirectional components. In all cases, therefore, the system is inherently stable. The inverse of the motional impedance matrix is

$$
{}^{-1} = \frac{1}{\Delta}\;
\begin{array}{c|c|c|c|}
 & \mathrm{a} & \mathrm{F} & \mathrm{s} \\
\hline
\mathrm{a} & (R_F+L_Fp)(F+Jp) & M\dot{\theta}(F+Jp)-M^2I^FI^a & MI^F(R_F+L_Fp) \\
\hline
\mathrm{F} & & \begin{array}{c}(R_a+L_ap)(F+Jp)\\ +M^2(I^F)^2\end{array} & \\
\hline
\mathrm{s} & -MI^F(R_F+L_Fp) & \begin{array}{c}-MI^a(R_a+L_ap)\\ -M^2\dot{\theta}I^F\end{array} & (R_F+L_Fp)(R_a+L_ap) \\
\hline
\end{array}
$$

If the supply voltages are constant and a sinusoidal disturbance of angular frequency h times the angular velocity $\dot{\theta}$ is applied to the shaft, we may continue the analysis of the oscillations in complex form although the steady-state conditions are represented by real quantities. If we write ω for $\dot{\theta}$, since the angular frequency of the disturbance is hω, we may replace p by jhω and since

$$
\begin{array}{c|c|}
\mathrm{a} & \Delta v_a \\
\hline
\mathrm{F} & \Delta v_F \\
\hline
\mathrm{s} & \Delta T_s
\end{array}
\;=\;
\begin{array}{c|c|}
\mathrm{a} & 0 \\
\hline
\mathrm{F} & 0 \\
\hline
\mathrm{s} & \Delta T_s
\end{array}
$$

we need only the last column of $\mathbf{Z}^{-1}$ to obtain the response in complex form as

$$
\begin{array}{c|c|}
\mathrm{a} & \Delta i^a \\
\hline
\mathrm{F} & \Delta i^F \\
\hline
\mathrm{s} & \Delta\dot{\vartheta}
\end{array}
\;=\;\frac{1}{\Delta}\;
\begin{array}{c|c|}
\mathrm{a} & MI^F(R_F+j\mathrm{h}X_F)\Delta T_s \\
\hline
\mathrm{F} & \\
\hline
\mathrm{s} & (R_F+j\mathrm{h}X_F)(R_a+j\mathrm{h}X_a)\Delta T_s
\end{array}
$$

where now

$$\Delta = (R_F + jhX_F)[\{R_aF + M^2(I^F)^2 - h^2X_a\omega J\} + jh\{X_aF + R_a\omega J\}]$$

Hence

$$\begin{array}{c|c|} a & \Delta i^a \\ F & \Delta i^F \\ s & \Delta\dot{\theta} \end{array} = \begin{array}{c|c|} a & MI^F\Delta T_s/[\{R_aF + M^2(I^F)^2 - h^2X_a\omega J\} + jh\{X_aF + R_a\omega J\}] \\ F & \\ s & (R_a + jhX_a)\Delta T_s/[\{R_aF + M^2(I^F)^2 - h^2X_a\omega J\} + jh\{X_aF + R_a\omega J\}] \end{array}$$

The field current thus remains constant but there are sinusoidal variations in the armature current and speed, the amplitude and phase of which are determined by this equation.

Balanced Induction Machine

The transient impedance matrix of a balanced induction machine in terms of axes stationary relative to the stator is

$\mathbf{Z} =$

	D	Q	p	q
D	$R_S + L_S p$		Mp	
Q		$R_S + L_S p$		Mp
d	Mp	$M\dot{\theta}$	$R_r + L_r p$	$L_r\dot{\theta}$
q	$-M\dot{\theta}$	Mp	$-L_r\dot{\theta}$	$R_r + L_r p$

If a transformation

$\mathbf{C} =$ †

	A	B	a	b
D	$\cos\phi$	$\sin\phi$		
Q	$\sin\phi$	$-\cos\phi$		
d			$\cos\phi$	$\sin\phi$
q			$\sin\phi$	$-\cos\phi$

† The use of the indices A, B, a, b here does not imply any connection with two- or three-phase axes.

is applied, the axes are transformed to new orthogonal axes at an angle ϕ to the D, Q, d, q system† as illustrated in Fig. 44. If ϕ is constant and equal to $\pi/2$, it is clear that no change is made to the matrix beyond replacing the indices D, Q, d, q by B, A, b, a respectively. If however, $\phi = \theta$, where $\dot{\theta}$ is, as usual, the angular velocity of the rotor,

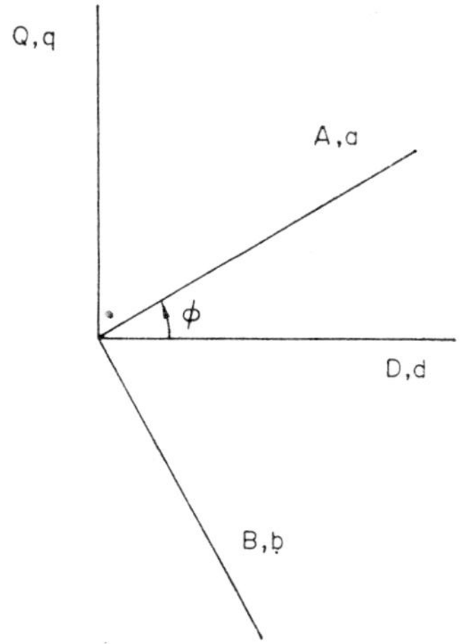

FIG. 44. Rotating axes.

the axes A, B, a, b are all stationary relative to the rotor. Yet again, if ϕ is varying with time so that $\dot{\phi} = \omega$, where ω is the angular frequency of the balanced excitation in D, Q, d, q axes, the new axes are all stationary relative to the rotating field, and are described as "synchronous" axes. They are obviously not at rest relative to any part of the physical structure except when the rotor happens to be revolving at synchronous speed. They are, nevertheless, "stationary", being at rest relative to one another. This would equally be true if $\dot{\phi}$ were chosen arbitrarily. We will, therefore, proceed without, as yet, any restriction on $\dot{\phi}$. Performing the first step of the transformation we get

† This may be verified by the technique used on p. 115.

ZC =

	A	B	a	b
D	$(R_S+L_Sp)\cos\phi$	$(R_S+L_Sp)\sin\phi$	$Mp\cos\phi$	$Mp\sin\phi$
Q	$(R_S+L_Sp)\sin\phi$	$-(R_S+L_Sp)\cos\phi$	$Mp\sin\phi$	$-Mp\cos\phi$
d	$M(p\cos\phi+\dot{\theta}\sin\phi)$	$M(p\sin\phi-\dot{\theta}\cos\phi)$	$(R_r+L_rp)\cos\phi$ $+L_r\dot{\theta}\sin\phi$	$(R_r+L_rp)\sin\phi$ $-L_r\dot{\theta}\cos\phi$
q	$M(p\sin\phi-\dot{\theta}\cos\phi)$	$-M(p\cos\phi+\dot{\theta}\sin\phi)$	$(R_r+L_rp)\sin\phi$ $-L_r\dot{\theta}\cos\phi$	$-(R_r+L_rp)\cos\phi$ $-L_r\dot{\theta}\sin\phi$

It has to be remembered that in this matrix the p's operate both on the functions of ϕ and on the currents i of the implied associated current matrix as products in a manner similar to that encountered on p. 59. Performing the differentiations,

ZC =

	A	B	a	b
D	$\cos\phi(R_S+L_Sp)$ $-\sin\phi L_S\dot{\phi}$	$\sin\phi(R_S+L_Sp)$ $+\cos\phi L_S\dot{\phi}$	$\cos\phi Mp$ $-\sin\phi M\dot{\phi}$	$\sin\phi Mp$ $+\cos\phi M\dot{\phi}$
Q	$\sin\phi(R_S+L_Sp)$ $+\cos\phi L_S\dot{\phi}$	$-\cos\phi(R_S+L_Sp)$ $+\sin\phi L_S\dot{\phi}$	$\sin\phi Mp$ $+\cos\phi M\phi$	$-\cos\phi Mp$ $+\sin\phi M\dot{\phi}$
d	$M\begin{pmatrix}\cos\phi p-\sin\phi\dot{\phi}\\ +\sin\phi\dot{\theta}\end{pmatrix}$	$M\begin{pmatrix}\sin\phi p+\cos\phi\dot{\phi}\\ -\cos\phi\dot{\theta}\end{pmatrix}$	$\cos\phi(R_r+L_rp)$ $L_r(-\sin\phi\dot{\phi}+\sin\phi\dot{\theta})$	$\sin\phi(R_r+L_rp)$ $L_r(\cos\phi\dot{\phi}-\cos\phi\dot{\theta})$
q	$M\begin{pmatrix}\sin\phi p+\cos\phi\dot{\phi}\\ -\cos\phi\dot{\theta}\end{pmatrix}$	$M\begin{pmatrix}-\cos\phi p+\sin\phi\dot{\phi}\\ -\sin\phi\dot{\theta}\end{pmatrix}$	$\sin\phi(R_r+L_rp)$ $L_r(\cos\phi\dot{\phi}-\cos\phi\dot{\theta})$	$-\cos\phi(R_r+L_rp)$ $L_r(\sin\phi\dot{\phi}-\sin\phi\dot{\theta})$

If this is premultiplied by $\mathbf{C}_t$,

$\mathbf{Z}' = \mathbf{C}_t\mathbf{Z}\mathbf{C}$

=

	A	B	a	b
A	R_S+L_Sp	$L_S\dot{\phi}$	Mp	$M\dot{\phi}$
B	$-L_S\dot{\phi}$	R_S+L_Sp	$-M\dot{\phi}$	Mp
a	Mp	$M(\dot{\phi}-\dot{\theta})$	R_r+L_rp	$L_r(\dot{\phi}-\dot{\theta})$
b	$-M(\dot{\phi}-\dot{\theta})$	Mp	$-L_r(\dot{\phi}-\dot{\theta})$	R_r+L_rp

If $\dot{\psi} = \dot{\phi}-\dot{\theta}$, the transient impedance matrix is

	A	B	a	b
A	R_S+L_Sp	$L_S\dot{\phi}$	Mp	$M\dot{\phi}$
B	$-L_S\dot{\phi}$	R_S+L_Sp	$-M\dot{\phi}$	Mp
a	Mp	$M\dot{\psi}$	R_r+L_rp	$L_r\dot{\psi}$
b	$-M\dot{\psi}$	Mp	$-L_r\dot{\psi}$	R_r+L_rp

Since $\dot{\phi}$ is the speed of the axes relative to the stator and $\dot{\theta}$ is the speed of the rotor relative to the stator, $\dot{\psi}$ is the speed of the axes relative to the rotor.

If the axes are synchronous, $\dot{\phi} = \omega$, and $\dot{\theta} = (1-s)\omega$, where ω is the angular frequency of the supply and s is the per unit slip. Hence $\dot{\psi} = \dot{\phi}-\dot{\theta} = \omega-(1-s)\omega = s\omega$.

The transient impedance matrix in terms of synchronous reference axes is therefore

$\mathbf{Z}' =$

	A	B	a	b
A	R_S+L_Sp	ωL_S	Mp	ωM
B	$-\omega L_S$	R_S+L_Sp	$-\omega M$	Mp
a	Mp	$s\omega M$	R_r+L_rp	$s\omega L_r$
b	$-s\omega M$	Mp	$-s\omega L_r$	R_r+L_rp

In order to find the steady-state currents i^A, i^B, i^a, i^b, let us assume that the balanced terminal voltages in D, Q, d, q axes are

D	$\hat{V}\cos\omega t$
Q	$\hat{V}\sin\omega t$
d	0
q	0

The time zero is thus determined by the positive maximum value of v_D.

Then

				D	Q	d	q		
A	v_A		A	$\cos\phi$	$\sin\phi$			D	$\hat{V}\cos\omega t$
B	v_B	=	B	$\sin\phi$	$-\cos\phi$			Q	$\hat{V}\sin\omega t$
a	v_a		a			$\cos\phi$	$\sin\phi$	d	0
b	v_b		b			$\sin\phi$	$-\cos\phi$	q	0

	A	$\hat{V}(\cos\phi\cos\omega t+\sin\phi\sin\omega t)$
=	B	$\hat{V}(\sin\phi\cos\omega t-\cos\phi\sin\omega t)$
	a	0
	b	0

	A	$\hat{V}\cos(\phi-\omega t)$
=	B	$\hat{V}\sin(\phi-\omega t)$
	a	0
	b	0

Now since $\dot{\phi} = \omega$, $\phi = \omega t +$ constant. Because the axes are not attached to either physical member, we can arbitrarily choose ϕ to be zero at time zero, i.e. $\phi = \omega t$. "Physically", therefore, the A axis is coincident with the D axis at time zero. Then $\phi - \omega t = 0$ and the voltage matrix is

A	$\hat{V}$
B	0
a	0
b	0

Since this is constant, the p of the transient impedance matrix may be replaced by zero for the determination of the steady-state currents, and if we write X for ωL and ωM,

$\mathbf{Z}' =$

	A	B	a	b
A	R_S	X_S		X_M
B	$-X_S$	R_S	$-X_M$	
a		sX_M	R_r	sX_r
b	$-sX_M$		$-sX_r$	R_r

of which the determinant is

$$\Delta = (R_S^2+X_S^2)(R_r^2+s^2X_r^2)+2sX_M^2(R_S R_r - sX_S X_r)+s^2X_M^4$$

By determining only the first, or A, column of the inverse of this $\mathbf{Z}'$ we get the steady-state currents as

A	I^A		A	$R_S(R_r^2+s^2X_r^2)+sR_rX_M^2$
B	I^B	$= \dfrac{\hat{V}}{\Delta}$	B	$X_S(R_r^2+s^2X_r^2)-s^2X_rX_M^2$
a	I^a		a	$-sX_M(X_SR_r+sR_SX_r)$
b	I^b		b	$sX_M\{R_SR_r-s(X_SX_r-X_M^2)\}$

From the first matrix for $\mathbf{Z}'$ on p. 263 the coefficient of $\dot{\theta}$ can be written down to obtain the $\mathbf{G}'$ matrix.

$\mathbf{G}' =$

	A	B	a	b
A				
B				
a		$-M$		$-L_r$
b	M		L_r	

Hence

$\mathbf{G}'_t =$

	A	B	a	b
A				M
B			$-M$	
a				L_r
b			$-L_r$	

and

$\mathbf{G}'+\mathbf{G}'_t =$

	A	B	a	b
A				M
B			$-M$	
a		$-M$		
b	M			

Therefore

$-\mathbf{i}'_t(\mathbf{G}'+\mathbf{G}'_t) = -$

A	B	a	b
I^A	I^B	I^a	I^b

	A	B	a	b
A				M
B			$-M$	
a		$-M$		
b	M			

$=$

A	B	a	b
$-MI^b$	MI^a	MI^B	$-MI^A$

and

$\mathbf{G'i'} =$

	A	B	a	b
A				
B				
a		$-M$		$-L_r$
b	M		L_r	

A	I^A
B	I^B
a	I^a
b	I^b

$=$

A	
B	
a	$-MI^B - L_r I^b$
b	$MI^A + L_r I^a$

The motional impedance matrix is therefore

	A	B	a	b	s
A	$R_S + L_S p$	X_S	Mp	X_M	
B	$-X_S$	$R_S + L_S p$	$-X_M$	Mp	
a	Mp	sX_M	$R_r + L_r p$	sX_r	$-MI^B - L_r I^b$
b	$-sX_M$	Mp	$-sX_r$	$R_r + L_r p$	$MI^A + L_r I^a$
s	$-MI^b$	MI^a	MI^B	$-MI^A$	$F + Jp$

If now the supply voltages are constant in the A, B, a, b axes and there is a torque disturbance at an anguar frequency $h\omega$, the p in the motional impedance matrix may be replaced by $jh\omega$, so that the matrix becomes

$\mathbf{Z} =$

	A	B	a	b	s
A	$R_S + jhX_S$	X_S	jhX_M	X_M	
B	$-X_S$	$R_S + jhX_S$	$-X_M$	jhX_M	
a	jhX_M	sX_M	$R_r + jhX_r$	sX_r	$-MI^B - L_r I^b$
b	$-sX_M$	jhX_M	$-sX_r$	$R_r + jhX_r$	$MI^A + L_r I^a$
s	$-MI^b$	MI^a	MI^B	$-MI^A$	$F + jh\omega J$

This is obviously as far as one would wish to go in general terms with the matrix written in full. However, under the given conditions, it is worthwhile to separate the electrical and mechanical axes into the two original matrix equations:

$$\Delta\mathbf{v} = \mathbf{Z}\,\Delta\mathbf{i} + \mathbf{Gi}\,\Delta\dot{\theta}$$

$$\Delta T_s = -\mathbf{i}_t(\mathbf{G}+\mathbf{G}_t)\,\Delta\mathbf{i} + (F+Jp)\,\Delta\dot{\theta}$$

From the first of these, $\Delta\mathbf{v}$ being a null matrix because the voltages are constant, we get

$$\mathbf{Z}\,\Delta\mathbf{i} = -\mathbf{Gi}\,\Delta\dot{\theta}$$

which, premultiplied by $\mathbf{Z}^{-1}$, gives

$$\Delta\mathbf{i} = -\mathbf{Z}^{-1}\mathbf{Gi}\,\Delta\dot{\theta}$$

Substituting this value of $\Delta\mathbf{i}$ in the second equation, gives

$$\begin{aligned}\Delta T_s &= \mathbf{i}_t(\mathbf{G}+\mathbf{G}_t)\mathbf{Z}^{-1}\mathbf{Gi}\,\Delta\dot{\theta} + (F+Jp)\,\Delta\dot{\theta} \\ &= [\mathbf{i}_t(\mathbf{G}+\mathbf{G}_t)\mathbf{Z}^{-1}\mathbf{Gi} + F + Jp]\,\Delta\dot{\theta}\end{aligned}$$

When p is replaced by $j\,\mathrm{h}\omega$, $\Delta\dot{\theta}$ follows easily, since the quantity in the square brackets is a scalar. The value of $\Delta\mathbf{i}$ then follows.

It is clear that the responses of speed and current in A, B, a, b axes will be of angular frequency hω. Since the electrical axes are rotating at an angular velocity ω relative to the D, Q, d, q axes, the angular frequencies in these axes are $(1+\mathrm{h})\omega$ and $(1-\mathrm{h})\omega$. These are the response frequencies in the stator windings. Similarly, since the A, B, a, b axes are rotating at a speed sω relative to the rotor, the actual rotor winding currents in α, β or a, b, c axes will be of angular frequencies $(\mathrm{s}+\mathrm{h})\omega$ and $(\mathrm{s}-\mathrm{h})\omega$.

Synchronous Machine

It is important to note that when there is no disturbance to the terminal voltages in the actual winding axes and the transformed voltages in a new set of axes are constant, it does not follow that the Δv in that new set of axes are all zero. This apparent inconsistency occurs when the two sets of axes are in non-uniform relative motion. It did not arise in the preceding analysis of the induction machine because the rotor circuit was there assumed to be closed, leading to zero rotor $\mathbf{v}$ and $\Delta\mathbf{v}$ in all sets of axes, whilst the motion of the axes relative to the stator was uniform. It does arise in the doubly-fed induction machine, i.e. one in which both stator and rotor have non-zero terminal voltages. It also arises in the synchronous machine, in which speed changes result in non-uniform motion of the armature axes a, b, c relative to the d, q axes used in the analysis.

It was shown on p. 212 that if the three-phase terminal voltages are

$$\begin{array}{c|c|}\hline \mathrm{a} & -\hat{V}\sin\omega t \\ \hline \mathrm{b} & -\hat{V}\sin(\omega t-2\pi/3) \\ \hline \mathrm{c} & -\hat{V}\sin(\omega t-4\pi/3) \\ \hline \end{array}$$

the stationary axes steady-state voltages are

$$\mathbf{v} = \begin{array}{c|c|}\hline \mathrm{d} & -\sqrt{\tfrac{3}{2}}\hat{V}\sin(\omega t-\theta) \\ \hline \mathrm{q} & -\sqrt{\tfrac{3}{2}}\hat{V}\cos(\omega t-\theta) \\ \hline \mathrm{F} & V_{\mathrm{F}} \\ \hline \end{array}$$

Hence

$$\mathbf{v}+\Delta\mathbf{v} = \begin{array}{c|c|}\hline \mathrm{d} & -\sqrt{\tfrac{3}{2}}\hat{V}\sin(\omega t-\theta-\Delta\theta) \\ \hline \mathrm{q} & -\sqrt{\tfrac{3}{2}}\hat{V}\cos(\omega t-\theta-\Delta\theta) \\ \hline \mathrm{F} & V_{\mathrm{F}} \\ \hline \end{array}$$

and

$$\Delta\mathbf{v} = (\mathbf{v}+\Delta\mathbf{v})-\mathbf{v} = \begin{array}{c|c|}\hline \mathrm{d} & -\sqrt{\tfrac{3}{2}}\hat{V}\{\sin(\omega t-\theta-\Delta\theta)-\sin(\omega t-\theta)\} \\ \hline \mathrm{q} & -\sqrt{\tfrac{3}{2}}\hat{V}\{\cos(\omega t-\theta-\Delta\theta)-\cos(\omega t-\theta)\} \\ \hline \mathrm{F} & \\ \hline \end{array}$$

$$= \begin{array}{c|c|}\hline \mathrm{d} & -\sqrt{\tfrac{3}{2}}\hat{V}2\cos(\omega t-\theta-\Delta\theta/2)\sin(-\Delta\theta/2) \\ \hline \mathrm{q} & -\sqrt{\tfrac{3}{2}}\hat{V}2\sin(\omega t-\theta-\Delta\theta/2)\sin(\Delta\theta/2) \\ \hline \mathrm{F} & \\ \hline \end{array}$$

As $\Delta\theta$ is small, $\sin(\Delta\theta/2)$ may be taken as $\Delta\theta/2$ and $\omega t-\theta-\Delta\theta/2$ may be taken as $\omega t-\theta = -\psi$. Then

$$\Delta\mathbf{v} = \begin{array}{c|c|} d & \sqrt{\tfrac{3}{2}}\hat{V}\cos\psi\,\Delta\theta \\ q & \sqrt{\tfrac{3}{2}}\hat{V}\sin\psi\,\Delta\theta \\ F & \end{array}$$

This $\Delta\mathbf{v}$ is additional to any which arises as a result of disturbances of the three-phase terminal voltage. If the frequency of the terminal voltage is constant, $\Delta\theta = \Delta\psi$, but there is little point in substituting for $\Delta\theta$ in $\Delta\mathbf{v}$.

This $\Delta\mathbf{v}$ thus appears as a component of the total $\Delta\mathbf{v}$ on the left-hand side of the equation on p. 256. Since, however, $\Delta\dot{\theta}$ is the second variable on the right-hand side of this equation, this $\Delta\mathbf{v}$ can be subtracted from both sides of the equation, if, on the right-hand side, it is expressed in terms of $\Delta\dot{\theta}$ instead of in terms of $\Delta\theta$ as

$$\begin{array}{c|c|} d & \sqrt{(\tfrac{3}{2})}\,\hat{V}\cos\psi\dfrac{1}{p} \\ q & \sqrt{(\tfrac{3}{2})}\,\hat{V}\sin\psi\dfrac{1}{p} \\ F & \end{array}\;\Delta\dot{\theta}$$

This column matrix can then be subtracted from the last column of $\mathbf{Z}$.

Alternatively, $\Delta\dot{\theta}$ may be replaced throughout by $\Delta\theta$, if the last column of the motional impedance matrix is post-multiplied by p. This avoids the need to use the inverse of p, but obscures the analogy between the mechanical and electrical equations.

On p. 213 the steady-state currents of a salient-pole synchronous machine without damper windings, for the terminal voltages defined as above, were shown to be

$$
\begin{array}{c|c|}
\text{d} & I^{\text{d}} \\
\hline
\text{q} & I^{\text{q}} \\
\hline
\text{F} & I^{\text{F}}
\end{array}
= \frac{1}{\Delta}
\begin{array}{c|c|}
\text{d} & \sqrt{\tfrac{3}{2}}\hat{V}R_{\text{F}}(R_{\text{a}}\sin\psi+X_{\text{q}}\cos\psi)-V_{\text{F}}X_{\text{q}}X_{\text{Fd}} \\
\hline
\text{q} & \sqrt{\tfrac{3}{2}}\hat{V}R_{\text{F}}(X_{\text{d}}\sin\psi-R_{\text{a}}\cos\psi)+V_{\text{F}}R_{\text{a}}X_{\text{Fd}} \\
\hline
\text{F} & V_{\text{F}}(R_{\text{a}}^2+X_{\text{d}}X_{\text{q}})
\end{array}
$$

which are constant under steady-state conditions.

Since

$$
\mathbf{G} =
\begin{array}{c|c|c|c|}
 & \text{d} & \text{q} & \text{F} \\
\hline
\text{d} & & L_{\text{q}} & \\
\hline
\text{q} & -L_{\text{d}} & & -L_{\text{Fd}} \\
\hline
\text{F} & & &
\end{array}
$$

$$
\mathbf{Gi} =
\begin{array}{c|c|c|c|}
 & \text{d} & \text{q} & \text{F} \\
\hline
\text{d} & & L_{\text{q}} & \\
\hline
\text{q} & -L_{\text{d}} & & -L_{\text{Fd}} \\
\hline
\text{F} & & &
\end{array}
\quad
\begin{array}{c|c|}
\text{d} & I^{\text{d}} \\
\hline
\text{q} & I^{\text{q}} \\
\hline
\text{F} & I^{\text{F}}
\end{array}
$$

$$
=
\begin{array}{c|c|}
\text{d} & L_{\text{q}}I^{\text{q}} \\
\hline
\text{q} & -L_{\text{d}}I^{\text{d}}-L_{\text{Fd}}I^{\text{F}} \\
\hline
\text{F} &
\end{array}
$$

$$
-\mathbf{i}_{\text{t}}(\mathbf{G}+\mathbf{G}_{\text{t}}) = -
\begin{array}{|c|c|c|}
\text{d} & \text{q} & \text{F} \\
\hline
I^{\text{d}} & I^{\text{q}} & I^{\text{F}}
\end{array}
\quad
\begin{array}{c|c|c|c|}
 & \text{d} & \text{q} & \text{F} \\
\hline
\text{d} & & L_{\text{q}}-L_{\text{d}} & \\
\hline
\text{q} & L_{\text{q}}-L_{\text{d}} & & -L_{\text{Fd}} \\
\hline
\text{F} & & -L_{\text{Fd}} &
\end{array}
$$

$$
=
\begin{array}{|c|c|c|}
\text{d} & \text{q} & \text{F} \\
\hline
I^{\text{q}}(L_{\text{d}}-L_{\text{q}}) & I^{\text{d}}(L_{\text{d}}-L_{\text{q}})+I^{\text{F}}L_{\text{Fd}} & I^{\text{q}}L_{\text{Fd}}
\end{array}
$$

The general motional impedance matrix is therefore

	d	q	F	s
d	$R_a+L_d p$	$L_q\dot{\theta}$	$L_{Fd}p$	$L_q I^q$
q	$-L_d\dot{\theta}$	$R_a+L_q p$	$-L_{Fd}\dot{\theta}$	$-L_d I^d - L_{Fd} I^F$
F	$L_{Fd}p$		$R_F+L_F p$	
s	$I^q(L_d-L_q)$	$I^d(L_d-L_q)+I^F L_{Fd}$	$I^q L_{Fd}$	$F+Jp$

in which I^d, I^q and I^F have the values given on p. 271.

If, however, we now subtract the voltage disturbances resulting from the transformation, we get a motional impedance matrix

	d	q	F	s
d	$R_a+L_d p$	$L_q\dot{\theta}$	$L_{Fd}p$	$L_q I^q - \sqrt{\tfrac{3}{2}}\,\hat{V}\cos\psi\,\dfrac{1}{p}$
q	$-L_d\dot{\theta}$	$R_a+L_q p$	$-L_{Fd}\dot{\theta}$	$-L_d I^q - L_{Fd} I^F - \sqrt{\tfrac{3}{2}}\,\hat{V}\sin\psi\,\dfrac{1}{p}$
F	$L_{Fd}p$		$R_F+L_F p$	
s	$I^q(L_d-L_q)$	$I^d(L_d-L_q)+I^F L_{Fd}$	$I^q L_{Fd}$	$F+Jp$

appropriate to the particular terminal conditions, and which relates the responses to the externally imposed disturbances.

The last column of this matrix may be multiplied by p to eliminate the $1/p$, provided that the matrix of changes of current and speed:

$$\begin{array}{c|c|} \text{d} & \Delta i_d \\ \text{q} & \Delta i_q \\ \text{F} & \Delta i_F \\ \text{s} & \Delta\dot{\theta} \end{array} = \begin{array}{c|c|} \text{d} & \Delta i_d \\ \text{q} & \Delta i_q \\ \text{F} & \Delta i_F \\ \text{s} & p\,\Delta\theta \end{array}$$

is replaced by a matrix of changes of current

and position:

d	Δi_d
q	Δi_q
F	Δi_F
s	$\Delta\theta$

Further analysis of this synchronous machine in general terms involves too much algebra to be pursued here, whilst that of the synchronous machine with dampers is worse, unless simplifying assumptions are made.

CHAPTER 13

Miscellaneous Machine Problems

The Metadyne Generator with its Quadrature Brushes displaced

We shall investigate the consequences of a small displacement of the quadrature brushes of a metadyne generator.† For simplicity we shall assume that the air-gap is uniform and that the stator windings are distributed and shall confine our attention to the no-load condition. Although the air-gap is uniform, the self-inductances of the arma-

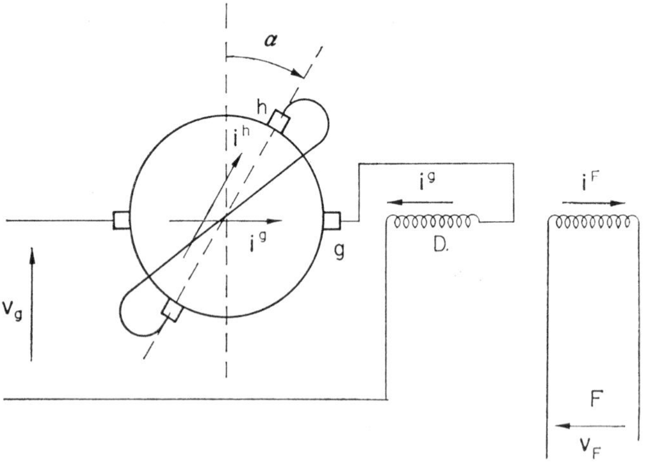

FIG. 45. Metadyne generator with quadrature brushes displaced.

† For a general treatment of the metadyne in matrix form see ref. 20.

ture on the direct and quadrature axes will not in general be equal. Commutating poles† are provided for the brushes on the output circuit but not for the other brushes. With the windings as designated below, L_d will be less than L_q for this reason.

Assuming that the quadrature brushes are displaced in the opposite direction to that of rotation, the arrangement is as shown in Fig. 45. The corresponding "primitive" machine is shown in Fig. 46.

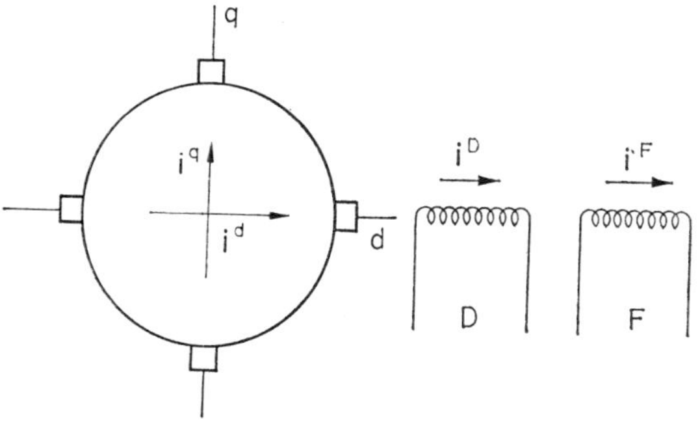

FIG. 46. Primitive machine.

It is important to note that the actual output brushes are designated g and not d although they are on the direct axis. This is necessary since the h circuit also has a component of m.m.f. along the d axis and hence both h and g contribute to d.

The brush-shift transformation is

$$\begin{array}{c|c|} d & i^d \\ \hline q & i^q \end{array} \;=\; \begin{array}{c|c|c|} & g & h \\ \hline d & 1 & \sin\alpha \\ \hline q & & \cos\alpha \end{array} \;\; \begin{array}{c|c|} g & i^g \\ \hline h & i^h \end{array}$$

† The effect of commutating poles will not be considered here but is treated generally in ref. 14.

The connection matrix is

$$
\begin{array}{c|c|}
 & \\
g & i^{g} \\
h & i^{h} \\
D & i^{D} \\
F & i^{F}
\end{array}
=
\begin{array}{c|c|c|c|}
 & g & h & F \\
g & 1 & & \\
h & & 1 & \\
D & -1 & & \\
F & & & 1
\end{array}
\begin{array}{c|c|}
 & \\
g & i^{g} \\
h & i^{h} \\
F & i^{F}
\end{array}
$$

The complete transformation is therefore

$$
\begin{array}{c|c|}
 & \\
d & i^{d} \\
q & i^{q} \\
D & i^{D} \\
F & i^{F}
\end{array}
=
\begin{array}{c|c|c|c|c|}
 & g & h & D & F \\
d & 1 & \sin\alpha & & \\
q & & \cos\alpha & & \\
D & & & 1 & \\
F & & & & 1
\end{array}
\begin{array}{c|c|c|c|}
 & g & h & F \\
g & 1 & & \\
h & & 1 & \\
D & -1 & & \\
F & & & 1
\end{array}
\begin{array}{c|c|}
 & \\
g & i^{g} \\
h & i^{h} \\
F & i^{F}
\end{array}
$$

$$
=
\begin{array}{c|c|c|c|}
 & g & h & F \\
d & 1 & \sin\alpha & \\
q & & \cos\alpha & \\
D & -1 & & \\
F & & & 1
\end{array}
\begin{array}{c|c|}
 & \\
g & i^{g} \\
h & i^{h} \\
F & i^{F}
\end{array}
$$

$$
\mathbf{C} =
\begin{array}{c|c|c|c|}
 & g & h & F \\
d & 1 & \sin\alpha & \\
q & & \cos\alpha & \\
D & -1 & & \\
F & & & 1
\end{array}
\qquad
\mathbf{C}_{t} =
\begin{array}{c|c|c|c|c|}
 & d & q & D & F \\
g & 1 & & -1 & \\
h & \sin\alpha & \cos\alpha & & \\
F & & & & 1
\end{array}
$$

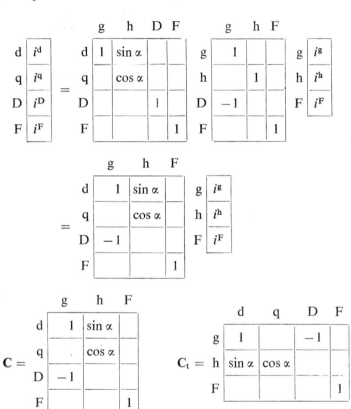

The impedance matrix of the primitive machine is

$\mathbf{Z} =$

	d	q	D	F
d	$R_a+L_d p$	$L_q\dot{\theta}$	$M_{dD}p$	$M_{dF}p$
q	$-L_d\dot{\theta}$	$R_a+L_q p$	$-M_{dD}\dot{\theta}$	$-M_{dF}\dot{\theta}$
D	$M_{dD}p$		$R_D+L_D p$	$M_{FD}p$
F	$M_{dF}p$		$M_{FD}p$	$R_F+L_F p$

Hence

$\mathbf{ZC} =$

	g	h	F
d	$R_a+L_d p-M_{dD}p$	$(R_a+L_d p)\sin\alpha+L_q\dot{\theta}\cos\alpha$	$M_{dF}p$
q	$-(L_d-M_{dD})\dot{\theta}$	$-L_d\dot{\theta}\sin\alpha+(R_a+L_q p)\cos\alpha$	$-M_{dF}\dot{\theta}$
D	$M_{dD}p-(R_D+L_D p)$	$M_{dD}p\sin\alpha$	$M_{FD}p$
F	$(M_{dF}-M_{FD})p$	$M_{dF}p\sin\alpha$	$R_F+L_F p$

and

$\mathbf{ZC} =$

	g	h	F
g	(R_a+R_D) $+(L_d+L_D-2M_{dD})p$	$R_a\sin\alpha+L_q\dot{\theta}\cos\alpha$ $+(L_d-M_{dD})p\sin\alpha$	$(M_{dF}-M_{FD})p$
h	$R_a\sin\alpha$ $+(L_d-M_{dD})p\sin\alpha$ $-(L_d-M_{dD})\dot{\theta}\cos\alpha$	$(R_a+L_d p)\sin^2\alpha$ $+(R_a+L_q p)\cos^2\alpha$ $+(L_q-L_d)\dot{\theta}\sin\alpha\cos\alpha$	$M_{dF}p\sin\alpha$ $-M_{dF}\dot{\theta}\cos\alpha$
F	$(M_{dF}-M_{FD})p$	$M_{dF}p\sin\alpha$	$R_F+L_F p$

Since $i^g = 0$ for the present investigation, for the currents we need consider only the voltage equation

$$\begin{array}{c|c|}\text{h} & v_h \\ \text{F} & v_F\end{array} = \begin{array}{c|c|c|} & \text{h} & \text{F} \\ \hline \text{h} & R_a+(L_d\sin^2\alpha+L_q\cos^2\alpha)p+(L_q-L_d)\dot{\theta}\sin\alpha\cos\alpha & M_{dF}p\sin\alpha-M_{dF}\dot{\theta}\cos\alpha \\ \hline \text{F} & M_{dF}p\sin\alpha & R_F+L_F p\end{array} \quad \begin{array}{c|c|}\text{h} & i^h \\ \text{F} & i^F\end{array}$$

The difference between unity and cos α for small values of α is slight. The principal effect of the displacement of the quadrature brushes is thus to change the apparent resistance of the h circuit from R_a to $R_a+(L_q-L_d)\,\dot{\theta}\sin\alpha\cos\alpha$. For positive values of $\dot{\theta}$ and α this represents an increase of resistance, since $L_q > L_d$. This will reduce both the amplification and also the time-constant of this circuit in exactly the same way as if an external resistance had been connected in series in the h circuit. The reduction of gain will affect steady-state and transient conditions alike.

The nature of the transient response is determined by the zeros of the determinant of the impedance matrix. If $\alpha = 0$, the determinant is $(R_F+L_Fp)(R_a+L_qp)$ and its zeros are $-R_F/L_F$ and $-R_a/L_q$ so that the response is then stable and non-oscillatory.

When, however, $\alpha \neq 0$, the determinant is

$$\begin{aligned}[L_F(L_d\sin^2\alpha+L_q\cos^2\alpha)-M_{dF}^2\sin^2\alpha]p^2 \\ +[L_F\{R_a+(L_q-L_d)\dot{\theta}\sin\alpha\cos\alpha\} \\ +R_F(L_d\sin^2\alpha+L_q\cos^2\alpha)+M_{dF}^2\dot{\theta}\sin\alpha\cos\alpha]p \\ +[R_F\{R_a+(L_q-L_d)\dot{\theta}\sin\alpha\cos\alpha\}]\end{aligned}$$

The algebraic form of the zeros of this quadratic are too cumbersome to consider. If the parameters were given, however, it would be simple to find the numerical values of the zeros and the nature of the transient. Here we can reach some conclusions as follows:

The coefficient of p^2 is positive for all values of α, both positive and negative because $L_FL_d > M_{Fd}^2$.

The coefficient of p and the constant term are positive for all values of $\alpha \geqslant 0$. If, however, α is increasingly negative, the coefficient of p will become smaller and then negative, followed by the constant term. This can be shown by considering the sign and order of magnitude of the coefficient of p when the constant term is zero. Consequently the response, originally non-oscillatory when $\alpha = 0$, will become damped oscillatory, undamped oscillatory, unstable oscillatory and finally unidirectionally unstable.

The practical importance of this analysis lies in the fact that when the orders of magnitude are known, it shows the necessity for very accurate brush setting. The actual commutating axis moves within the width of the brush[21] with changing quadrature current and other changes of conditions, so that some of the effects considered here may be experienced even with the brush centre-line exactly on the neutral axis.

The output voltage can be found from

$$v_g = \{R_a \sin\alpha + (L_d - M_{dD})p \sin\alpha + L_q\dot{\theta}\cos\alpha\}i^h + (M_{dF} - M_{FD})pi^D$$

since $i^g = 0$.

The effect when the machine is on load can be determined in a similar manner, but involves too much algebra in the general case for consideration here.

The Ferraris–Arno Phase Converter

The three-phase induction motor connected to a single-phase line-to-line supply is covered by the analysis on p. 186, where it is shown that with the A-phase open circuit

$$\begin{array}{c|c|} \mathrm{P} & i^P \\ \mathrm{N} & i^N \\ \mathrm{O} & i^O \end{array} = \begin{array}{c|c|} \mathrm{P} & ji^B \\ \mathrm{N} & -ji^B \\ \mathrm{O} & 0 \end{array}$$

By putting $Z_U = 0$ in the impedance matrix on p. 182 we obtain the steady-state voltage equation of the balanced induction machine in terms of P, N axes as

$$\begin{array}{c|c|} & \\ \mathrm{P} & V_P \\ \mathrm{N} & V_N \end{array} = \begin{array}{c|c|c|} & \mathrm{P} & \mathrm{N} \\ \mathrm{P} & R_S + jX_S + \dfrac{sX_M^2}{R_r + jsX_r} & \\ \mathrm{N} & & R_S + jX_S + \dfrac{(2-s)X_M^2}{R_r + j(2-s)X_r} \end{array} \begin{array}{c|c|} & \\ \mathrm{P} & I^P \\ \mathrm{N} & I^N \end{array}$$

This may be written as

$$\begin{array}{c|c|} & \\ \text{P} & V_P \\ \hline \text{N} & V_N \end{array} = \begin{array}{c|c|c|} & \text{P} & \text{N} \\ \text{P} & Z_{PP} & \\ \hline \text{N} & & Z_{NN} \end{array} \begin{array}{c|c|} & \\ \text{P} & I^P \\ \hline \text{N} & I^N \end{array} \quad \text{say.}$$

Hence

$$\frac{V_N}{V_P} = \frac{Z_{NN}I^N}{Z_{PP}I^P} = \frac{Z_{NN}(-jI^B)}{Z_{PP}(jI^B)} = -\frac{Z_{NN}}{Z_{PP}}$$

$$= -\frac{R_S + jX_S + \dfrac{(2-s)X_M^2}{R_r + j(2-s)X_r}}{R_S + jX_S + \dfrac{sX_M^2}{R_r + jsX_r}}$$

This ratio cannot be zero except for some value of slip s greater than 2, and even then only fortuitously if the constants were in the appropriate ratio. However, the ratio is small as s tends to zero and consequently if the motor is running light there is across its terminals an almost balanced three-phase voltage even though only two terminals are connected to a supply. Provided that some means exists for starting, therefore, the machine can be used as a phase converter, producing an approximately balanced three-phase voltage from a single-phase supply. Other three-phase motors can then be started by the normal three-phase motor methods. When other motors are connected to its terminals, the value of the negative-sequence component of the converter terminal voltage will depend upon the ratio of the effective negative-sequence impedance to the effective positive-sequence impedance of all the machines in parallel at the particular slips at which they are operating. Since the motors will in general be on load, their slips will be greater than that of the phase converter and consequently the latter will reduce the proportion of negative sequence-current in the other machines. It is interesting to note that if a passive load were connected across the converter there would be an appreciable increase in the negative-sequence voltage, since the negative-sequence and positive-sequence impedances of a passive load are equal.

If an impedance Z consisting of a resistance and capitance in series is connected between the A-phase terminal and the B-phase terminal,

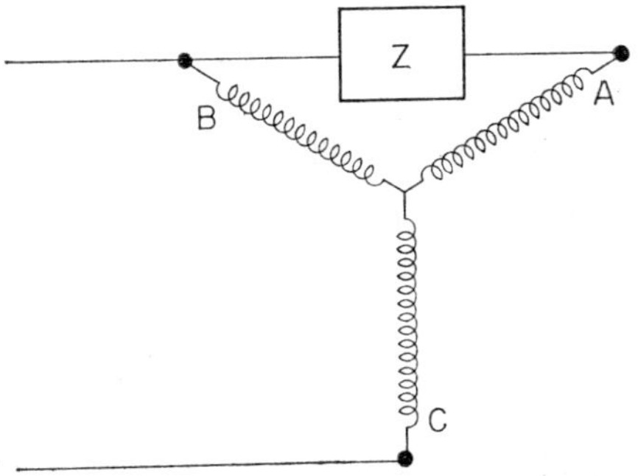

FIG. 47. Stator winding of modified Ferraris–Arno phase converter.

as shown in Fig. 47, it is possible for the converter to produce a perfectly balanced output voltage. The value of Z is related, however, to the loads, both mechanical and electrical, connected to the frequency changer, and consequently can only give perfect balance at one load if fixed values of resistance and capacitance are used.

The impedance matrix of this external impedance is

	A	B	C
A	Z		
B			
C			

This is transformed to P, N, O by

$$\frac{1}{\sqrt{3}}\begin{array}{c|ccc|} & \text{A} & \text{B} & \text{C} \\ \hline \text{P} & 1 & h & h^2 \\ \text{N} & 1 & h^2 & h \\ \text{O} & 1 & 1 & 1 \\ \hline \end{array}\quad \begin{array}{c|ccc|} & \text{A} & \text{B} & \text{C} \\ \hline \text{A} & Z & & \\ \text{B} & & & \\ \text{C} & & & \\ \hline \end{array}\quad \frac{1}{\sqrt{3}}\begin{array}{c|ccc|} & \text{P} & \text{N} & \text{O} \\ \hline \text{A} & 1 & 1 & 1 \\ \text{B} & h^2 & h & 1 \\ \text{C} & h & h^2 & 1 \\ \hline \end{array}$$

$$= \frac{1}{3}\begin{array}{c|ccc|} & \text{P} & \text{N} & \text{O} \\ \hline \text{P} & Z & Z & Z \\ \text{N} & Z & Z & Z \\ \text{O} & Z & Z & Z \\ \hline \end{array}$$

When this is connected in series with the machine we can ignore the zero-sequence row and column since will there not be any zero-sequence current. The effective impedance† is then

	P	N
P	$Z_{PP}+Z/3$	$Z/3$
N	$Z/3$	$Z_{NN}+Z/3$

The voltage applied to the system is defined in terms of the line voltages only by

$$V_{AB}=0, \quad V_{BC}=V, \quad V_{CA}=-V$$

If we define V_{AB} as V_A-V_B, we have

$$\begin{array}{c|c|}\text{AB} & 0\\ \text{BC} & V\\ \text{CA} & -V\end{array} = \begin{array}{c|c|}\text{AB} & V_A-V_B\\ \text{BC} & V_B-V_C\\ \text{CA} & V_C-V_A\end{array} = \begin{array}{c|ccc|}& \text{A} & \text{B} & \text{C}\\ \text{AB} & 1 & -1 & \\ \text{BC} & & 1 & -1\\ \text{CA} & -1 & & 1\end{array}\;\begin{array}{c|c|}\text{A} & V_A\\ \text{B} & V_B\\ \text{C} & V_C\end{array}$$

$$= \begin{array}{c|ccc|}& \text{A} & \text{B} & \text{C}\\ \text{AB} & 1 & -1 & \\ \text{BC} & & 1 & -1\\ \text{CA} & -1 & & 1\end{array}\;\frac{1}{\sqrt{3}}\;\begin{array}{c|ccc|}& \text{P} & \text{N} & \text{O}\\ \text{A} & 1 & 1 & 1\\ \text{B} & h^2 & h & 1\\ \text{C} & h & h^2 & 1\end{array}\;\begin{array}{c|c|}\text{P} & V_P\\ \text{N} & V_N\\ \text{O} & V_0\end{array}$$

$$= \frac{1}{\sqrt{3}}\;\begin{array}{c|ccc|}& \text{P} & \text{N} & \text{O}\\ \text{AB} & 1-h^2 & 1-h & 0\\ \text{BC} & h^2-h & h-h^2 & 0\\ \text{CA} & h-1 & h^2-1 & 0\end{array}\;\begin{array}{c|c|}\text{P} & V_P\\ \text{N} & V_N\\ \text{O} & V_0\end{array}$$

† Note that the interconnection has here been done in P, N axes and not in the actual winding axes. This is permissible only because the identical transformation (A, B, C to P, N, O) has been applied to the two systems to be connected.

This transformation matrix is singular, corresponding to the obvious fact that it is not possible to determine the zero-sequence component of the phase voltages from the line voltages. However, we do not require to find the zero-sequence component of voltage, at least so far, and we can omit the corresponding column. This leaves us with a three-by-two matrix which is also singular, but now it is singular because one of the equations is redundant and not because the data is insufficient. Hence we proceed with

$$\begin{array}{c|c|}\mathrm{AB} & 0 \\ \mathrm{BC} & V\end{array} = \frac{1}{\sqrt{3}} \begin{array}{c|c|c|}& \mathrm{P} & \mathrm{N} \\ \hline \mathrm{AB} & 1-h^2 & 1-h \\ \mathrm{BC} & h^2-h & h-h^2\end{array} \begin{array}{c|c|}\mathrm{P} & V_{\mathrm{P}} \\ \mathrm{N} & V_{\mathrm{N}}\end{array}$$

The determinant of this matrix is $(h-h^2)$ and the inverse is

$$\frac{1}{\sqrt{3}(h-h^2)} \begin{array}{c|c|c|}& \mathrm{AB} & \mathrm{BC} \\ \hline \mathrm{P} & h-h^2 & h-1 \\ \mathrm{N} & h-h^2 & 1-h^2\end{array} = \frac{1}{\sqrt{3}} \begin{array}{c|c|c|}& \mathrm{AB} & \mathrm{BC} \\ \hline \mathrm{P} & 1 & -h^2 \\ \mathrm{N} & 1 & -h\end{array}$$

Hence

$$\begin{array}{c|c|}\mathrm{P} & V_{\mathrm{P}} \\ \mathrm{N} & V_{\mathrm{N}}\end{array} = \frac{1}{\sqrt{3}} \begin{array}{c|c|c|}& \mathrm{AB} & \mathrm{BC} \\ \hline \mathrm{P} & 1 & -h^2 \\ \mathrm{N} & 1 & -h\end{array} \begin{array}{c|c|}\mathrm{AB} & 0 \\ \mathrm{BC} & V\end{array} = \frac{1}{\sqrt{3}} \begin{array}{c|c|}\mathrm{P} & -h^2V \\ \mathrm{N} & -hV\end{array}$$

The voltage equation is

$$\begin{array}{c|c|}\mathrm{P} & -h^2V/\sqrt{3} \\ \mathrm{N} & -hV/\sqrt{3}\end{array} = \begin{array}{c|c|c|}& \mathrm{P} & \mathrm{N} \\ \hline \mathrm{P} & Z_{\mathrm{PP}}+Z/3 & Z/3 \\ \mathrm{N} & Z/3 & Z_{\mathrm{NN}}+Z/3\end{array} \begin{array}{c|c|}\mathrm{P} & I^{\mathrm{P}} \\ \mathrm{N} & I^{\mathrm{N}}\end{array}$$

For balanced currents, I^N will be zero, and

$$-h^2V/\sqrt{3} = (Z_{PP}+Z/3)I^P$$
$$-hV/\sqrt{3} = (Z/3)I^P$$

Dividing,

$$h = (Z_{PP}+Z/3)/(Z/3)$$
$$hZ/3 = Z_{PP}+Z/3$$
$$(h-1)Z/3 = Z_{PP}$$
$$Z/3 = Z_{PP}/(h-1)$$
$$Z = (h^2-1)Z_{PP}$$
$$= -(\tfrac{3}{2}+j\sqrt{3}/2)Z_{PP}$$

If Z_{PP} be represented by $R_{PP}+jX_{PP}$

$$Z = -(\tfrac{3}{2}+j\sqrt{3}/2)(R_{PP}+jX_{PP})$$
$$= (-3R_{PP}/2+\sqrt{3}X_{PP}/2)-j(\sqrt{3}R_{PP}/2+3X_{PP}/2)$$

The resistance component of Z is therefore positive only when $\sqrt{3}X_{PP}/2$ is greater than $3R_{PP}/2$, the other component is always capacitive. This method of balancing can be used therefore only when the power factor of the converter and any connected load is less than 0·5, which makes the method unattractive.

This result could have been obtained more directly, but having used this approach we can proceed to find the sequence components of the converter output for loads other than that for which it is balanced.

Inverting the impedance matrix

	P	N
P	$Z_{PP}+Z/3$	$Z/3$
N	$Z/3$	$Z_{NN}+Z/3$

gives

$$\frac{1}{(Z_{PP}+Z/3)(Z_{NN}+Z/3)-(Z/3)^2}$$

	P	N
P	$Z_{NN}+Z/3$	$-Z/3$
N	$-Z/3$	$Z_{PP}+Z/3$

Hence

$$\begin{array}{|c|}\hline I^{P} \\ \hline I^{N} \\ \hline \end{array}$$

$$= \frac{1}{(Z_{PP}+Z/3)\,(Z_{NN}+Z/3)-(Z/3)^2} \quad \begin{array}{c|c|c|} & P & N \\ \hline P & Z_{NN}+Z/3 & -Z/3 \\ \hline N & -Z/3 & Z_{PP}+Z/3 \\ \hline \end{array} \quad \begin{array}{c|c|} \hline P & -h^2V/\sqrt{3} \\ \hline N & -hV/\sqrt{3} \\ \hline \end{array}$$

$$= \frac{-hV/\sqrt{3}}{(Z_{PP}+Z/3)\,(Z_{NN}+Z/3)-(Z/3)^2} \quad \begin{array}{c|c|} \hline P & h(Z_{NN}+Z/3)-Z/3 \\ \hline N & -hZ/3+(Z_{FP}+Z/3) \\ \hline \end{array}$$

The phase converter terminal voltages are given by

$$\begin{array}{c|c|} \hline P & V_P \\ \hline N & V_N \\ \hline \end{array} = \begin{array}{c|c|c|} & P & N \\ \hline P & Z_{PP} & \\ \hline N & & Z_{NN} \\ \hline \end{array} \quad \begin{array}{c|c|} \hline P & I^{P} \\ \hline N & I^{N} \\ \hline \end{array} = \begin{array}{c|c|} \hline P & Z_{PP}I^{P} \\ \hline N & Z_{NN}I^{N} \\ \hline \end{array}$$

$$= \frac{-hV/\sqrt{3}}{(Z_{PP}+Z/3)\,(Z_{NN}+Z/3)-(Z/3)^2} \quad \begin{array}{c|c|} \hline P & Z_{PP}\{hZ_{NN}+(h-1)Z/3\} \\ \hline N & Z_{NN}\{Z_{PP}+(1-h)Z/3\} \\ \hline \end{array}$$

Bearing in mind the relative complexity of Z_{PP} and Z_{NN}, there is little point in proceeding in algebraic terms; if, however, the parameters are known, this is a relatively simple expression for the ratio of the positive-sequence and negative-sequence voltages.

This problem is the subject of ref. 22.

The Polyphase Induction Machine with a Single-phase Secondary Circuit

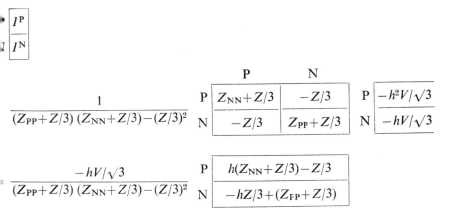

It is possible to analyse the performance of a polyphase machine with an unbalanced polyphase winding on the secondary, but the simpler case of the machine with a single-phase winding on the secondary will be treated here as an example.†

Since the secondary winding is unbalanced it is essential to take the reference axes stationary relative to the secondary winding. In order to utilize our normal impedance matrix which has the reference axes stationary relative to the stator, it will be assumed here that the rotor is the primary. There will of course be no difference in the form of the equations if the rotor is taken as the secondary and the reference axes as stationary relative to the rotor.

The arrangement of the windings is shown diagrammatically in Fig. 48.

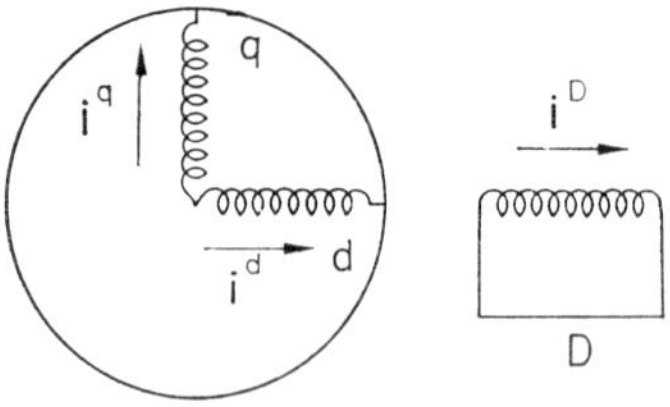

FIG. 48. Polyphase induction machine with single-phase secondary circuit.

The impedance matrix, written down according to the usual rules in terms of stationary reference axes, is

		d	q	D
	d	R_r+L_rp	$L_r\dot{\theta}$	Mp
$\mathbf{Z}$ =	q	$-L_r\dot{\theta}$	R_r+L_rp	$-M\dot{\theta}$
	D	Mp		R_S+L_Sp

The determinant of this matrix is

$$\Delta = (R_S+L_Sp)\,\{(R_r+L_rp)^2+(L_r\dot{\theta})^2\}-Mp\{(R_r+L_rp)Mp+L_rM\dot{\theta}^2\}$$

† It is not possible to analyse in a routine manner a single-phase induction motor with a single-phase secondary winding in terms of stationary reference axes since neither member then has a balanced winding.

and its inverse is

$$
{}^{-1} = \frac{1}{\Delta}\;
\begin{array}{c|c|c|c|}
 & \text{d} & \text{q} & \text{D} \\
\hline
\text{d} & (R_r+L_rp)\,(R_S+L_Sp) & -(R_S+L_Sp)L_r\dot{\theta} & -(R_r+L_rp)Mp-L_rM\dot{\theta}^2 \\
\hline
\text{q} & \begin{matrix}(R_S+L_Sp)\,(L_r\dot{\theta})\\ -M^2\dot{\theta}p\end{matrix} & \begin{matrix}(R_r+L_rp)\,(R_S+L_Sp)\\ -M^2p^2\end{matrix} & (R_r+L_rp)M\dot{\theta}-L_rM\dot{\theta}p \\
\hline
\text{D} & -(R_r+L_rp)Mp & L_rM\dot{\theta}p & (R_r+L_rp)^2+(L_r\dot{\theta})^2 \\
\hline
\end{array}
$$

The terminal voltages of the rotor windings a, b, c are balanced and may be defined as

$$
\begin{array}{c|c|}
\text{a} & v_a \\
\text{b} & v_b \\
\text{c} & v_c \\
\end{array}
=
\begin{array}{c|c|}
\text{a} & \hat{V}\cos \omega t \\
\text{b} & \hat{V}\cos(\omega t-2\pi/3) \\
\text{c} & \hat{V}\cos(\omega t-4\pi/3) \\
\end{array}
$$

The stator winding being short-circuited, $v_D = 0$ and the terminal voltages in the d, q, D reference axes are therefore

$$
\begin{array}{c|c|}
\text{d} & (\sqrt{3/2})\hat{V}\cos(\theta-\omega t) \\
\text{q} & (\sqrt{3/2})\hat{V}\sin(\theta-\omega t) \\
\text{D} & 0 \\
\end{array}
= \sqrt{\tfrac{3}{2}}\hat{V}\;
\begin{array}{c|c|}
\text{d} & \cos(\theta-\omega t) \\
\text{q} & \sin(\theta-\omega t) \\
\text{D} & 0 \\
\end{array}
$$

Now if v is the per-unit speed, $\dot{\theta} = \mathrm{v}\omega$, and integrating $\theta = \mathrm{v}\omega t+\psi$ or $\theta - \omega t = -(1-\mathrm{v})\omega t+\psi = -s\omega t+\psi$, where s is the per-unit slip. Hence

$$
\begin{array}{c|c|}
\text{d} & v_d \\
\text{q} & v_q \\
\text{D} & v_D \\
\end{array}
= \sqrt{\tfrac{3}{2}}\hat{V}\;
\begin{array}{c|c|}
\text{d} & \cos(-s\omega t+\psi) \\
\text{q} & \sin(-s\omega t+\psi) \\
\text{D} & 0 \\
\end{array}
$$

$$
= \sqrt{\tfrac{3}{2}}\hat{V}\;
\begin{array}{c|c|}
\text{d} & \cos(s\omega t-\psi) \\
\text{q} & -\sin(s\omega t-\psi) \\
\text{D} & 0 \\
\end{array}
$$

Thus v_d and v_q are sinusoidal voltages of angular frequency $s\omega$, with v_q leading v_d by $\pi/2$, showing that the rotor field is rotating clockwise at slip speed relative to the stator.

If we are concerned only with steady-state conditions we must note that in d, q, D terms, the only exciting functions are v_d and v_q which are of angular frequency $s\omega$ and that consequently the p of the impedance matrix can be replaced by $js\omega$ to obtain the steady-state impedance matrix. Since we have already inverted this matrix, the same substitution can be made directly in the inverse. Again, noting that $v_D = 0$, we shall not need to determine the elements of the last column of $\mathbf{Z}^{-1}$. Writing $(1-s)\omega$ for $\dot{\theta}$, X_r for ωL_r, X_S for ωL_S, X_m for ωM, gives

$$\mathbf{Z}^{-1} = \frac{1}{\Delta}\begin{array}{c|c|c|c|} & \text{d} & \text{q} & \text{D} \\ \hline \text{d} & (R_r+jsX_r)(R_S+jsX_S) & -(R_S+jsX_S)(1-s)X_r & --- \\ \hline \text{q} & \begin{array}{c}(R_S+jsX_S)(1-s)X_r \\ -js(1-s)X_m^2\end{array} & \begin{array}{c}(R_r+jsX_r)(R_S+jsX_S) \\ +s^2X_m^2\end{array} & --- \\ \hline \text{D} & -(R_r+jsX_r)(jsX_m) & js(1-s)X_rX_m & --- \\ \hline \end{array}$$

In complex terms the terminal voltages are

$$\begin{array}{c|c|} \text{d} & v_d \\ \hline \text{q} & v_q \\ \hline \text{D} & v_D \end{array} = \begin{array}{c|c|} \text{d} & V_d \\ \hline \text{q} & jV_d \\ \hline \text{D} & 0 \end{array}$$

where $V_d = \frac{\sqrt{3}}{2}\hat{V} = \sqrt{\tfrac{3}{2}}V$

The currents are therefore

$$\begin{array}{c|c|} \text{d} & I^d \\ \hline \text{q} & I^q \\ \hline \text{D} & I^D \end{array} = \frac{1}{\Delta}\begin{array}{c|c|c|c|} & \text{d} & \text{q} & \text{D} \\ \hline \text{d} & (R_r+jsX_r)(R_S+jsX_S) & -(R_S+jsX_S)(1-s)X_r & --- \\ \hline \text{q} & \begin{array}{c}(R_S+jsX_S)(1-s)X_r \\ -js(1-s)X_m^2\end{array} & \begin{array}{c}(R_r+jsX_r)(R_S+jsX_S) \\ +s^2X_m^2\end{array} & --- \\ \hline \text{D} & -(R_r+jsX_r)(jsX_m) & js(1-s)X_rX_m & --- \\ \hline \end{array} \begin{array}{c|c|} \text{d} & V_d \\ \hline \text{q} & jV_d \\ \hline \text{D} & 0 \end{array}$$

$$
= \frac{V_d}{\Delta}\;
\begin{array}{c|c|}
\text{d} & (R_S+jsX_S)\,\{R_r+jsX_r-j(1-s)X_r\} \\ \hline
\text{q} & j(R_S+jsX_S)\,\{R_r+jsX_r-j(1-s)X_r\}+jX_m^2\{s^2-s(1-s)\} \\ \hline
\text{D} & X_rX_m\{s^2-s(1-s)\}-jsX_mR_r
\end{array}
$$

$$
= \frac{V_d}{\Delta}\;
\begin{array}{c|c|}
\text{d} & (R_S+jsX_S)\,\{R_r+j(2s-1)X_r\} \\ \hline
\text{q} & j(R_S+jsX_S)\,\{R_r+j(2s-1)X_r\}+js(2s-1)X_m^2 \\ \hline
\text{D} & -jsX_m\{R_r+j(2s-1)X_r\}
\end{array}
$$

where Δ

$$
\begin{aligned}
&= (R_S+jsX_S)\,\{(R_r+jsX_r)^2+(1-s)^2X_r^2\} \\
&\quad -jsX_m\{(R_r+jsX_r)\,(jsX_m)+(1-s)^2X_rX_m\} \\
&= (R_S+jsX_S)\,[\{R_r^2+(1-2s)X_r^2\}+j2sR_rX_r]-jsX_m^2[(1-2s)X_r+jsR_r] \\
&= [R_S\{R_r^2+(1-2s)X_r^2\}-2s^2R_rX_rX_S+s^2R_rX_m^2] \\
&\quad +j[sX_S\{R_r^2+(1-2s)X_r^2\}+2sR_rR_SX_r-s(1-2s)X_rX_m^2]
\end{aligned}
$$

The currents I^d, I^q, and I^D will, of course, be of angular frequency $s\omega$. Since $\theta = v\omega t+\psi = (1-s)\omega t+\psi$, when the actual rotor currents are determined there will be terms involving the products of trigonometric functions of $(1-s)\omega t$ and of $s\omega t$. It follows that the currents i^a, i^b and i^c will have components of fundamental frequency and of angular frequency $(1-2s)\omega t$. The current in the secondary winding, I^D, is, of course, wholly of angular frequency $s\omega$ provided that the supply voltage is sinusoidal at angular frequency ω.

Torque

From the impedance matrix on p. 286 we can write down

$$
G = \begin{array}{c|c|c|c|}
 & \text{d} & \text{q} & \text{D} \\ \hline
\text{d} & & L_r & \\ \hline
\text{q} & -L_r & & -M \\ \hline
\text{D} & & & \\ \hline
\end{array}
= \frac{1}{\omega}\;
\begin{array}{c|c|c|c|}
 & \text{d} & \text{q} & \text{D} \\ \hline
\text{d} & & X_r & \\ \hline
\text{q} & -X_r & & -X_m \\ \hline
\text{D} & & & \\ \hline
\end{array}
$$

The two-pole steady-state torque T is therefore

$$\operatorname{Re}\frac{1}{\omega}\begin{array}{c}\begin{array}{ccc} \mathrm{d} & \mathrm{q} & \mathrm{D}\end{array}\\ \begin{bmatrix} I^{d*} & I^{q*} & I^{D*}\end{bmatrix}\end{array} \begin{array}{c|ccc|} & \mathrm{d} & \mathrm{q} & \mathrm{D}\\ \hline \mathrm{d} & & X_r & \\ \mathrm{q} & -X_r & & -X_m\\ \mathrm{D} & & & \end{array} \begin{array}{c|c|} \mathrm{d} & I^d\\ \mathrm{q} & I^q\\ \mathrm{D} & I^D\end{array}$$

$$= \operatorname{Re}\frac{1}{\omega}\begin{array}{c}\begin{array}{ccc} \mathrm{d} & \mathrm{q} & \mathrm{D}\end{array}\\ \begin{bmatrix} I^{d*} & I^{q*} & I^{D*}\end{bmatrix}\end{array} \begin{array}{c|c|} \mathrm{d} & X_r I^q\\ \mathrm{q} & -X_r I^d - X_m I^D\\ \mathrm{D} & \end{array}$$

$$= \operatorname{Re}\frac{1}{\omega}\{I^{d*}X_r I^q - I^{q*}X_r I^d - I^{q*}X_m I^D\}$$

Now $I^{d*}X_r I^q$ and $I^{q*}X_r I^d$ are conjugates and hence their difference is wholly imaginary and the torque expression reduces to

$$\operatorname{Re}\frac{1}{\omega}\{-I^{q*}X_m I^D\}$$

The torque is therefore

$$\operatorname{Re}\frac{1}{\omega}\,\frac{-V_d^2 X_m}{\Delta^*\Delta}\,[-j(R_S - jsX_S)\{R_r - j(2s-1)X_r\}$$
$$-js(2s-1)X_m^2]\,[-jsX_m\{R_r + j(2s-1)X_r\}]$$

$$= \operatorname{Re}\frac{1}{\omega}\,\frac{sV_d^2X_m^2}{\Delta^*\Delta}\,[(R_S - jsX_S)\{R_r - j(2s-1)X_r\}$$
$$+s(2s-1)X_m^2]\,[R_r + j(2s-1)X_r]$$

$$= \frac{1}{\omega}\,\frac{sV_d^2X_m^2}{\Delta^*\Delta}\,[R_S\{R_r^2 + (2s-1)^2X_r^2\} + s(2s-1)X_m^2R_r]^\dagger$$

$$= \frac{1}{\omega}\,\frac{sV_d^2X_m^2}{\Delta^*\Delta}\,[2(2R_SX_r^2 + R_rX_m^2)s^2 - (4R_SX_r^2 + R_rX_m^2)s + R_S(R_r^2 + X_r^2)]$$

† Note: $\Delta^*\Delta$ is real.

The torque of the two-pole three-phase machine is thus

$$\frac{1}{\omega}\,\frac{3}{2}\,\frac{sV^2X_m^2}{\Delta\Delta^*}\,[2(2R_SX_r^2+R_rX_m^2)s^2-(4R_SX_r^2+R_rX_m^2)s+R_S(R_r^2+X_r^2)]$$

where V is the r.m.s. phase voltage.

The torque will be zero when s = 0 and when

$$2(2R_SX_r^2+R_rX_m^2)s^2-(4R_SX_r^2+R_rX_m^2)s+R_S(R_r^2+X_r^2) = 0$$

Solving this quadratic for s we get

$$s = \frac{4R_SX_r^2+R_rX_m^2 \pm R_r\sqrt{(X_m^4-8R_rR_SX_m^2-16R_S^2X_r^2)}}{4(2R_SX_r^2+R_rX_m^2)}$$

Although there is no exact algebraic value for the square root, a value could, of course, be found in a numerical case. For the present we can obtain an approximate solution on the assumption that $X_m \gg R_r$ or R_S, in which case the value of the square root is very little less than X_m^2, say $X_m^2-\delta$. Then the values of s are

$$\frac{4R_SX_r^2+R_rX_m^2 \pm R_r(X_m^2-\delta)}{4(2R_SX_r^2+R_rX_m^2)}$$

$$= \frac{4R_SX_r^2+2R_rX_m^2-R_r\delta}{4(2R_SX_r^2+R_rX_m^2)} \quad \text{or} \quad \frac{4R_SX_r^2+R_r\delta}{4(2R_SX_r^2+R_rX_m^2)}$$

The first of these is slightly under $\frac{1}{2}$. The value of the second one can be estimated as follows. When referred to the same number of turns R_S and R_r are of the same order of magnitude, so also are X_r and X_m which differ by the rotor leakage reactance. Hence $R_SX_r^2$ and $R_rX_m^2$ are of the same order of magnitude and the second root is about $\frac{1}{3}$.

It is easy to ascertain that the torque is positive for s = 1 and also when s is small but positive. The general shape of the torque-speed curve is therefore as shown in Fig. 49.

This problem is treated theoretically and experimentally in ref. 23.

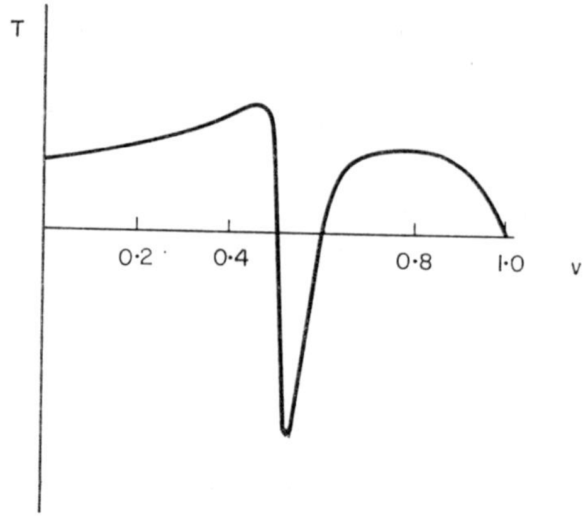

FIG. 49. Torque-speed curve of induction machine with single-phase secondary circuit.

Power Selsyns

Power selsyns are machines of the slip-ring induction type and are used in pairs. Both primary windings are supplied from the same source and the secondary windings are connected in series opposition as shown in Fig. 50. The primary supply is balanced polyphase for large machines but may be single phase for small ones. The characteristic of

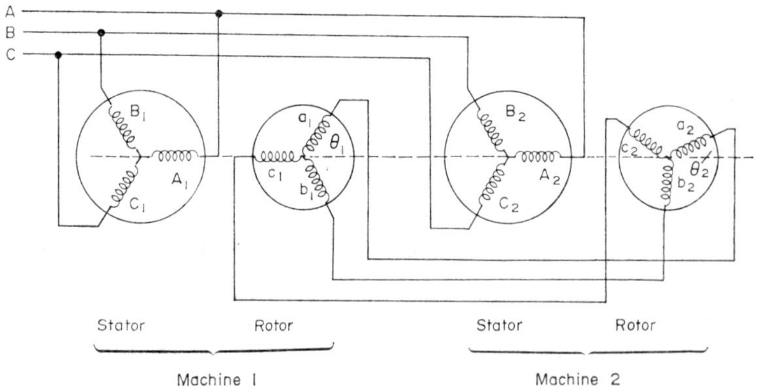

FIG. 50. Three-phase selsyns.

the system is that the torques oppose any misalignment of the shafts. The arrangement is therefore used to maintain synchronism (except for a "load" angle) between two otherwise independent drives, without any direct mechanical connection.

The system is chosen here to illustrate the technique of interconnection of machines. It is essential to express the interconnection in terms of the actual winding axes or of windings which are at rest relative to the actual winding axes. Thus, in the present case, assuming that the rotors are three-phase wound, the interconnection can be expressed in terms of a, b, c or α, β axes, but *not* in terms of d, q axes. This is because the rotor angle θ will not be the same for the two machines, and hence, although the transformations from a, b, c to α, β are identical, those to d, q are not.

If the machines are designated 1 and 2, from the matrix on p. 59 the impedance matrix $\mathbf{Z}$ of the combination before interconnection, in terms of two-phase axes, is as given on p. 294.

If the α_1, β_1 windings are connected to the α_2, β_2 windings respectively, the interconnection matrix is

$\mathbf{C}_1 =$

	D_1	Q_1	α_1	β_1	D_2	Q_2
D_1	1					
Q_1		1				
α_1			1			
β_1				1		
D_2					1	
Q_2						1
α_2			-1			
β_2				-1		

The matrix is designated 1 because it is the first of several, not because it has any particular relation to machine 1, although it is expressed

$\mathbf{Z} =$

	$D_1(A_1')$	$Q_1(B_1')$	α_1	β_1	$D_2(A_2')$	$Q_2(B_2')$	α_2	β_2
$D_1(A_1')$	R_S+L_Sp		$Mp\cos\theta_1$	$Mp\sin\theta_1$				
$Q_1(B_1')$		R_S+L_Sp	$Mp\sin\theta_1$	$-Mp\cos\theta_1$				
α_1	$Mp\cos\theta_1$	$Mp\sin\theta_1$	R_r+L_rp					
β_1	$Mp\sin\theta_1$	$-Mp\cos\theta_1$		R_r+L_rp				
$D_2(A_2')$					R_S+L_Sp		$Mp\cos\theta_2$	$Mp\sin\theta_2$
$Q_2(B_2')$						R_S+L_Sp	$Mp\sin\theta_2$	$-Mp\cos\theta_2$
α_2					$Mp\cos\theta_2$	$Mp\sin\theta_2$	R_r+L_rp	
β_2					$Mp\sin\theta_2$	$-Mp\cos\theta_2$		R_r+L_rp

in terms of the currents of machine 1 in preference to those of machine 2.

The impedance matrix of the system after connection is

= $\mathbf{C}_{1t}\mathbf{Z}\mathbf{C}_1$ =

D_1	Q_1	α_1	β_1	D_2	Q_2
R_S+L_Sp		$Mp\cos\theta_1$	$Mp\sin\theta_1$		
	R_S+L_Sp	$Mp\sin\theta_1$	$-Mp\cos\theta_1$		
$Mp\cos\theta_1$	$Mp\sin\theta_1$	$2(R_r+L_rp)$		$-Mp\cos\theta_2$	$-Mp\sin\theta_2$
$Mp\sin\theta_1$	$-Mp\cos\theta_1$		$2(R_r+L_rp)$	$-Mp\sin\theta_2$	$Mp\cos\theta_2$
		$-Mp\cos\theta_2$	$-Mp\sin\theta_2$	R_S+L_Sp	
		$-Mp\sin\theta_2$	$Mp\cos\theta_2$		R_S+L_Sp

This matrix contains functions of the time-varying θ_1 and θ_2 and the system must be transformed to a stationary reference frame if Laplace transforms are to be used to obtain a solution of the differential equations in a routine manner. Since D_1, Q_1 and D_2, Q_2 axes are all stationary relative to one another, it seems appropriate to transform α, β to d, q in terms of one, say the first, machine. The transformation is therefore

$\mathbf{C}_2$ =

	D_1	Q_1	d_1	q_1	D_2	Q_2
D_1	1					
Q_1		1				
α_1			$\cos\theta_1$	$\sin\theta_1$		
β_1			$\sin\theta_1$	$-\cos\theta_1$		
D_2					1	
Q_2						1

Since $\mathbf{C}_2$ also contains time-varying functions, it is necessary to form the product $\mathbf{Z}'\mathbf{C}_2\mathbf{i}'$, perform the differentiations and then to premultiply by $\mathbf{C}_{2t}$. In practice, only the product $\mathbf{Z}'\mathbf{C}_2$ will be written down, the implied presence of the current matrix will, however, be taken into account when differentiating.

The completed transformation obtained by premultiplying this last matrix $\mathbf{Z}'\mathbf{C}_2$ by $\mathbf{C}_{2t}$ is therefore $\mathbf{C}_{2t}\mathbf{Z}'\mathbf{C}_2 = \mathbf{Z}''$ as given on p. 299.

For this matrix to represent a differential equation with constant coefficients, it is clear that $(\theta_1 - \theta_2)$ must be constant, i.e. the angle between the rotors must be constant. It follows that $(\dot{\theta}_1 - \dot{\theta}_2) = 0$. In other words, the present analysis is inherently restricted to the case when the shafts are rotating at the same speed and consequently with a constant angle between them. This is, of course, the steady-state mechanical condition. Let $(\theta_1 - \theta_2) = \alpha$ and $\dot{\theta}_1 = \dot{\theta}_2 = \dot{\theta}$, then the impedance matrix with the above restriction written in is

$\mathbf{Z}'' =$

	D_1	Q_1	d_1	q_1	D_2	Q_2
D_1	$R_S + L_S p$		Mp			
Q_1		$R_S + L_S p$		Mp		
d_1	Mp	$M\dot{\theta}$	$2(R_r + L_r p)$	$2L_r\dot{\theta}$	$-M\cos\alpha\,.p$ $-M\sin\alpha\,.\dot{\theta}$	$M\sin\alpha\,.$ $-M\cos\alpha\,.$
q_1	$-M\dot{\theta}$	Mp	$-2L_r\dot{\theta}$	$2(R_r + L_r p)$	$-M\sin\alpha\,.p$ $+M\cos\alpha\,.\dot{\theta}$	$-M\cos\alpha\,.$ $-M\sin\alpha\,.$
D_2			$-M\cos\alpha\,.p$	$-M\sin\alpha\,.p$	$R_S + L_S p$	
Q_2			$M\sin\alpha\,.p$	$-M\cos\alpha\,.p$		$R_S + L_S p$

$\mathbf{Z'C_2}$ =

	D$_1$	Q$_1$	d$_1$	q$_1$	D$_2$	Q$_2$
D$_1$	R_S+L_Sp		Mp			
Q$_1$		R_S+L_Sp		Mp		
α_1	$Mp\cos\theta_1$	$Mp\sin\theta_1$	$2(R_r+L_rp)\cos\theta_1$	$2(R_r+L_rp)\sin\theta_1$	$-Mp\cos\theta_2$	$-Mp\sin\theta_2$
β_1	$Mp\sin\theta_1$	$-Mp\cos\theta_1$	$2(R_r+L_rp)\sin\theta_1$	$-2(R_r+L_rp)\cos\theta_1$	$-Mp\sin\theta_2$	$Mp\cos\theta_2$
D$_2$			$-Mp\cos(\theta_1-\theta_2)$	$-Mp\sin(\theta_1-\theta_2)$	R_S+L_Sp	
Q$_2$			$Mp\sin(\theta_1-\theta_2)$	$-Mp\cos(\theta_1-\theta_2)$		R_S+L_Sp

		D_1	Q_1	d_1	q_1	D_2	Q_2
	D_1	R_S+L_Sp		Mp			
	Q_1		R_S+L_Sp		Mp		
	α_1	$M\cos\theta_1 p$ $-M\sin\theta_1\dot{\theta}_1$	$M\sin\theta_1 p$ $+M\cos\theta_1\dot{\theta}_1$	$2\cos\theta_1(R_r+L_rp)$ $-2L_r\sin\theta_1\dot{\theta}_1$	$2\sin\theta_1(R_r+L_rp)$ $+2L_r\cos\theta_1\dot{\theta}_1$	$-M\cos\theta_2 p$ $+M\sin\theta_2\dot{\theta}_2$	$-M\sin\theta_2 p$ $-M\cos\theta_2\dot{\theta}_2$
=	β_1	$M\sin\theta_1 p$ $+M\cos\theta_1\dot{\theta}_1$	$-M\cos\theta_1 p$ $+M\sin\theta_1\dot{\theta}_1$	$2\sin\theta_1(R_r+L_rp)$ $+2L_r\cos\theta_1\dot{\theta}_1$	$-2\cos\theta_1(R_r+L_rp)$ $+2L_r\sin\theta_1\dot{\theta}_1$	$-M\sin\theta_2 p$ $-M\cos\theta_2\dot{\theta}_2$	$M\cos\theta_2 p$ $-M\sin\theta_2\dot{\theta}_2$
	D_2			$-M\cos(\theta_1-\theta_2)p$ $+M\sin(\theta_1-\theta_2)(\dot{\theta}_1-\dot{\theta}_2)$	$-M\sin(\theta_1-\theta_2)p$ $-M\cos(\theta_1-\theta_2)(\dot{\theta}_1-\dot{\theta}_2)$	R_S+L_Sp	
	Q_2			$M\sin(\theta_1-\theta_2)p$ $+M\cos(\theta_1-\theta_2)(\dot{\theta}_1-\dot{\theta}_2)$	$-M\cos(\theta_1-\theta_2)p$ $+M\sin(\theta_1-\theta_2)(\dot{\theta}_1-\dot{\theta}_2)$		R_S+L_Sp

$\mathbf{Z}'' =$

	D_1	Q_1	d_1	q_1	D_2	Q_2
D_1	$R_S + L_S p$		Mp			
Q_1		$R_S + L_S p$		Mp		
d_1	Mp	$M\dot{\theta}_1$	$2(R_r + L_r p)$	$2L_r\dot{\theta}_1$	$-M\cos(\theta_1-\theta_2)p$ $-M\sin(\theta_1-\theta_2)\dot{\theta}_2$	$M\sin(\theta_1-\theta_2)p$ $-M(\cos(\theta_1-\theta_2)\dot{\theta}_2$
q_1	$-M\dot{\theta}_1$	Mp	$-2L_r\dot{\theta}_1$	$2(R_r + L_r p)$	$-M\sin(\theta_1-\theta_2)p$ $+M\cos(\theta_1-\theta_2)\dot{\theta}_2$	$-M\cos(\theta_1-\theta_2)p$ $-M\sin(\theta_1-\theta_2)\dot{\theta}_2$
D_2			$-M\cos(\theta_1-\theta_2)p$ $+M\sin(\theta_1-\theta_2)(\dot{\theta}_1-\dot{\theta}_2)$	$-M\sin(\theta_1-\theta_2)p$ $-M\cos(\theta_1-\theta_2)(\dot{\theta}_1-\dot{\theta}_2)$	$R_S + L_S p$	
Q_2			$M\sin(\theta_1-\theta_2)p$ $+M\cos(\theta_1-\theta_2)(\dot{\theta}_1-\dot{\theta}_2)$	$-M\cos(\theta_1-\theta_2)p$ $+M\sin(\theta_1-\theta_2)(\dot{\theta}_1-\dot{\theta}_2)$		$R_S + L_S p$

Since the excitation on the $D_1(A_1')$, $Q_1(B_1')$, and $D_2(A_2')$, $Q_2(B_2')$ axes is all of angular frequency ω, the steady-state electrical condition can be expressed by replacing p by $j\omega$ and if we also put $\dot{\theta} = \mathrm{v}\omega$ and express the results in terms of reactance, we have

$\mathbf{Z}'' =$

	D_1	Q_1	d_1	q_1	D_2	Q_2
D_1	R_S+jX_S		jX_M			
Q_1		R_S+jX_S		jX_M		
d_1	jX_M	$\mathrm{v}X_M$	$2(R_r+jX_r)$	$2\mathrm{v}X_r$	$-jX_M \cos\alpha$ $-\mathrm{v}X_M \sin\alpha$	jX_M sin $-\mathrm{v}X_M$ cos
q_1	$-\mathrm{v}X_M$	jX_M	$-2\mathrm{v}X_r$	$2(R_r+jX_r)$	$-jX_M \sin\alpha$ $+\mathrm{v}X_M \cos\alpha$	$-jX_M$ cos $-\mathrm{v}X_M$ sin
D_2			$-jX_M \cos\alpha$	$-jX_M \sin\alpha$	R_S+jX_S	
Q_2			$jX_M \sin\alpha$	$-jX_M \cos\alpha$		R_S+jX_S

It is obviously impossible to proceed without reducing this complicated matrix to a more manageable form and previous induction motor experience suggests transformation to P, N, f, b axes. The transformation matrix for this will be

$\mathbf{C}_3 = \dfrac{1}{\sqrt{2}}$

	P_1	N_1	f	b	P_2	N_2
$D_1(A_1')$	1	1				
$Q_1(B_1')$	$-j$	j				
d_1			1	1		
q_1			$-j$	$-j$		
$D_2(A_2')$					1	1
$Q_2(B_2')$					$-j$	j

so that

$$C^{*}_{3t} = \frac{1}{\sqrt{2}}$$

	$D_1(A'_1)$	$Q_1(B'_1)$	d_1	q_1	$D_2(A'_2)$	$Q_2(B'_2)$
P_1	1	j				
N_1	1	$-j$				
f			1	j		
b			1	$-j$		
P_2					1	j
N_2					1	$-j$

and the transformed impedance matrix $\mathbf{Z}'''$ is as given on p. 302.

The benefit of this transformation is revealed by rearranging the order of the rows and columns. At the same time $(1-v)$ and $(1+v)$ will be replaced by s and $(2-s)$ respectively, to get the impedance matrix given on p. 303.

If the supply voltages are balanced, we need proceed only with the P_1, f, P_2 axes, since they form a system which is independent of the N_1, b, N_2 axes and all the terminal voltages of the latter are zero.

By inverting the P_1, f, P_2 sub-matrix, the positive-sequence currents are obtained from the equation

$$\begin{array}{c|c|} P_1 & I^{P_1} \\ \hline f & I^{f} \\ \hline P_2 & I^{P_2} \end{array} = \frac{1}{\Delta} \begin{array}{c|c|c|c|} & P_1 & f & P_2 \\ \hline P_1 & 2(R_S+jX_S)(R_r+jsX_r)+sX_M^2 & - & sX_M^2 e^{j\alpha} \\ \hline f & -jsX_M(R_S+jX_S) & - & jsX_M(R_S+jX_S)e^{j\alpha} \\ \hline P_2 & sX_M^2 e^{-j\alpha} & - & 2(R_S+jX_S)(R_r+jsX_r)+sX_m^2 \end{array} \begin{array}{c|c|} P_1 & V_P \\ \hline f & 0 \\ \hline P_2 & V_P \end{array}$$

in which the f column of the admittance matrix, being irrelevant, has not been calculated and where

$$\begin{aligned} \Delta &= (R_S+jX_S)\{2(R_S+jX_S)(R_r+jsX_r)+sX_M^2\}-jX_M(jsX_M)(R_S+jX_S) \\ &= 2(R_S+jX_S)\{(R_S+jX_S)(R_r+jsX_r)+sX_M^2\} \end{aligned}$$

$\mathbf{Z}''' = \mathbf{C}_{3t}\mathbf{Z}''\mathbf{C}_3 =$

	P_1	N_1	f	b	P_2	N_2
P_1	R_S+jX_S		jX_M			
N_1		R_S+jX_S		jX_M		
f	$j(1-v)X_M$		$2\{R_r+j(1-v)X_r\}$		$-j(1-v)X_M e^{j\alpha}$	
b		$j(1+v)X_M$		$2\{R_r+j(1+v)X_r\}$		$-jX_M(1+v)e^{-j\alpha}$
P_2			$-jX_M e^{-j\alpha}$		R_S+jX_S	
N_2				$-jX_M e^{j\alpha}$		R_S+jX_S

$\mathbf{Z}'' =$

	P₁	f	P₂	N₁	b	N₂
P₁	R_S+jX_S	jX_M				
f	jsX_M	$2\{R_r+jsX_r\}$	$-jsX_Me^{j\alpha}$			
P₂		$-jX_Me^{-j\alpha}$	R_S+jX_S			
N₁				R_S+jX_S	jX_M	
b				$j(2-s)X_M$	$2\{R_r+j(2-s)X_r\}$	$-j(2-s)X_Me^{-j\alpha}$
N₂					$-jX_Me^{j\alpha}$	R_S+jX_S

From p. 172, the two-pole torque of machine 1 is

$$T_1 = \text{Re}\,\frac{1}{\omega}\{-jX_M I^{P_1} I^{f*}\}$$

$$= \text{Re}\,\frac{V_P^2}{\omega}(-jX_M)\,\frac{\{2(R_S+jX_S)(R_r+jsX_r)+sX_M^2(1+e^{j\alpha})\}\{jsX_M(R_S-jX_S)(1-e^{-j\alpha}\ldots}{\Delta\Delta^*}$$

$$= \text{Re}\,\frac{V_P^2}{\omega\Delta\Delta^*}\,sX_M^2\{2(R_S+jX_S)(R_S-jX_S)(R_r+jsX_r)(1-e^{-j\alpha})$$

$$+sX_M^2(R_S-jX_S)(e^{j\alpha}-e^{-j\alpha})\}$$

$$= \frac{V_P^2}{\omega\Delta\Delta^*}\,sX_M^2\{2(R_S^2+X_S^2)[R_r(1-\cos\alpha)-sX_r\sin\alpha]+sX_M^2(2X_S\sin\alpha)\}$$

$$= \frac{2V_P^2}{\omega\Delta\Delta^*}\,sX_M^2\{(R_S^2+X_S^2)R_r(1-\cos\alpha)-sX_r(R_S^2+X_S^2-X_S X_M^2/X_r)\sin\alpha\}$$

in which $V_P = \sqrt{2}V$ for the two-phase machine and $V_P = \sqrt{3}V$ for the three-phase machine.

The torque of the second machine can be similarly calculated bearing in mind that I^f above is the negative of the corresponding current in the second machine. Alternatively, it follows directly from the expression just calculated if α is replaced by $-\alpha$.

If α is positive but not large, the second term in the braces is the bigger, and, irrespective of the sign of s, T_1 is negative and clockwise. If $s < 1$, the machines are running counterclockwise, and, since $\theta_1 > \theta_2$, T_1 is a retarding torque in the leading machine 1. If $s > 1$, the machines are running clockwise and, since $\theta_1 > \theta_2$, T_1 is an accelerating torque on the lagging machine 1. Similarly, T_2 accelerates machine 2 in the direction of rotation when it is lagging behind machine 1 and retards it when it is leading machine 1. The torques therefore tend to maintain synchronism.

If $s = 0$, the torques on both machines are zero, irrespective of the sign or value of α. The system cannot be used to maintain synchronism at or through synchronous speed unless running clockwise, i.e. against the stator fields with $s = 2$.

The analysis of the single-phase selsyns is identical with that of the polyphase system as far as the derivation of the impedance matrix in P_1, f, P_2, N_1, b, N_2 axes. At this point, for the single-phase selsyns it is necessary to write the constraints that $I^{N_1} = -I^{P_1}$ and $I^{N_2} = -I^{P_2}$ in the form of a further transformation matrix which reduces the impedance matrix to a four-by-four in P_1, f, P_2, b axes. By inverting this, one can find all the currents and put the values in the torque expression given on p. 172.

The Single-phase Performance of the Synchronous Generator with a Uniform Air-gap and no Damping Circuits

Since the single-phase condition is a particular case of unbalanced operation, the single-phase performance of a polyphase synchronous generator, and therefore also that of a single-phase generator, can be analysed in symmetrical component axes. It has already been shown that the negative-sequence impedance of the synchronous machine depends on the current wave-shape and that a possibly significant harmonic component is neglected in analysis in these terms. This will also occur with analysis directly in terms of single-phase or other unbalanced conditions. For example, if a single-phase sinusoidal load current is assumed and the machine terminal voltage deduced, this terminal voltage will contain harmonics. It is improbable that any normal load would have a sinusoidal current when its terminal voltage contains harmonic. In practice both current and voltage will contain harmonics. In some cases it may be justifiable to ignore these harmonics, but, when it is not, a more accurate solution is required.

If the load is treated as an impedance and its matrix combined with the transient impedance matrix of the alternator, as was done for a *balanced* load in pp. 202–6, the equations which result cannot be transformed to a set with constant coefficients by the matrices on pp. 116 and 117 or by any other transformation. Accordingly, the problem cannot be solved by the routine use of Laplace transforms. The complete solution involves considerable algebra and only an indication of the technique is given here. For simplicity, a machine without dampers will be considered. Since there is no advantage in using any other than the actual winding axes, we will therefore return to the voltage equations of p. 59 and omit one of the stator windings Q and one of the armature windings. Experience shows that the solution is more convenient in two-phase terms taking $i^\alpha = 0$, but it would be equally valid to take $i^\beta = 0$. The voltage equation in matrix form is

				β	F		
β	v_β	=	β	R_a+L_dp	$Mp \sin\theta$	β	i^β
F	v_F		F	$Mp\sin\theta$	R_F+L_Fp	F	i^F

where F denotes the field winding. In this matrix equation the p's of the $Mp \sin\theta$ elements of the impedance matrix operate on the products of the $\sin\theta$ and the appropriate current. There is little advantage in using matrices therefore and we can write

$$v_\beta = (R_a+L_dp)i^\beta+Mp(\sin\theta i^F)$$
$$v_F = Mp(\sin\theta i^\beta)+(R_F+L_Fp)i^F$$

Even if we are interested only in steady-state conditions we may not assume that i^F is a direct current—in fact it is not. We cannot "invert" any matrix since this is appropriate only to the solution of problems in which the equations have constant coefficients. We can, however, employ Laplace transforms after a little preliminary consideration:

$$\sin\theta i = \frac{\varepsilon^{j\theta}-\varepsilon^{-j\theta}}{j2}\, i = -\frac{j}{2}\{\varepsilon^{j(\omega t+\psi)}-\varepsilon^{-j(\omega t+\psi)}\}i$$
$$= -\frac{j}{2}\{\varepsilon^{j\psi}\varepsilon^{j\omega t}i-\varepsilon^{-j\psi}\varepsilon^{-j\omega t}i\}$$

if $\theta = \omega t+\psi$.

Since ψ is a constant and since the Laplace transform of $\varepsilon^{j\omega t}\, i$ is $\bar{i}(s-j\omega)$, the transform of $\sin\theta\, i$ is

$$-\frac{j}{2}\{\varepsilon^{j\psi}\bar{i}(s-j\omega)-\varepsilon^{-j\psi}\bar{i}(s+j\omega)\}$$

and that of $p(\sin\theta\, i)$ is

$$-\frac{j}{2}s\{\varepsilon^{j\psi}\bar{i}(s-j\omega)-\varepsilon^{-j\psi}\bar{i}(s+j\omega)\}+\frac{j}{2}\{\varepsilon^{j\psi}-\varepsilon^{-j\psi}\}i_o$$

where i_o is the value of i at zero time.

If R_a and L_d include any external impedance in the armature circuit, the voltage transform equations with a constant field winding voltage V_F are

$$0 = (R_a+L_d s)\, \bar{\imath}^{\beta}(s)$$
$$+M\left[-\frac{j}{2}s\{\varepsilon^{j\varphi}\bar{\imath}^{F}(s-j\omega)-\varepsilon^{-j\varphi}\bar{\imath}^{F}(s+j\omega)\}+\frac{j}{2}\{\varepsilon^{j\varphi}-\varepsilon^{-j\varphi}\}V_F/R_F\right]$$

$$\frac{V_F}{s} = M\left[-\frac{j}{2}s\{\varepsilon^{j\varphi}\bar{\imath}^{\beta}(s-j\omega)-\varepsilon^{-j\varphi}\bar{\imath}^{\beta}(s+j\omega)\}\right]+(R_F+L_F s)\, \bar{\imath}^{F}(s)$$

since at time zero $i_o^F = V_F/R_F$ and $i_o^\beta = 0$.

From the first equation

$$\bar{\imath}^{\beta}(s) = -\frac{M}{R_a+L_d s}$$
$$\times\left[-\frac{j}{2}s\{\varepsilon^{j\varphi}\bar{\imath}^{F}(s-j\omega)-\varepsilon^{-j\varphi}\bar{\imath}^{F}(s+j\omega)\}+\frac{j}{2}\{\varepsilon^{j\varphi}-\varepsilon^{-j\varphi}\}V_F/R_F\right]$$

and from the second equation

$$\bar{\imath}^{F}(s) = \frac{V_F}{s(R_F+L_F s)}-\frac{M}{R_F+L_F s}$$
$$\times\left[-\frac{j}{2}s\{\varepsilon^{j\varphi}\bar{\imath}^{\beta}(s-j\omega)-\varepsilon^{-j\varphi}\bar{\imath}^{\beta}(s+j\omega)\}\right]$$

We can thus substitute for $\bar{\imath}^\beta$ in the second equation or for $\bar{\imath}^F$ in the first equation by writing $(s-j\omega)$ or $(s+j\omega)$ for (s) where necessary, thus

$$\bar{\imath}^{F}(s-j\omega) = \frac{V_F}{(s-j\omega)\{R_F+L_F(s-j\omega)\}}-\frac{M}{R_F+L_F(s-j\omega)}$$
$$\times\left[-\frac{j}{2}(s-j\omega)\{\varepsilon^{j\varphi}\bar{\imath}^{\beta}(s-j2\omega)-\varepsilon^{-j\varphi}\bar{\imath}^{\beta}(s)\}\right]$$

and

$$\bar{\imath}^{F}(s+j\omega) = \frac{V_F}{(s+j\omega)\{R_F+L_F(s+j\omega)\}}-\frac{M}{R_F+L_F(s+j\omega)}$$
$$\times\left[-\frac{j}{2}(s+j\omega)\{\varepsilon^{j\varphi}\bar{\imath}^{\beta}(s)-\varepsilon^{-j\varphi}\bar{\imath}^{\beta}(s+j2\omega)\}\right]$$

The first voltage transform equation thus becomes

$$
\begin{aligned}
0 = {} & (R_a + L_d s)\, \bar{i}^{\beta}(s) \\
& - \frac{jMs\varepsilon^{j\varphi}}{2\{R_F + L_F(s-j\omega)\}} \\
& \times \left[\frac{V_F}{s-j\omega} + \frac{jM(s-j\omega)}{2} \{\varepsilon^{j\varphi}\bar{i}^{\beta}(s-j2\omega) - \varepsilon^{-j\varphi}\bar{i}^{\beta}(s)\} \right] \\
& + \frac{jMs\varepsilon^{-j\varphi}}{2\{R_F + L_F(s+j\omega)\}} \\
& \times \left[\frac{V_F}{s+j\omega} + \frac{jM(s+j\omega)}{2} \{\varepsilon^{j\varphi}\bar{i}^{\beta}(s) - \varepsilon^{-j\varphi}\bar{i}^{\beta}(s+j2\omega)\} \right] \\
& + \frac{jM}{2} \{\varepsilon^{j\varphi} - \varepsilon^{-j\varphi}\} \frac{V_F}{R_F}
\end{aligned}
$$

The terms of this equation would all be known in a given case except for the functions $\bar{i}^{\beta}(s)$, $\bar{i}^{\beta}(s-j2\omega)$ and $\bar{i}^{\beta}(s+j2\omega)$, which are identical functions of the three variables s, $s-j2\omega$, $s+j2\omega$. Such an equation is a difference equation and is susceptible to solution either approximately or numerically. When $\bar{i}^{\beta}(s)$ is known, i^{β} follows and similarly for i^{F}. For a full solution the reader is referred to refs. 12 and 24.

It will be noted that matrix algebra has contributed nothing to the solution of this problem, which has been included to show that, in such a case, the formulation of the equations may be simple, but the routine techniques for solution fail, although it is not necessary to abandon *all* hope.

CHAPTER 14

Conclusion

THIS chapter is necessary only to put matrix analysis of electrical machinery into perspective.

Firstly, electrical machines form a group within a broader class which can be described as electromechanical devices and includes also such things as contactors and meters. The fundamental equations of all such apparatus are the same. Rotating machines alone have been treated in detail here since they form an important group worthy of detailed attention and also because they have one thing in common which most of the others lack, namely a steady-state constant speed condition. This makes them a better introduction to electromechanical systems than the seemingly simpler devices which have only a limited movement.

Secondly, no book on matrix analysis of electrical machines would be complete without a reference to tensor analysis. Tensor analysis is a much deeper fundamental study than matrix analysis, employing a very concise and powerful notation and is closely associated with the study of geometry and dynamics. The tensor approach to electrical machinery is usually based on the Lagrange Dynamical Equation, but to use this it is necessary to know the expression for the energy stored in a magnetic field. This approach is thus as dependent on the circuit equation as is the treatment in this book. Moreover, when it is applied to a particular electrical machine, it is indistinguishable from the matrix analysis in this book, since the matrix array is the most convenient arrangement for parameters and variables.

To embark first upon tensor analysis of electromechanical systems can be a very trying experience since, in addition to the matrix algebra, it is necessary to study the dynamics in tensor form before any application can be made. It is preferable, therefore, to study matrix analysis and some of its application to actual machines before studying tensor analysis. The latter then provides not only the mathematical philosophy behind the matrix analysis, but also enables one to go beyond the scope of this book with a confidence which would otherwise be quite unjustified.

Appendix A

Restriction on Rotor Windings

We will show that it is necessary for the rotor windings to be balanced, even when the air-gap is uniform, if they are to be transformed to stationary windings to obtain equations with constant coefficients.

Let us assume that, with the same notation as on p. 50,

$$L_{\alpha D} = M_{\alpha D} \cos \theta \qquad L_{\beta D} = M_{\beta D} \sin \theta$$

$$L_{\alpha Q} = M_{\alpha Q} \sin \theta \qquad L_{\beta Q} = -M_{\beta Q} \cos \theta$$

where $M_{\alpha D}$, $M_{\beta D}$, $M_{\alpha Q}$ and $M_{\beta Q}$ are constant but not necessarily equal.

The third and fourth voltage equations of p. 58 now become

$$v_D = p(M_{\alpha D} \cos \theta i^{\alpha} + M_{\beta D} \sin \theta i^{\beta}) + R_{DD} i^D + L_{DD} p i^D$$

$$v_Q = p(M_{\alpha Q} \sin \theta i^{\alpha} - M_{\beta Q} \cos \theta i^{\beta}) + R_{QQ} i^Q + L_{QQ} p i^Q$$

The transformation required is thus

$$M_{dD} i^d = M_{\alpha D} \cos \theta i^{\alpha} + M_{\beta D} \sin \theta i^{\beta}$$

$$M_{qQ} i^q = M_{\alpha Q} \sin \theta i^{\alpha} - M_{\beta Q} \cos \theta i^{\beta}$$

where M_{dD} and M_{qQ} are constant inductances which are necessary to maintain dimensional consistency. In matrix form the transformation is

$$\begin{array}{c|c|} d & i^d \\ \hline q & i^q \end{array} = \begin{array}{c|c|c|} & \alpha & \beta \\ \hline d & \dfrac{M_{\alpha D}}{M_{dD}} \cos \theta & \dfrac{M_{\beta D}}{M_{dD}} \sin \theta \\ \hline q & \dfrac{M_{\alpha Q}}{M_{qQ}} \sin \theta & -\dfrac{M_{\beta Q}}{M_{qQ}} \cos \theta \end{array} \begin{array}{c|c|} \alpha & i^{\alpha} \\ \hline \beta & i^{\beta} \end{array}$$

The equation expressing i^α and i^β in terms of i^d and i^q is

$$\begin{array}{c|c|} \alpha & i^\alpha \\ \hline \beta & i^\beta \end{array} = \frac{1}{\Delta} \begin{array}{c|c|c|} & \text{d} & \text{q} \\ \alpha & -\dfrac{M_{\beta Q}}{M_{qQ}}\cos\theta & -\dfrac{M_{\beta D}}{M_{dD}}\sin\theta \\ \hline \beta & -\dfrac{M_{\alpha Q}}{M_{qQ}}\sin\theta & \dfrac{M_{\alpha D}}{M_{dD}}\cos\theta \end{array} \begin{array}{c|c|} \text{d} & i^d \\ \hline \text{q} & i^q \end{array}$$

The determinant

$$\Delta = -\frac{M_{\alpha D}M_{\beta Q}}{M_{dD}M_{qQ}}\cos^2\theta - \frac{M_{\alpha Q}M_{\beta D}}{M_{dD}M_{qQ}}\sin^2\theta,$$

thus appears in the denominators of i^α and i^β expressed in terms of i^d and i^q. Since the first two voltage equations include the time differentials of i^α and i^β, these equations will inevitably be complicated functions of θ unless Δ is independent of θ, i.e. unless

$$\frac{M_{\alpha D}M_{\beta Q}}{M_{dD}M_{qQ}} = \frac{M_{\alpha Q}M_{\beta D}}{M_{dD}M_{qQ}} = -\Delta$$

so that we must assume that $M_{\alpha D}M_{\beta Q} = M_{\alpha Q}M_{\beta D}$.

Again v_d, v_q in terms of v_α, v_β are given by

$$\begin{array}{c|c|} \text{d} & v_d \\ \hline \text{q} & v_q \end{array} = \frac{1}{\Delta} \begin{array}{c|c|c|} & \alpha & \beta \\ \text{d} & -\dfrac{M_{\beta Q}}{M_{qQ}}\cos\theta & -\dfrac{M_{\alpha Q}}{M_{qQ}}\sin\theta \\ \hline \text{q} & -\dfrac{M_{\beta D}}{M_{dD}}\sin\theta & \dfrac{M_{\alpha D}}{M_{dD}}\cos\theta \end{array} \begin{array}{c|c|} \alpha & v_\alpha \\ \hline \beta & v_\beta \end{array}$$

Hence $$v_d = \frac{1}{\Delta}\left[-\frac{M_{\beta Q}}{M_{qQ}}\cos\theta v_\alpha - \frac{M_{\alpha Q}}{M_{qQ}}\sin\theta v_\beta\right]$$

It is not necessary to expand the whole of this expression and we will consider only the resistance voltage drops which are:

$$\frac{1}{\Delta}\left[-\frac{M_{\beta Q}}{M_{qQ}}\cos\theta R_{\alpha\alpha}i^{\alpha}-\frac{M_{\alpha Q}}{M_{qQ}}\sin\theta R_{\beta\beta}i^{\beta}\right]$$

$$=\frac{1}{\Delta}\left[-\frac{M_{\beta Q}}{M_{qQ}}\cos\theta R_{\alpha\alpha}\frac{1}{\Delta}\left\{-\frac{M_{\beta Q}}{M_{qQ}}\cos\theta i^{d}-\frac{M_{\beta D}}{M_{dD}}\sin\theta i^{q}\right\}\right.$$

$$\left.-\frac{M_{\alpha Q}}{M_{qQ}}\sin\theta R_{\beta\beta}\frac{1}{\Delta}\left\{-\frac{M_{\alpha Q}}{M_{qQ}}\sin\theta i^{d}+\frac{M_{\alpha D}}{M_{dD}}\cos\theta i^{q}\right\}\right]$$

$$=\frac{1}{\Delta^2}\left[\left\{\frac{M^2_{\beta Q}}{M^2_{qQ}}R_{\alpha\alpha}\cos^2\theta+\frac{M^2_{\alpha Q}}{M^2_{qQ}}R_{\beta\beta}\sin^2\theta\right\}i^{d}\right.$$

$$\left.+\left\{\frac{M_{\beta Q}M_{\beta D}}{M_{qQ}M_{dD}}R_{\alpha\alpha}-\frac{M_{\alpha Q}M_{\alpha D}}{M_{qQ}M_{dD}}R_{\beta\beta}\right\}\sin\theta\cos\theta i^{q}\right]$$

For this to be independent of θ it will be necessary for

$$\frac{M^2_{\beta Q}}{M^2_{qQ}}R_{\alpha\alpha}=\frac{M^2_{\alpha Q}}{M^2_{qQ}}R_{\beta\beta}\quad\text{and}\quad\frac{M_{\beta Q}M_{\beta D}}{M_{qQ}M_{dD}}R_{\alpha\alpha}=\frac{M_{\alpha Q}M_{\alpha D}}{M_{qQ}M_{dD}}R_{\beta\beta}$$

From the first of these conditions $M^2_{\beta Q}R_{\alpha\alpha}=M^2_{\alpha Q}R_{\beta\beta}$ and from the second, $M_{\beta Q}M_{\beta D}R_{\alpha\alpha}=M_{\alpha Q}M_{\alpha D}R_{\beta\beta}$.

One set of relations which would comply with these and the earlier requirement is that

$$R_{\alpha\alpha}=R_{\beta\beta},\quad M_{\alpha Q}=M_{\beta Q},\quad M_{\alpha D}=M_{\beta D}$$

which are true if the α and β windings are balanced.

A second more general set of relations is

$$R_{\alpha\alpha}=n^2R_{\beta\beta},\quad M_{\alpha Q}=nM_{\beta Q},\quad M_{\alpha D}=nM_{\beta D}$$

which also meet the earlier condition.

It is apparent that this second set of relations is the same as the first set except that the turns ratio is n instead of unity. This will increase $M_{\alpha Q}$ and $M_{\alpha D}$ in the turns ratio and the resistance in the square of the turns ratio, assuming that the total volume of the conductor is unchanged, i.e. that the space factor is the same. There is little practical application for the wider condition, hence effectively it is necessary for the rotor windings to be balanced. A similar result is obtained if the induced voltage terms are considered in place of the resistance drops.

Appendix B

Torque under Saturated Conditions

If hysteresis is neglected, in a device with a single coil the flux-linkage of that coil is a function of the current in it and of the relative position of the moving element relative to the stationary element. It may, therefore, be written as $\psi(\theta, i)$. In Fig. 51 the graph *OEA* represents the flux-linkage ψ as a function of the current i for a fixed relative position θ_1, and the graph *OFB* represents the flux-linkage for a fixed position θ_2.

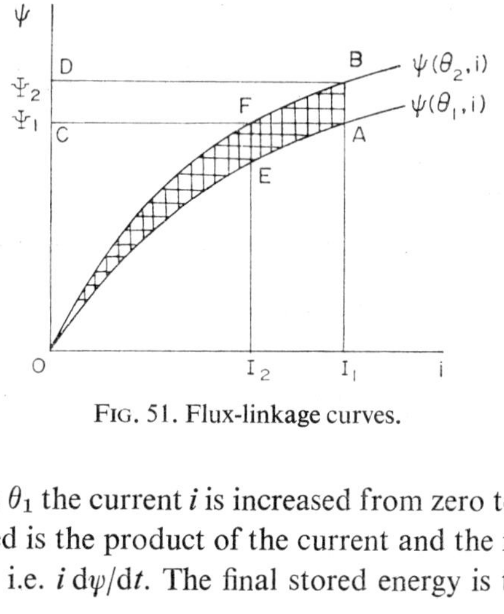

FIG. 51. Flux-linkage curves.

If in position θ_1 the current i is increased from zero to I_1 in time τ_1, the power stored is the product of the current and the rate of change of flux-linkage, i.e. $i\,d\psi/dt$. The final stored energy is then

$$\int_0^{\tau_1} i(d\psi/dt)\,dt = \int_0^{\Psi_1} i(\theta_1)\,d\psi = \text{area}\,OEAFCO$$

where Ψ_1 is the final flux-linkage at time τ_1 with the current I_1.

If now the position is changed from θ_1 to θ_2 at time τ_2, while the current is kept constant at I_1, the input power is $I_1\, \mathrm{d}\psi/\mathrm{d}t$ and the additional energy input is

$$\int_{\tau_1}^{\tau_2} I_1(\mathrm{d}\psi/\mathrm{d}t)\, \mathrm{d}t = \int_{\Psi_1}^{\Psi_2} I_1\, \mathrm{d}\psi = I_1(\Psi_2 - \Psi_1)$$
$$= \text{area } ABDC$$

The total energy input to the system is now

$$\int_0^{\Psi_1} i(\theta_1)\, \mathrm{d}\psi + I_1(\Psi_2 - \Psi_1) = \text{area } OEABDO$$

But in position θ_2 with a current I_1 the stored energy is

$$\int_0^{\Psi_2} i(\theta_2)\, \mathrm{d}\psi = \text{area } OFBDO$$

The amount by which the total energy input exceeds the final stored energy in position θ_2 must have been converted into mechanical energy as an output. Hence the mechanical output energy is

$$\int_0^{\Psi_1} i(\theta_1)\, \mathrm{d}\psi + I_1(\Psi_2 - \Psi_1) - \int_0^{\Psi_2} i(\theta_2)\, \mathrm{d}\psi = \text{area } OEABFO$$

Now integration by parts gives

$$\int_0^{\Psi_1} i(\theta_1)\, \mathrm{d}\psi = I_1 \Psi_1 - \int_0^{I_1} \psi(\theta_1)\, \mathrm{d}i$$

and

$$\int_0^{\Psi_2} i(\theta_2)\, \mathrm{d}\psi = I_1 \Psi_2 - \int_0^{I_1} \psi(\theta_2)\, \mathrm{d}i$$

Substituting these values gives the mechanical output energy as

$$\int_0^{I_1} \psi(\theta_2)\, \mathrm{d}i - \int_0^{I_1} \psi(\theta_1)\, \mathrm{d}i = \int_0^{I_1} \{\psi(\theta_2) - \psi(\theta_1)\}\, \mathrm{d}i$$

Now suppose that $\theta_1 = \theta$ and $\theta_2 = \theta+\Delta\theta$, then the mechanical output energy is

$$\int_0^{I_1} \{\psi(\theta+\Delta\theta)-\psi(\theta)\}\, \mathrm{d}i$$

and the mean torque is

$$\frac{1}{\Delta\theta}\int_0^{I_1} \{\psi(\theta+\Delta\theta)-\psi(\theta)\}\, \mathrm{d}i = \int_0^{I_1} \frac{\psi(\theta+\Delta\theta)-\psi(\theta)}{\Delta\theta}\, \mathrm{d}i$$

which in the limit as $\Delta\theta \to 0$, becomes

$$\int_0^{I_1} \frac{\partial\psi}{\partial\theta}\, \mathrm{d}i$$

As an alternative, suppose that the movement $\Delta\theta$ had taken place not at constant current, but at constant flux-linkage, that is from A to F instead of from A to B in Fig. 51. Constant flux-linkage implies zero rate of change of flux-linkage and hence there is no electrical power input or output (other than I^2R) during the motion. The mechanical output energy is simply the stored energy in position θ_1 with current I_1 less the stored enegy in position θ_2 with the corresponding current I_2. The latter is

$\int_0^{\Psi_1} i(\theta_2)\, \mathrm{d}\psi$ so that the mechanical output energy is

$$\int_0^{\Psi_1} i(\theta_1)\, \mathrm{d}\psi - \int_0^{\Psi_1} i(\theta_2)\, \mathrm{d}\psi$$
$$= \int_0^{\Psi_1} \{i(\theta_1)-i(\theta_2)\}\, \mathrm{d}\psi$$

If again $\theta_1 = \theta$ and $\theta_2 = \theta+\Delta\theta$, the mean torque is

$$\frac{1}{\Delta\theta}\int_0^{\Psi_1} \{i(\theta)-i(\theta+\Delta\theta)\}\, \mathrm{d}\psi$$
$$= -\int_0^{\Psi_1} \frac{i(\theta+\Delta\theta)-i(\theta)}{\Delta\theta}\, \mathrm{d}\psi$$

which in the limit as $\Delta\theta \to 0$, becomes

$$-\int_0^{\Psi_1} \frac{\partial i}{\partial \theta}\, d\psi$$

Although this is a different expression from that obtained before, the two expressions are equal and this will be shown by a slightly different mathematical treatment.

In the following analysis I and Ψ denote the values of current and flux-linkage at which the torque is to be determined, while i and ψ denote the values during the period in which the current and flux-linkage are brought up from zero to I and Ψ. Now, however, neither I nor Ψ are necessarily constant during the movement $\Delta\theta$. This means that the terms $\partial\Psi/\partial\theta$ and $\partial I/\partial\theta$, which arise, are not necessarily zero.

With terminal voltage V and current I the input energy in a time Δt is $VI\,\Delta t$. Of this $RI^2\,\Delta t$ will be dissipated in the resistances, ΔW will be stored, and $T\,\Delta\theta$ converted to mechanical energy. Since $V = RI + d\Psi/dt$,

$$VI\,\Delta t = RI^2\,\Delta t + I(d\Psi/dt)\,\Delta t$$

Hence

$$T\,\Delta\theta + \Delta W = I(d\Psi/dt)\,\Delta t = I\,\Delta\Psi$$

and

$$T = I(\Delta\Psi/\Delta\theta) - \Delta W/\Delta\theta$$

In the limit as $\Delta\theta \to 0$, this becomes

$$T = I\frac{\partial\Psi}{\partial\theta} - \frac{\partial W}{\partial\theta}$$

Now the stored energy $\quad W = \int_0^{\Psi} i\, d\psi$

Hence

$$T = I\frac{\partial\Psi}{\partial\theta} - \frac{\partial}{\partial\theta}\int_0^{\Psi} i\, d\psi$$

By the rules of differentiation under the integral sign†

$$\frac{\partial}{\partial\theta}\int_0^{\Psi} i\,\mathrm{d}\psi = \int_0^{\Psi}\frac{\partial i}{\partial\theta}\,\mathrm{d}\psi + i(\Psi)\frac{\partial\Psi}{\partial\theta}$$
$$= \int_0^{\Psi}\frac{\partial i}{\partial\theta}\,\mathrm{d}\psi + I\frac{\partial\Psi}{\partial\theta}$$

and hence

$$T = I\frac{\partial\Psi}{\partial\theta} - \int_0^{\Psi}\frac{\partial i}{\partial\theta}\,\mathrm{d}\psi - I\frac{\partial\Psi}{\partial\theta}$$
$$= -\int_0^{\Psi}\frac{\partial i}{\partial\theta}\,\mathrm{d}\psi$$

Alternatively, $\int_0^{\Psi} i\,\mathrm{d}\psi$ may be integrated by parts to get

$$\int_0^{\Psi} i\,\mathrm{d}\psi = I\Psi - \int_0^{I}\psi\,\mathrm{d}i$$

Again, by the rules for differentiation under the integral sign,

$$\frac{\partial}{\partial\theta}\int_0^{I}\psi\,\mathrm{d}i = \int_0^{I}\frac{\partial\psi}{\partial\theta}\,\mathrm{d}i + \psi(I)\frac{\partial I}{\partial\theta}$$
$$= \int_0^{I}\frac{\partial\psi}{\partial\theta}\,\mathrm{d}i + \Psi\frac{\partial I}{\partial\theta}$$

Hence

$$\frac{\partial}{\partial\theta}\int_0^{\Psi} i\,\mathrm{d}\psi = \frac{\partial}{\partial\theta}(I\Psi) - \frac{\partial}{\partial\theta}\int_0^{I}\psi\,\mathrm{d}i$$
$$= I\frac{\partial\Psi}{\partial\theta} + \Psi\frac{\partial I}{\partial\theta} - \int_0^{I}\frac{\partial\psi}{\partial\theta}\,\mathrm{d}i - \Psi\frac{\partial I}{\partial\theta}$$
$$= I\frac{\partial\Psi}{\partial\theta} - \int_0^{I}\frac{\partial\psi}{\partial\theta}\,\mathrm{d}i$$

† See, for example, Pipes, *Applied Mathematics for Engineers and Physicists*, p. 320.

Hence

$$T = I\frac{\partial \Psi}{\partial \theta} - \frac{\partial}{\partial \theta}\int_0^{\Psi} i\,\mathrm{d}\psi$$

$$= I\frac{\partial \Psi}{\partial \theta} - I\frac{\partial \Psi}{\partial \theta} + \int_0^I \frac{\partial \psi}{\partial \theta}\,\mathrm{d}i$$

$$= \int_0^I \frac{\partial \psi}{\partial \theta}\,\mathrm{d}i$$

Thus the two expressions $\int_0^I \frac{\partial \psi}{\partial \theta}\,\mathrm{d}i$ and $-\int_0^{\Psi} \frac{\partial i}{\partial \theta}\,\mathrm{d}\psi$ are equal and both give the torque irrespective of how i and Ψ are considered to behave during the infinitesimal movement $\Delta\theta$.

Just as $\int_0^{\Psi} i\,\mathrm{d}\psi$ represents energy, so also does $\int_0^I \psi\,\mathrm{d}i$, which is called the co-energy. The torque may thus be expressed in terms of the partial derivative with respect to θ of either the stored magnetic energy or of the co-energy.

Torque in a Multi-circuit Device

In a device involving a number of circuits, the flux-linkage of each depends on the relative position θ and on all the currents. By similar reasoning it is possible to show that the torque in such a device involving n circuits is given by

$$\sum_{x=1,2,\ldots,n} \int_0^{I_x} \frac{\partial \psi_x(\theta, i^1, i^2, \ldots i^n)}{\partial \theta}\,\mathrm{d}i_x$$

In this case it is necessary to consider the integrability of $i^1\,\mathrm{d}\psi_1 + i^2\,\mathrm{d}\psi_2 + \ldots + i^n\,\mathrm{d}\psi_n$. It is interesting to note that the principle of the conservation of energy, expressed in terms appropriate to this case, satisfies the mathematical condition for integrability. This shows in mathematical form what is clear from physical considerations, namely that we cannot use the preceding technique to determine torque taking hysteresis into account. If hysteresis is neglected, conservation of energy applies and stored energy and torque are state functions depending only on the state of the device and independent of how it was brought to that state. If hysteresis is present, for a given current the flux-linkage depends on whether the current has been increased or decreased to the given value. The stored energy and torque do not then depend solely on the state of the device, but also on how that state was reached. Without this detailed information, the torque cannot be determined by any means.

Calculation of Torque

In some cases of a single-circuit device where the flux-linkage can be suitably expressed, it may be possible to perform the mathematical operations required to obtain a general expression for the torque. In general, however, only numerical solutions are possible. Where a number of circuits is involved, although the amount of data required is considerable, it is not necessary to know the values of all flux-linkages as functions of position for all possible combinations of currents. Since the torque is a state function, we may consider that all currents are brought up from zero to the specific values in the same time or, alternatively, that they are brought up in succession, so that as each is brought up the others are constant at either the specified value or zero.

Only in exceptional circumstances could this technique be employed for electrical machines. Fortunately in most machines the variation in permeance of the magnetic circuits depends primarily on the variation of length of the part of the circuit in air, and there is little or no change in the conditions in the iron. The system can be treated without great errors as if it were linear. This is equivalent to regarding the part of the circuit in iron as replaced by an equivalent path in air. Practically, this means that the inductances used in the torque expressions derived in Chapter 6 should not be the unsaturated values, but should be adjusted to allow for saturation.†

It is left to the reader to express the flux-linkages in terms of Li in the expressions for torque deduced in this Appendix and show that this leads to the same expression as given in Chapter 6.

† See ref. 25.

Appendix C

Definitions of Systems of Axes

This appendix is included in the hope of assisting the reader to proceed to the study of other literature on the subject.

Any reader not already aware that the definition of, and relationship between, the two sets of axes α, β and d, q is arbitrary, will surely be convinced by a very brief study of Fig. 52.

The following list gives some of the writers who have used the various systems of axes shown in Fig. 52.

(i) Kron, G., Non-Riemannian Dynamics of Rotating Electrical Machinery (*Journal of Mathematics and Physics*) 1934. and subsequent writings.
Gibbs, W. J., *Tensors in Electrical Machine Theory*, 1952.
Lynn, J. W., *Tensors in Electrical Engineering*, 1963.
Jones, C. V., *The Unified Theory of Electrical Machines*, 1967.

(ii) Gibbs, W. J., The Modern Approach to Electrical Machine Analysis (*The Engineer*) 1951.
White, D. C. and Woodson, H. H., *Electromechanical Energy Conversion*, 1959.

(iii) Gibbs, W. J., *Electric Machine Analysis using Matrices* 1962.
Hancock, N. N., *Matrix Analysis of Electrical Machinery*, 1964.
Gibbs, W. J., *Electric Machine Analysis using Tensors*, 1967.

(iv) Messerle, H. K., *Dynamic Circuit Theory*, 1965.

(v) Jevons, M., *Electrical Machine Theory*, 1966.

(vi) Jones, C. V., *The Unified Theory of Electrical Machines*, 1967.

(vii) O'Kelly, D. and Simmons, S., *Introduction to Generalized Electrical Machine Theory*, 1968.

It would appear that a number of writers, including the present author, did not care to follow Kron in making θ a positive angle in the *clockwise* direction, but, apart from this, it is difficult to see any general basis for the particular choices. The systems shown in (ii) and (v) are related to that shown in (i) by a reflection and rotation through $\pi/2$, and accordingly they lead to the same transformation matrix and impedances as does the Kron system of (i). The system shown in (vii) similarly results from a reflection and rotation of that shown in (iv) and consequently they have the same transformation matrix, which, however, differs from that of Kron in having the opposite sign for θ. As a result the elements of the impedance matrix associated with the rotational voltages are of opposite sign to those of Kron. The systems shown in (iii) and (vi) are also related to each other by reflection and rotation and have the same transformation matrix, which, although different from all the others, leads to the same impedance matrices as does that of Kron. That reversal of the directions of both the q axis and θ would do this was pointed out by Kron (*Tensor Analysis for Electrical Engineers*, 1942).

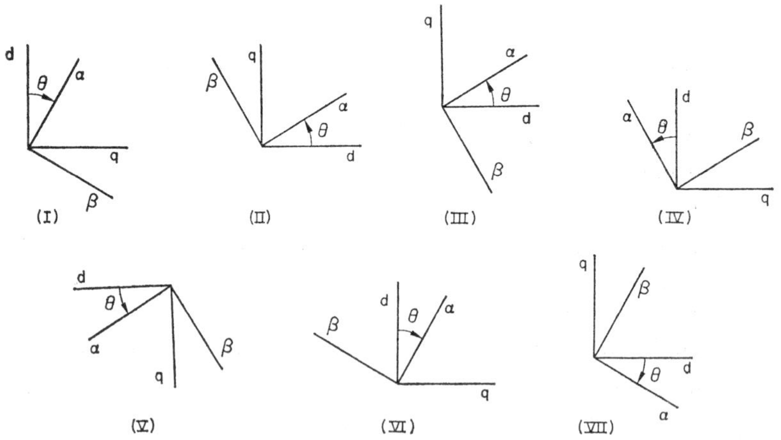

FIG. 52. α, β and d, q axes relationships.

Dr. W. J. Gibbs and the author adopted the system shown in (iii) in the first instance to obtain v_α leading v_β when a machine is operated as a revolving armature synchronous machine with $\dot{\theta}$ positive. A consequence is, however, that when the machine is operated as an induction machine with the stator as the primary, if v_D leads v_Q, then, although i^α leads i^β at super-synchronous speeds, it lags behind i^β at the more important sub-synchronous speeds. There is no single system free from all objections in respect of rotor phase-sequence and Jones avoided this particular objection by using different definitions with different transformation matrices for synchronous and induction machines. The present author prefers to use one system throughout on the grounds that the phase-sequence of the currents in the secondary winding of an induction machine is rarely of any importance, and in all systems reverses as the machine passes through synchronous speed. A great advantage of the systems shown in (iii) and (vi) is that the transformation matrix is both orthogonal and symmetric, whereas the others are only orthogonal. The form of the transpose and inverse are thus both identical with that of the matrix itself, so that no thought is required to determine where the negative sign should be. Gibbs has explained (*Electric Machine Analysis using Tensors*, p. 35) that whereas in the other systems α, β are related to d, q by a simple or "proper" rotation through an angle θ, in this system there is both rotation *and* reflection or an "improper" rotation. It is this double operation which causes the matrix to be its own inverse.

The relationship between the three-phase axes a, b, c and the two-phase axes α, β is also arbitrary. Although all writers appear to be agreed that a transformation which permits direct comparison of the α and a phases is to be preferred, some omit the factor $\sqrt{\frac{2}{3}}$. This leads to $v_\alpha = v_a$ and $i^\alpha = i^a$, but, of course, not to invariant *total* power. Whilst possibly a computational convenience, this is contrary to the whole philosophy of tensor analysis which underlies matrix analysis.

Similarly many writers retain Fortescue's original definition of symmetrical components given on p. 106 in preference to the orthogonal relationships used in this book. This again causes the total power to change under the transformation and is also fundamentally unacceptable, although it may be more convenient for some particular purposes.

Appendix D

Trigonometric Formulae

$$\sin A \cos B = \tfrac{1}{2}\{\sin(A+B)+\sin(A-B)\}$$

$$\sin A \sin B = \tfrac{1}{2}\{-\cos(A+B)+\cos(A-B)\}$$

$$\cos A \cos B = \tfrac{1}{2}\{\cos(A+B)+\cos(A-B)\}$$

$$\sin A+\sin B = 2\sin\frac{(A+B)}{2}\cos\frac{(A-B)}{2}$$

$$\sin A-\sin B = 2\cos\frac{(A+B)}{2}\sin\frac{(A-B)}{2}$$

$$\cos A+\cos B = 2\cos\frac{(A+B)}{2}\cos\frac{(A-B)}{2}$$

$$\cos A-\cos B = -2\sin\frac{(A+B)}{2}\sin\frac{(A-B)}{2}$$

$$\sin(A+B) = \sin A\cos B+\cos A\sin B$$

$$\sin(A-B) = \sin A\cos B-\cos A\sin B$$

$$\cos(A+B) = \cos A\cos B-\sin A\sin B$$

$$\cos(A-B) = \cos A\cos B+\sin A\sin B$$

$$\cos^2 A = \tfrac{1}{2}(1+\cos 2A)$$

$$\sin^2 A = \tfrac{1}{2}(1-\cos 2A)$$

$$\sin A\cos A = \tfrac{1}{2}\sin 2A$$

Laplace Transforms

Time function	Transform
A	A/s
$\varepsilon^{-\alpha t}$	$1/(s+\alpha)$
$\sin \omega t$	$\omega/(s^2+\omega^2)$
$\cos \omega t$	$s/(s^2+\omega^2)$
$\sin(\omega t+\phi)$	$(s \sin \phi+\omega \cos \phi)/(s^2+\omega^2)$
$\varepsilon^{-\alpha t} \sin \omega t$	$\omega/\{(s+\alpha)^2+\omega^2\}$
$\varepsilon^{-\alpha t} \cos \omega t$	$(s+\alpha)/\{(s+\alpha)^2+\omega^2\}$
$\frac{\mathrm{d}}{\mathrm{d}t} \mathrm{f}(t)$	$s\,\mathrm{f}(s)-\underset{t\to 0}{\mathrm{f}(t)}$
$\int_{-\infty}^{t} \mathrm{f}(t)\,\mathrm{d}t$	$\frac{1}{s}\mathrm{f}(s)+\frac{1}{s}\int_{-\infty}^{0+} \mathrm{f}(t)\,\mathrm{d}t$

Exercises

1. Multiply (a) $\begin{bmatrix} 1 & 2 & 3 \\ 8 & 9 & 4 \\ 7 & 6 & 5 \end{bmatrix} \begin{bmatrix} 1 \\ 3 \\ 2 \end{bmatrix}$ (b) $\begin{bmatrix} 1 & 2 & 3 \end{bmatrix} \begin{bmatrix} 1 & 2 & 3 \\ 8 & 9 & 4 \\ 7 & 6 & 5 \end{bmatrix}$

2. Multiply (a) $\begin{bmatrix} 1 & 2 & 3 \\ 8 & 9 & 4 \\ 7 & 6 & 5 \end{bmatrix} \begin{bmatrix} 1 & 2 & 3 \\ 3 & 1 & 2 \\ 2 & 3 & 1 \end{bmatrix}$

   (b) $\begin{bmatrix} 1 & 2 & 3 \\ 3 & 1 & 2 \\ 2 & 3 & 1 \end{bmatrix} \begin{bmatrix} 1 & 2 & 3 \\ 8 & 9 & 4 \\ 7 & 6 & 5 \end{bmatrix}$

3. Invert $\begin{bmatrix} A & B \\ C & D \end{bmatrix}$ (and memorize the result).

4. Invert $\begin{bmatrix} 1 & 2 & 3 \\ 8 & 9 & 4 \\ 7 & 6 & 5 \end{bmatrix}$ (a) by the routine procedure; (b) by partitioning.

5. Invert $\begin{bmatrix} 1 & 2 & 3 \\ 3 & 1 & 2 \\ 2 & 3 & 1 \end{bmatrix}$ (a) by the routine procedure; (b) by transformation to diagonal form, using the **C** of p. 28.

6. Invert the results of Exercise 2, and compare the inverses with the products (in reverse order) of the results of Exercises 4 and 5.

7. A two-by-two symmetric matrix is pre-multiplied by a skew-symmetric matrix and post-multiplied by another skew-symmetric matrix. What is the form of the product? Is the result the same for three-by-three matrices?

8. Three impedances Z_{11}, Z_{22}, Z_{33} with mutual impedances $X_{12} =$

X_{21}, $X_{13} = X_{31}$, $X_{23} = X_{32}$, are connected with impedance 1 in series with the parallel combination of impedances 2 and 3. The terminal voltage of the network is V. Determine the currents in impedances 1 and 2.

9. If the impedance matrix of a four-phase machine (one with an angle $\pi/2$ between adjacent phase windings) is a four-by-four circulant matrix, what is the form of the impedance matrix of the two-phase machine formed by connecting opposite phases of the four-phase machine in series opposition?

10. A network consists solely of a number of parallel branches with mutual inductance between them. If $\mathbf{v} = \mathbf{Zi}$ expresses the branch voltages in terms of the branch currents and $\mathbf{v}' = \mathbf{Z}'\mathbf{i}'$ expresses the terminal voltage of the network in terms of its current, determine the value of $\mathbf{Z}'$.

11. A transformer has a primary to secondary turns ratio of 10/1. With the secondary winding open circuit, the primary winding impedance is $1+j110$ ohms. With the primary winding open circuit, the secondary winding impedance is $0{\cdot}01+j1{\cdot}1$ ohms. The mutual inductance between the two windings corresponds to a reactance $j10$ ohms. The output from the secondary winding is 9·9 V, 10 A at unity power factor. Taking the output current as "real", determine the primary current and voltage in complex form, (a) from the equivalent circuit in terms of actual values, and (b) from the equivalent circuit in which all quantities are referred to the primary winding.

12. Transform the impedance matrix of p. 63 to axes stationary relative to the rotor.

13. Transform the impedance matrix of p. 59 to axes stationary relative to the rotor.

14. A two-by-two impedance matrix $\mathbf{Z}$ is transformed to $\mathbf{Z}'$ by the $\mathbf{C}$ of p. 111. What is the form of $\mathbf{Z}'$ when $\mathbf{Z}$ is (a) scalar, (b) diagonal, (c) symmetric, (d) skew-symmetric, and (e) hermitian?

15. The three-phase to symmetrical component transformation (p. 108) makes a circulant matrix diagonal (see p. 109). Is there any more general form of three-by-three matrix which this transformation will make symmetric?

16. In two-phase axes an impedance matrix is of the form

	α	β	o
α	$Z_{\alpha\alpha}$	$Z_{\alpha\beta}$	
β	$Z_{\beta\alpha}$	$Z_{\beta\beta}$	
o			Z_{oo}

where $Z_{\alpha\beta} \neq Z_{\beta\alpha}$. Determine the impedance matrix in terms of three-phase axes a, b, c. What is the form of this matrix if $Z_{\alpha\alpha} = Z_{\beta\beta}$ and $Z_{\beta\alpha} = -Z_{\alpha\beta}$?

17. The terminal conditions of a network with a given impedance matrix are in the form of terminal voltages of some branches, currents in some branches, voltage and current of some branches and neither voltage nor current in the remaining branches. Assuming that there is just sufficient data, develop a procedure for determining all the unknown voltages and currents.

18. A metadyne (Fig. 11) with a uniform air-gap is driven at constant speed with a constant d.c. voltage applied to the d brushes, a variable resistance being connected across the q brushes. The stator winding is unexcited. Determine the steady-state currents in the two circuits and the voltage between the q brushes. What driving torque will be required? Comment on the magnitude of the q current if the armature winding resistance is negligible.

19. A machine with a uniform air-gap has no stator windings but the commutator-type armature winding is connected to slip-rings to form a balanced three-phase winding and also to a commutator, the brushes of which are also spaced to form a balanced three-phase system. Determine the frequency and the magnitude of the terminal voltage of the other brushes when the armature is driven at an angular velocity $\dot{\theta}$ and (a) the slip-ring brushes, or (b) the commutator brushes, are supplied with balanced voltage V at an angular frequency ω.

If the commutator brushes are connected to a balanced load when the supply is connected to the slip-ring brushes, determine the driving torque required.

20. A d.c. machine has a uniform air-gap, two field windings on the direct axis and brushes on both the d and q axes (Metadyne). The brushes on the q axis are short-circuited, whilst those on the d axis are connected in series opposition with field winding 1 to form the output circuit. Determine (a) the steady-state output voltage in terms of the output current when a constant voltage V_2 is applied to field winding 2, and (b) the transient e.m.f. in the output circuit in terms of the instantaneous terminal voltage of field winding 2 (transfer function).

Assume that the distributions are sinusoidal and that $L_d = M_{d1}$ and $M_{12} = M_{d2}$.

21. A d.c. series motor is operating with a current I_o when there is a sudden short circuit at its terminals. Determine the instantaneous value of the motor current after the short circuit. Neglect changes of saturation.

22. A d.c. compound motor with a cumulative series winding is on load with an armature current I^a and shunt field current I^Z, when there is a sudden short circuit at the terminals. Assuming that the speed remains effectively constant during the decay of the currents, determine the transient armature current.

23. Determine the form of the impedance matrix obtained by partitioning the impedance matrix given on p. 162 and eliminating the rotor axes d, q.

24. Show that a balanced induction machine connected to a balanced supply system can generate active power but not reactive power.

25. If the machine of Exercise 19 had also had a stator winding, what would have been the frequency and magnitude of its open-circuit voltage under the two conditions (a) and (b)?

26. A commutator machine with a uniform air-gap has balanced D and Q stator windings and two sets of commutator brushes at right angles to one another. Each stator winding is connected in series with one set of brushes to one phase of the balanced two-phase supply. Determine the steady-state torque in terms of the speed and displacement of the brushes from the neutral position.

27. A commutator machine with a uniform air-gap has balanced three-phase stator windings which are connected to a balanced supply.

The three symmetrically spaced brushes are connected to the same supply through a step-down transformer of ratio $n/1$. Determine the steady-state torque when the brushes are displaced from the neutral position.

28. A balanced, star-connected, three-phase induction machine has $R_S = R_r = 1$, $X_S = X_r = 110$ and $X_M = 100$. Plot the speed–torque curve for (a) a balanced supply voltage of 400 V (line), 50 Hz, and (b) a single-phase supply voltage of 400, line-to-line.

29. A resistance R_X is connected in series with one phase of the primary winding of a star-connected, balanced three-phase induction machine. Determine the impedance matrix of the system in symmetrical component axes and thence deduce an expression for the starting torque in terms of the balanced supply voltage. Calculate the torque for the machine of Exercise 28 when $R_X = 30$.

30. Taking the axes of an "inverted" three-phase induction motor as stationary relative to the secondary (stator), deduce an expression for the torque in terms of the voltage applied to the primary (rotor).

31. A balanced polyphase induction machine has two separate cages on the rotor, there being mutual coupling between them. Determine (a) its impedance matrix, and (b) the torque with balanced supply voltage.

32. Deduce an expression for the instantaneous torque of a single-phase induction motor and from it determine whether the torque is positive over the whole cycle or whether it is at times negative.

33. Two terminals of a balanced, star-connected, three-phase induction machine are connected to a single-phase source of voltage V and the third terminal is connected to one of the others through a capacitor. Derive an expression for the starting torque.

34. Both stator and rotor windings of a balanced three-phase induction machine are supplied from balanced three-phase sources, the angular frequencies being ω_S and ω_r respectively. Deduce an expression for the steady-state torque when the machine is operating synchronously with $\dot{\theta} = \omega_S + \omega_r$.

35. A salient-pole machine has a balanced three-phase armature winding which is connected to a balanced a.c. supply. There is no

winding on the salient member (reluctance motor). Determine the torque at synchronous speed. Compare your result with the expression on p. 214 for $E = 0$. Can this machine act as a generator and if so is there any restriction on its operation?

36. Calculate v_b corresponding to the v_a of p. 192.

37. Determine the torque required to drive a salient-pole synchronous machine on steady-state short circuit.

38. The field winding of a three-phase, salient-pole, synchronous machine without damper windings is open circuited and a single-phase voltage applied to the star-connected armature winding with the machine at rest (a) between the B and C terminals with A unconnected, (b) between the A terminal and B and C connected together, and (c) between the A terminal and the star point. Determine the ratio of voltage to current for all connections when (i) $\theta = 0$, and (ii) when $\theta = \pi/2$.

39. Determine the ratios of voltage to current for the machine of Exercise 39 under the same armature conditions but with the field winding short-circuited.

40. An induction-type machine has balanced three-phase windings on both stator and rotor. It is used as a synchronous machine with a d.c. supply connected to the stator winding. Determine the difference, if any, in the transient impedance matrices when the d.c. is supplied (a) to the terminals B and C with A unconnected, and (b) to the terminal A and to the terminals B and C connected together.

41. Derive an expression for the instantaneous torque of a three-phase reluctance motor under steady-state conditions with $\omega > \dot{\theta} > 0$.

42. A salient-pole synchronous machine without dampers is connected to a balanced a.c. supply but is rotating at an angular velocity $\dot{\theta} < \omega$. Determine the steady-state current and mean torque (a) with the field winding short-circuited, and (b) with a constant voltage V_F applied to the field winding.

43. The induction machine equivalent circuits (pp. 170, 184, and 187) were derived for steady-state a.c. conditions. Under what transient conditions, if any, are they valid?

44. A balanced two-phase induction machine is connected to two equal capacitors, one across each stator phase, but not to any supply.

Determine the consequences of driving the machine at an angular velocity $\dot{\theta}$.

45. Calculate i^b corresponding to i^a of p. 233.

46. Calculate v_b corresponding to v_a of p. 239, and compare the value of negative-sequence impedance obtained therefrom with that of p. 239.

47. Calculate i^b corresponding to i^a of p. 242, and compare the value of negative-sequence impedance obtained therefrom with that of p. 243.

48. A synchronous machine without dampers has $R_a = 1$, $R_F = 0.1$, $X_d = 110$, $X_q = 80$, $X_F = 125$ and $X_{Fd} = 100$. Determine the roots of the equation $\Delta = 0$ and compare them with the values obtained from the approximate expressions of p. 226.

49. Investigate the stability of a d.c. series motor following a torque impulse. Neglect friction.

50. Investigate the possibility of obtaining a general solution for the stability of a d.c. compound motor. Neglect friction.

51. By neglecting the resistances and friction, determine an approximate value for the natural frequency of oscillation of a reluctance motor on no load.

52. Deduce the motional impedance matrices of an induction machine for the two connections described in Exercise 40.

53. The field winding of a separately-excited d.c. motor is supplied at constant voltage and the armature from a series generator driven at constant speed. Investigate the stability of the motor.

54. A d.c. generator has a distributed compensating winding, a separately-excited field winding and a uniform air-gap. The brushes are displaced slightly from the neutral position. Determine the steady-state and sudden short-circuit currents.

55. Plot the speed–torque curve for the machine of Exercise 28 when the stator is supplied with balanced voltage of 400 V (line) 50 Hz but with one phase of the secondary winding open circuited.

56. An impedance $R_X + jX_X$ is connected in series with one phase of the secondary winding of a star-connected three-phase induction motor. Determine the steady-state torque in terms of the balanced supply voltage.

Hints and Answers to Exercises

1. (a) $\begin{bmatrix} 13 \\ 43 \\ 35 \end{bmatrix}$ (b) $[38 \quad 38 \quad 26]$

2. (a) $\begin{bmatrix} 13 & 13 & 10 \\ 43 & 37 & 46 \\ 35 & 35 & 38 \end{bmatrix}$ (b) $\begin{bmatrix} 38 & 38 & 26 \\ 25 & 27 & 23 \\ 33 & 37 & 23 \end{bmatrix}$

3. $\dfrac{1}{(\mathrm{AD}-\mathrm{BC})}\begin{bmatrix} \mathrm{D} & -\mathrm{B} \\ -\mathrm{C} & \mathrm{A} \end{bmatrix}$

4. $\frac{1}{48}\begin{bmatrix} -21 & -8 & 19 \\ 12 & 16 & -20 \\ 15 & -8 & 7 \end{bmatrix}$

5. $\frac{1}{18}\begin{bmatrix} -5 & 7 & 1 \\ 1 & -5 & 7 \\ 7 & 1 & -5 \end{bmatrix}$

6. (a) $\dfrac{1}{144}\begin{bmatrix} 34 & 24 & -38 \\ 4 & -24 & 28 \\ -35 & 0 & 13 \end{bmatrix}$

   (b) $\dfrac{1}{432}\begin{bmatrix} 115 & -44 & -86 \\ -92 & -8 & 112 \\ -17 & 76 & -38 \end{bmatrix}$

7. (a) Symmetric; (b) no recognizable form.

8. $i^1 = V(Z_{22}-2X_{23}+Z_{33})/\Delta,$

   $i^2 = V(Z_{33}+X_{13}-X_{12}-X_{23})/\Delta,$

   where $\Delta = (Z_{11}+2X_{13}+Z_{33})(Z_{33}-2X_{23}+Z_{22}) - (X_{12}+X_{23}-X_{13}-Z_{33})^2.$

9.

A	B
−B	A

i.e. scalar plus skew-symmetric.

10. Hint: Define a connection matrix by $\mathbf{v} = \mathbf{K}\mathbf{v}'$, where $\mathbf{v}$ is the branch voltages and $\mathbf{v}'$ is a one-by-one matrix whose sole element is the terminal voltage. Deduce the transformation for invariant current and hence find $\mathbf{Z}'$ by substituting in $\mathbf{i} = \mathbf{Z}^{-1}\mathbf{v}$. $\mathbf{Z}' = [\mathbf{K}_t\mathbf{Z}^{-1}\mathbf{K}]^{-1}$, i.e. the inverse of the sum of the admittances of the branches.

11. $I^P = 1{\cdot}1 - j1 \qquad V_P = 111{\cdot}1 + j20$

12. Hint: Don't forget the differentiations before pre-multiplying by C_t.

	α	β	A′	B′
α	$R_r + L_r p$		Mp	
β		$R_r + L_r p$		Mp
A′	Mp	$M\dot{\theta}$	$R_S + L_S p$	$L_S\dot{\theta}$
B′	$-M\dot{\theta}$	Mp	$-L_S\dot{\theta}$	$R_S + L_S p$

13. As 12.

14.

Z	scalar	diagonal	symmetric	skew-symmetric	hermitian
Z′	scalar	symmetric	hermitian	diagonal	hermitian

15. Hint: transform a general 3×3 matrix and equate the appropriate elements of the transformed matrix *or* apply the inverse transformation to a symmetric 3×3 matrix.

A	B	C
C	D	E
B	F	D

that is

P		
		Q
	R	

plus any circulant.

16.

		a	b	c
	a	$Z_{\alpha\alpha}+\frac{1}{2}Z_{oo}$	$-\frac{1}{2}Z_{\alpha\alpha}+\frac{\sqrt{3}}{2}Z_{\alpha\beta}$ $+\frac{1}{2}Z_{oo}$	$-\frac{1}{2}Z_{\alpha\alpha}-\frac{\sqrt{3}}{2}Z_{\alpha\beta}$ $+\frac{1}{2}Z_{oo}$
$\frac{2}{3}$	b	$-\frac{1}{2}Z_{\alpha\alpha}+\frac{\sqrt{3}}{2}Z_{\beta\alpha}$ $+\frac{1}{2}Z_{oo}$	$\frac{1}{4}Z_{\alpha\alpha}-\frac{\sqrt{3}}{4}Z_{\alpha\beta}$ $-\frac{\sqrt{3}}{4}Z_{\beta\alpha}+\frac{3}{4}Z_{\beta\beta}$ $+\frac{1}{2}Z_{oo}$	$\frac{1}{4}Z_{\alpha\alpha}+\frac{\sqrt{3}}{4}Z_{\alpha\beta}$ $-\frac{\sqrt{3}}{4}Z_{\beta\alpha}-\frac{3}{4}Z_{\beta\beta}$ $+\frac{1}{2}Z_{oo}$
	c	$-\frac{1}{2}Z_{\alpha\alpha}-\frac{\sqrt{3}}{2}Z_{\beta\alpha}$ $+\frac{1}{2}Z_{oo}$	$\frac{1}{4}Z_{\alpha\alpha}-\frac{\sqrt{3}}{4}Z_{\alpha\beta}$ $+\frac{\sqrt{3}}{4}Z_{\beta\alpha}-\frac{3}{4}Z_{\beta\beta}$ $+\frac{1}{2}Z_{oo}$	$\frac{1}{4}Z_{\alpha\alpha}+\frac{\sqrt{3}}{4}Z_{\alpha\beta}$ $+\frac{\sqrt{3}}{4}Z_{\beta\alpha}+\frac{3}{4}Z_{\beta\beta}$ $+\frac{1}{2}Z_{oo}$

which is a circulant, if $Z_{\alpha\alpha} = Z_{\beta\beta}$ and $Z_{\beta\alpha} = -Z_{\alpha\beta}$.

17. Express the matrices as compound matrices in which the sub-matrices are 1 in which both voltage and current are known, 2 in which currents only are known, 3 in which voltages only are known and 4 in which neither are known. From voltage equations 1 and 3, by transferring the voltages due to currents 1 and 2 to the left-hand side, we can find currents 3 and 4. The unknown voltages follow from $\mathbf{v} = \mathbf{Zi}$.

18.
$$I^d = V(R+R_a)/\{R_a(R+R_a)+L_a^2\dot{\theta}^2\},$$
$$I^q = VL_a\dot{\theta}/\{R_a(R+R_a)+L_a^2\dot{\theta}^2\}$$

Voltage between q brushes $VRL_a\dot{\theta}/\{R_a(R+R_a)+L_a^2\dot{\theta}^2\}$. If the armature winding resistance is negligible, I^q is constant irrespective of the value of R.

19. (a) and (b) Frequency: $\dot{\theta}\pm\omega$. Magnitude: same. No torque.

20. (a) $V_o = (R_a+R_1)I^o+(L_aM_{d2}\dot{\theta}^2/R_a)(V_2/R_2)$.

(b) $V_2L_aM_{d2}\dot{\theta}^2/\{(R_a+L_ap)(R_2+L_2p)\}$.

21.
$$I_o \exp\{-(R_a+R_F-M_{dF}\dot{\theta})/(L_q+L_F)\}t.$$

22.
$$i^a = I^a\left\{\frac{\alpha-\gamma}{\alpha-\beta}\,\varepsilon^{-\alpha t}-\frac{\beta-\gamma}{\alpha-\beta}\,\varepsilon^{-\beta t}\right\},$$

where $\alpha, \beta = -\{R_a L_Z + R_Z L_a - L_Z M_{dY}\dot{\theta} + M_{YZ} M_{dZ}\dot{\theta}\}$

$$\frac{\pm\sqrt{[(R_a L_Z + R_Z L_a - L_Z M_{dY}\dot{\theta} + M_{YZ} M_{dZ}\dot{\theta})^2 - 4R_Z(L_Z L_a - M_{YZ}^2)(R_a - M_{dY}\dot{\theta})]}}{2(L_Z L_a - M_{YZ}^2)}$$

and

$$\gamma = \frac{R_Z L_a + M_{YZ} M_{dZ}\dot{\theta} + (R_Z M_{YZ} + L_Z M_{dZ}\dot{\theta})(R_a - M_{dY}\dot{\theta})/(R_Z + M_{dZ}\dot{\theta})}{(L_Z L_a - M_{YZ}^2)}$$

in which a denotes the armature circuit, Y the series field winding and Z the shunt field winding.

23. Scalar plus skew-symmetric.

24. Hint: Express I^A in terms of V_A.

With a real denominator, the imaginary part of the numerator of I^A is $-s^2 X_r(X_r X_S - X_M^2) - R_r^2 X_S$. Since $X_r X_S > X_M^2$, this is always negative, i.e. as a motor the current always lags and as a generator the current always leads.

The machine cannot be a source of reactive power.

25. (a) Frequency: $(\dot{\theta} - \omega)$, Magnitude: $\dfrac{VM(\dot{\theta} - \omega)}{\sqrt{\{R_r^2 + \omega^2 L_r^2\}}}$.

(b) Frequency: ω, Magnitude: $\dfrac{VM\omega}{\sqrt{\{R_r^2 + (\dot{\theta} - \omega)^2 L_r^2\}}}$.

26.

$$-\frac{2V^2 X_M}{\omega} \frac{\sin\alpha}{\{R_S + R_r - (1-s)X_M \sin\alpha\}^2 + \{X_S + sX_r + (1+s)X_M \cos\alpha\}^2}.$$

27.

$$-\frac{3V^2 X_M}{\omega} \frac{[\{R_S R_r + s(X_S X_r - X_M^2)\}\sin\alpha + \{R_r X_S - sR_S X_r\}\cos\alpha]/n + (R_S/n^2 - sR_r)X_M}{\{R_S R_r - s(X_S X_r - X_M^2)\}^2 + \{R_r X_S + sR_S X_r\}^2}.$$

28.

Speed	Torque (a)	Torque (b)
0	1·1434	0
0·1	1·2690	0·0577
0·2	1·4255	0·1190
0·3	1·6257	0·1882
0·4	1·8909	0·2716
0·5	2·2583	0·3795
0·6	2·7997	0·5319
0·7	3·6720	0·7736
0·8	5·2781	1·2313
0·9	8·7325	2·4274
0·95	10·5417	4·6978
1·0	0	−0·01262

29. Hint: Transform external resistance from A, B, C axes to P, N axes and add to machine impedance matrix in P, N axes.

$$\frac{V_P^2 X_M^2 R_r}{\omega}\;\frac{(R_S R_r + R_X R_r/3 - X_S X_r + X_M^2)^2 + (R_r X_S + R_S X_r + R_X X_r/3)^2 - (R_r^2 + X_r^2)(R_X/3)^2}{[(R_S R_r + 2R_X R_r/3 - X_S X_r + X_M^2)^2 + (R_r X_S + R_S X_r + 2R_X X_r/3)^2] \times [(R_S R_r - X_S X_r + X_M^2)^2 + (R_r X_S + R_S X_r)^2]}$$

Starting torque: 0·55.

30.

$$\frac{1}{\omega}\;\frac{3V^2 X_M^2 s R_S}{(R_S R_r - sX_S X_r + sX_M^2)^2 + (R_S X_r + sR_r X_S)^2}$$

Hint: There is a large complicated common factor of the numerator and denominator of the initial expression.

31. The positive-sequence impedance sub-matrix is

	P	f_1	f_2
P	$R_S + jX_S$	jX_{M1}	jX_{M2}
f_1	jsX_{M1}	$R_{r1} + jsX_{r1}$	jsX_{M3}
f_2	jsX_{M2}	jsX_{M3}	$R_{r2} + jsX_{r2}$

where r1, r2, denote the rotor winding resistances; M1, M2 and M3 denote the mutual couplings of stator and rotor winding 1, stator and rotor winding 2 and the two rotor windings respectively.

$$\frac{sV_P^2}{\omega}\frac{\begin{bmatrix} s^2(R_{r1}X_{r2}+R_{r2}X_{r1})(X_{M1}^2X_{r2}-2X_{M1}X_{M2}X_{M3}+X_{M2}^2X_{r1}) \\ +(R_{r2}X_{M1}^2+R_{r1}X_{M2}^2)(R_{r1}R_{r2}-s^2X_{r1}X_{r2}+s^2X_{M3}^2) \end{bmatrix}}{\begin{array}{c}\{R_S(R_{r1}R_{r2}-s^2X_{r1}X_{r2}+s^2X_{M3}^2)-sX_S(R_{r1}X_{r2}+R_{r2}X_{r1}) \\ +sX_{M1}^2R_{r2}+sX_{M2}^2R_{r1}\}^2 \\ +\{X_S(R_{r1}R_{r2}-s^2X_{r1}X_{r2}+s^2X_{M3}^2)+sR_S(R_{r1}X_{r2}+R_{r2}X_{r1}) \\ +s^2(X_{M1}^2X_{r2}+X_{M2}^2X_{r1}-2X_{M1}X_{M2}X_{M3})\}^2\end{array}}$$

32.

$$\frac{V^2X_M^2R_r(1-s)}{\omega}\times\frac{\{R_r^2-s(2-s)X_r^2\}+\sqrt{[\{R_r^2-s(2-s)X_r^2\}^2+\{2R_rX_r\}^2]}\sin(2\omega t+\psi)}{\begin{array}{c}[R_SR_r^2-s(2-s)R_SX_r^2-R_r(2X_SX_r-X_M^2)]^2 \\ +[2R_SR_rX_r+R_r^2X_S-s(2-s)X_r(X_SX_r-X_M^2)]^2\end{array}}$$

where ψ is an angle dependent on s and the machine constants. The magnitude of the alternating component exceeds that of the mean component and the torque thus reverses during part of the cycle.

33. Hint: The impedance matrix in P, N axes can be found as in Exercise 29. The line voltages being known, their symmetrical components can be found and the phase voltage components (omitting zero-sequence) from them.

$$\frac{2V^2X_M^2X_CR_r}{3\sqrt{3}}\frac{R_r(R_SR_r+X_M^2)+R_SX_r^2}{\begin{array}{c}[(R_SR_r-X_SX_r+X_M^2)^2+(R_SX_r+R_rX_S)^2]^2 \\ +[(R_rR_S-X_SX_r+X_M^2)^2+(R_SX_r+R_rX_S)^2] \\ \times[(2X_C/3)^2(R_r^2+X_r^2)-4(X_C/3)\{X_S(R_r^2+X_r^2-X_rX_M^2)\}]\end{array}}$$

34. Hint: Use synchronous axes so that both excitations are d.c.

$$\frac{3X_M}{\omega}\frac{\begin{array}{c}V_S^2sX_MR_r-V_r^2X_MR_S-V_SV_r[(R_rX_S-sR_SX_r)\cos\delta \\ +\{R_rR_S+s(X_SX_r-X_M^2)\}\sin\delta]\end{array}}{\{R_rR_S-s(X_SX_r-X_M^2)\}^2+\{R_rX_S+sR_SX_r\}^2}$$

35.

$$-\frac{3V^2}{2\omega}\frac{(R_a^2+X_d^2)^{1/2}(R_a^2+X_q^2)^{1/2}(X_d-X_q)}{(R_a^2+X_dX_q)^2}[\sin(2\psi-\alpha-\beta)+\sin(\alpha-\beta)]$$

where $\psi = \theta-\omega t$, $\alpha = \arctan R_a/X_q$ and $\beta = \arctan R_a/X_d$.

For $v_a = \hat{V}\sin\omega t$,

$$i^a = \frac{\hat{V}}{R_a^2+X_dX_q}\left[\left\{R_a-\frac{1}{2}(X_q-X_d)\sin 2\psi\right\}\sin\omega t - \{X_d\sin^2\psi+X_q\cos^2\psi\}\cos\omega t\right]$$

The sine (in-phase) component of the current thus changes sign with ψ, but the quadrature component always lags by $\pi/2$ as a motor current and leads by $\pi/2$ as a generated current. Thus irrespective of the sign of ψ, the machine cannot be a source of reactive power.

36.

$$\begin{aligned} v_b = \sqrt{\tfrac{2}{3}}\,[R_a\sqrt{\{(I^d)^2+(I^q)^2\}}\sin(\theta-2\pi/3+\psi) \\ +X_d\sqrt{\{(I^d)^2+(I^q)^2\}}\cos(\theta-2\pi/3+\psi) \\ -X_{Fd}I^F\sin(\theta-2\pi/3)] \end{aligned}$$

37.

$$-\frac{1}{\omega}\frac{R_aV_F^2X_{Fd}^2[R_a^2+X_q^2]}{R_F^2(R_a^2+X_dX_q)^2}$$

38. (a) (i) $2(R_a+L_qp)$. (ii) $2(R_a+L_dp)$.
(b) (i) $\frac{3}{2}(R_a+L_dp)$. (ii) $\frac{3}{2}(R_a+L_qp)$.
(c) (i) $R_a+\{(2L_d+L_o)/3\}p$ (ii) $R_a+\{(2L_q+L_o)/3\}p$.

39. (a) (i) $2(R_a+L_qp)$. (ii) $2[R_a+\{L_d-L_{Fd}^2p/(R_F+L_Fp)\}p]$.
(b) (i) $\frac{3}{2}[R_a+\{L_d-L_{Fd}^2p/(R_F+L_Fp)\}p]$.
(ii) $\frac{3}{2}(R_a+L_qp)$.
(c) (i) $R_a+[\frac{2}{3}\{L_d-L_{Fd}^2p/(R_F+L_Fp)\}+\frac{1}{3}L_o]p$.
(ii) $R_a+\{(2L_q+L_o)/3\}p$.

40. (a)

	d	q	S
d	R_r+L_rp	$L_r\dot{\theta}$	$\sqrt{2}M\dot{\theta}$
q	$-L_r\dot{\theta}$	R_r+L_rp	$\sqrt{2}Mp$
S		$\sqrt{2}Mp$	$2(R_S+L_Sp)$

(b)

	d	q	S
d	R_r+L_rp	$\{L_r-M^2p/(R_S+L_Sp)\}\dot{\theta}$	$\sqrt{\tfrac{3}{2}}\,Mp$
q	$-L_r\dot{\theta}$	$R_r+\{L_r-M^2p/(R_S+L_Sp)\}p$	$-\sqrt{\tfrac{3}{2}}\,M\dot{\theta}$
S	$\sqrt{\tfrac{3}{2}}Mp$		$\tfrac{3}{2}(R_S+L_Sp)$

41.

$$\frac{3}{2}\,\frac{V^2(X_d-X_q)}{\omega}\,\frac{(2s-1)\,R_a(X_d-X_q)-[\{(2s-1)\,R_a(X_d+X_q)\}^2+\{R_a^2-(2s-1)^2X_dX_q\}^2]^{1/2}\sin(2s\omega t+\psi)}{\{R_a^2-(2s-1)X_dX_q\}^2+\{sR_a(X_d+X_q)\}^2}$$

42. Hint: For the *mean* torque, the components due to armature and field voltages may be determined separately.

$$\frac{3V^2}{2\omega}\,\frac{\begin{array}{c}(2s-1)\,R_a(R_F^2+s^2X_F^2)\,(X_d-X_q)^2+sR_FX_{Fd}^2\{R_a^2+(2s-1)^2X_q^2\}\\ +s^2(2s-1)\,R_aX_{Fd}^2\{X_{Fd}^2-2X_F(X_d-X_q)\}\end{array}}{\begin{array}{c}(R_F^2+s^2X_F^2)\,[\{R_a^2-(2s-1)X_dX_q\}^2+s^2R_a^2(X_d+X_q)^2]\\ +s^2X_{Fd}^2[\{s^2R_a^2+(2s-1)^2X_q^2\}\{X_{Fd}^2-2X_FX_d\}\\ -2R_aR_F\{R_a^2+(2s-1)X_q^2\}+2R_a^2(s-1)^2X_qX_F]\end{array}}$$

$$-\frac{V_F^2}{\omega}\,\frac{(1-s)\,R_aX_{Fd}^2\{R_a^2+(1-s)^2X_q^2\}}{R_F^2\{R_a^2+(1-s)^2X_dX_q\}^2}$$

43. Hint: Compare the determinant of the impedance matrix of the equivalent circuit with that of the machine given on p. 220.

When $\dot{\theta}=0$, i.e. when the machine is at rest.

44. *Either* assume steady-state a.c. conditions and investigate the limitations on $\dot{\theta}$ and/or on C *or* consider the determinant of the transient impedance matrix. For the P, f sub-matrix, the determinant

will be a cubic in p. Neglecting all resistances gives approximate values for the frequencies (imaginary parts of roots). Reduction of the roots by the simplest of these leads to an approximate value of the corresponding real part. Division by the corresponding factor leads to a quadratic (with complex coefficients), the roots of which can be written down to obtain the complete approximate solution.

The machine will self-excite at an angular frequency approximately equal to $\dot{\theta}$, for values of $\dot{\theta}$ between $1/\sqrt{(L_S C)}$ and $1/\sqrt{\{(L_S - M^2/L_r)C\}}$.

45.

$$-\sqrt{\frac{2}{3}}\,\frac{V_F}{R_F}\,\frac{L_{Fd}\omega^2}{L_F L_q L_d'}\left[\begin{array}{l}F\cos(\omega t-2\pi/3+\psi)-K\sin(\omega t-2\pi/3+\psi)\\ +\{G\cos(\omega t-2\pi/3+\psi)-N\sin(\omega t-2\pi/3+\psi)\}\varepsilon^{-\alpha t}\\ +\{H/2[\cos(2\omega t+\psi-2\pi/3)+\cos(\psi-2\pi/3)]\\ +J/2[\sin(2\omega t+\psi-2\pi/3)-\sin(\psi-2\pi/3)]\\ -P/2[\sin(2\omega t+\psi-2\pi/3)+\sin(\psi-2\pi/3)]\\ -Q/2[-\cos(2\omega t+\psi-2\pi/3)+\cos(\psi-2\pi/3)]\}\varepsilon^{-\beta t}\end{array}\right]$$

46.

$$R_a\hat{I}\sin(\omega t+2\pi/3)+\frac{X_d'+X_q}{2}\,\hat{I}\cos(\omega t+2\pi/3)$$
$$+3\,\frac{X_d'-X_q}{2}\,\hat{I}\cos\left(3\omega t+2\psi-\frac{2\pi}{3}\right).$$

Same.

47.

$$\frac{\hat{V}}{2X_d'}\{\cos(2\omega t+2\psi-2\pi/3)+\cos(2\pi/3)$$
$$-\cos(3\omega t+2\psi-2\pi/3)-\cos(\omega t+2\pi/3)\}$$
$$+\frac{\hat{V}}{2X_q'}\{-\cos(2\omega t+2\psi-2\pi/3)+\cos(2\pi/3)$$
$$+\cos(3\omega t+2\psi-2\pi/3)-\cos(\omega t+2\pi/3)\}.$$

Same.

48. Hint: Use Newton's approximation starting with the values given on p. 226.

	α/ω	β/ω	γ/ω
Approximate values:	$= 0{\cdot}0029\dot{3}$,	$= 0{\cdot}02291\dot{6}$,	$= 1$
Accurate values:	0·0029326532	0·022917007	0·99991[illegible]
Errors:	0·0232%	0·0015%	0·00898[illegible]

49. Hint: Check the sign of $\dot{\theta}$, i.e. of the steady-state torque. The zeros of the determinant of the motional impedance matrix are

$$\frac{-(R-M\dot{\theta})J \pm \sqrt{[(R-M\dot{\theta})^2 J^2 - 8LJV^2M^2/(R-M\dot{\theta})^2]}}{2LJ}$$

Since $\dot{\theta}$ is negative and $M\dot{\theta} > R$, and since $8LJV^2M^2/(R-M\dot{\theta})^2$ is necessarily positive, the magnitude of the square root will be less than that of $(R-M\dot{\theta})J$ and the real part of the zeros of Δ will always be negative. The system is thus positively damped, but will oscillate if $J < 8LM^2V^2/(R-M\dot{\theta})^4$.

50. The determinant of the motional impedance matrix will not factorize and, without information on the relative magnitudes of the parameters, it is not possible to make justifiable assumptions. Only specific numerical cases could be investigated with certainty.

51.

$$\frac{\sqrt{3}V}{2\pi}\sqrt{\left\{\frac{(X_d - X_q)}{JX_dX_q\omega}\right\}}.$$

52.

	d	q	S	s
d	$R_r + L_r p$	$L_r\dot{\theta}$	$\sqrt{2}M\dot{\theta}$	$L_r I^q + \sqrt{2}MI^S - \sqrt{\tfrac{3}{2}}\,\hat{V}\cos\psi/p$
q	$-L_r\dot{\theta}$	$R_r + L_r p$	$\sqrt{2}Mp$	$-L_r I^d - \sqrt{\tfrac{3}{2}}\,\hat{V}\sin\psi/p$
S		$\sqrt{2}Mp$	$2(R_S + L_S p)$	
s	$-\sqrt{2}MI^S$		$-\sqrt{2}MI^d$	$F + Jp$

d	p	S	s
R_r+L_rp	$\left(L_r-\dfrac{M^2p}{R_S+L_Sp}\right)\dot{\theta}$	$\sqrt{\tfrac{3}{2}}\,Mp$	$\left(L_r-\dfrac{M^2p}{R_S+L_Sp}\right)I^q$ $-\sqrt{\tfrac{3}{2}}\hat{V}\cos\psi/p$
$-L_r\dot{\theta}$	$R_r+\left(L_r-\dfrac{M^2p}{R_S+L_Sp}\right)p$	$-\sqrt{\tfrac{3}{2}}\,M\dot{\theta}$	$-L_rI^d-\sqrt{\tfrac{3}{2}}\,MI^S$ $-\sqrt{\tfrac{3}{2}}\hat{V}\sin\psi/p$
$\sqrt{\tfrac{3}{2}}Mp$		$\tfrac{3}{2}(R_S+L_Sp)$	
$-\dfrac{M^2p}{R_S+L_Sp}I^q$	$-\dfrac{M^2p}{R_S+L_Sp}I^d-\sqrt{\tfrac{3}{2}}\,MI^S$	$-\sqrt{\tfrac{3}{2}}\,MI^q$	$F+Jp$

53. The system will oscillate if $J < 4LG_2^2(I^{F2})^2/(G_1\dot{\theta}_1-R)^2$, where R and L are the total armature circuit resistance and inductance, and the index 1 relates to the generator and 2 to the motor.
Note: The system usually oscillates, but the amplitude increases very rapidly so that the oscillation becomes non-linear and resembles that of a relaxation oscillator.

54.

$$\text{Steady-state:}\quad I^o = \frac{V_F M_{Fd}\cos\alpha\,\dot{\theta}}{R_F(R_a+R_K+M_{Kq}\sin\alpha\,\dot{\theta})}.$$

Transient:

$$i^o = \frac{V_F M_{Fd}\cos\alpha\,\dot{\theta}}{R_F R'_a}\left[1-\frac{R'_a/L''_a-\gamma}{\beta-\gamma}\varepsilon^{-\beta t}-\frac{R'_a/L''_a-\beta}{\beta-\gamma}\varepsilon^{-\gamma t}\right],$$

$$\text{where}\quad R'_a = R_a+R_K+M_{Kq}\sin\alpha\,\dot{\theta},$$
$$L'_a = L_a+L_K-2M_{Kq}\cos\alpha,$$
$$L''_a = L'_a-M^2_{Fd}\sin^2\alpha/L_F,$$

and β and γ are the roots of

$$s^2+\frac{R'_aL'_F}{L_FL''_a}s+\frac{R_FR'_a}{L_FL''_a}=0,\quad \text{where}$$

$$L'_F = L_F+R_FL'_a/R'_a-(M^2_{Fd}/R'_a)\sin\alpha\cos\alpha\,\dot{\theta}$$

55.

Speed	Torque
0	1·0441
0·1	1·2227
0·2	1·4920
0·3	1·9670
0·4	3·1621
0·5	0·1009
0·6	−0·9262
0·7	0·7158
0·8	1·9683
0·9	4·0775
0·95	5·2631
1·0	0

56.

$$\frac{\mathrm{s}X_M^2V^2}{\omega} \times \frac{\begin{bmatrix} \{R_r^2+(2-\mathrm{s}1)^2X_r^2\}\,\{R_D(R_Q^2+\mathrm{s}^2X_Q^2)+R_Q(R_D^2+\mathrm{s}^2X_D^2)\} \\ +\mathrm{s}(2\mathrm{s}-1)\,X_M^2\{R_r(R_D+R_Q)^2-2\mathrm{s}(2\mathrm{s}-1)\,X_r(R_DX_Q+R_QX_D) \\ +\mathrm{s}^2R_r(X_D-X_Q)^2\}+\mathrm{s}^2(2\mathrm{s}-1)^2X_M^4(R_D+R_Q) \end{bmatrix}}{\begin{bmatrix} \{R_r^2-(2\mathrm{s}-1)\,X_r^2\}(R_DR_Q-\mathrm{s}^2X_DX_Q)-2\mathrm{s}^2R_rX_r(R_DX_Q+R_DX_Q) \\ +\mathrm{s}^2R_rX_M^2(R_D+R_Q)-\mathrm{s}^2(2\mathrm{s}-1)\,X_rX_M^2(X_D+X_Q)+\mathrm{s}^2(2\mathrm{s}-1)X_M^4 \end{bmatrix}^2 + \begin{bmatrix} \mathrm{s}\{R_r^2-(2\mathrm{s}-1)X_r^2\}(R_DX_Q+R_QX_D)+2\mathrm{s}R_rX_r(R_DR_Q-\mathrm{s}^2X_DX_Q) \\ +\mathrm{s}^3R_rX_M^2(X_D+X_Q)+\mathrm{s}(2\mathrm{s}-1)X_rX_M^2(R_D+R_Q) \end{bmatrix}^2}$$

References

1. KRON, G., Non-Riemannian dynamics of rotating electrical machinery, *J. Math. Phys.* **13,** 103–94 (1934).
2. KRON, G., The application of tensors to the analysis of rotating electrical machinery, *General Electric Review*, 1935–8. Largely reprinted in book form (1938).
3. HINDMARSH, J., *Electrical Machines and their Applications*, 2nd edition, Pergamon, 1970.
4. KRON, G., Tensor analysis of networks, *Orthogonal Networks*, Wiley, pp. 407–35.
5. SLEMON, G. R., Equivalent circuits for transformers and machines including non-linear effects, Monograph No. 68, *Proc. IEE* **100,** Part IV, 129–43 (1953).
6. WIESEMAN, R. W., Graphical determination of magnetic fields: Practical application to salient-pole synchronous-machine design, *Trans. AIEE* **46,** 141–54 (1927).
7. JONES, C. V., *The Unified Theory of Electrical Machines*, Butterworth.
8. GIBBS, W. J., *Electric Machine Analysis using Tensors*, Pitman.
9. MORRILL, W. J., The revolving field theory of the capacitor motor, *Trans. AIEE* **48,** 614–29 (1929).
10. CONCORDIA, C., *Synchronous Machines*, Wiley.
11. KILGORE, L. A., Calculation of synchronous machine constants, *Trans. AIEE* **50,** 1201–14 (1931).
12. CHING, Y. K. and ADKINS, B., Transient theory of synchronous generators under unbalanced conditions, Monograph No. 85, *Proc. IEE* **101,** Part IV, 166–82 (1954).
13. ROUTH, E. J., *Dynamics of a System of Rigid Bodies*, Part II, 6th edition, Macmillan, 1905, Ch. VI. (Reprinted as a paperback by Dover, 1961.)
14. JONES, C. V., An analysis of commutation for the unified-machine theory, Monograph No. 302, *Proc. IEE* **105,** Part C, 476–88 (1958).
15. PRESCOTT, J. C. and EL-KHARASHI, A. K., A method of measuring self-inductances applicable to large electrical machines, *Proc. IEE* **106,** Part A, 169–73 (1959).
16. HANCOCK, N. N., *Electric Power Utilization*, Pitman.
17. BROWN, J. E. and BUTLER, O. I. The zero-sequence parameters and performance of three-phase induction motors, Monograph No. 92, *Proc. IEE* **101,** Part IV, 219–24 (1954).
18. WRIGHT, S. H., Determination of synchronous-machine constants by test, *Trans. AIEE* **50,** 1331–50 (1931).
19. GIBBS, W. J., *Electric Machine Analysis using Matrices*, Pitman.

20. RIAZ, M., Transient analysis of the metadyne generator, *Trans. AIEE* **72,** Part III, 52–62 (1953).
21. ADKINS, B., Rotating amplifiers, Section 1, *Theory of Cross-field Generators,* (ed. by M. G. Say), Newnes, p. 25.
22. BROWN, J. E. and RUSSELL, R. L., Symmetrical component analysis applied to phase convertors of the Ferraris–Arno type, *Proc. IEE* **105,** Part A, 538–44 (1958).
23. BARTON, T. H. and DOXEY, B. C., The operation of three-phase induction motors with unsymmetrical impedance in the secondary circuit, *Proc. IEE* **102,** Part A, 71–79 (1955).
24. SHENTON, J. C. and SHENTON, L. R., *Harmonics in Synchronous Machines, RAE (Farnborough) Technical Memoranda: Single-phase Line-to-line Resistive Load*, No. EL 1843, June 1960; *Three-phase Short Circuit* $R_a > 0$, $R_b = R_c = 0$, No. EL 1847, November 1960; *Three-phase Short Circuit* $R_a > 0$, $R_b = R_c \neq 0$, No. EL 1855, September 1961.
25. DOHERTY, R. E. and PARK, R. H., Mechanical force between electric circuits, *Trans. AIEE* **45,** 240–52 (1926).

INDEX